Concert Halls by Nagata Acoustics

Yasuhisa Toyota • Motoo Komoda • Daniel Beckmann
Marc Quiquerez • Erik Bergal

Concert Halls by Nagata Acoustics

Thirty Years of Acoustical Design for Music Venues and Vineyard-Style Auditoria

Springer

ASA PRESS

Yasuhisa Toyota
Nagata Acoustics
Los Angeles, CA, USA

Motoo Komoda
Nagata Acoustics
Los Angeles, CA, USA

Daniel Beckmann
Nagata Acoustics
Los Angeles, CA, USA

Marc Quiquerez
Nagata Acoustics
Paris, France

Erik Bergal
Nagata Acoustics
Los Angeles, CA, USA

ISBN 978-3-030-42449-7 ISBN 978-3-030-42450-3 (eBook)
https://doi.org/10.1007/978-3-030-42450-3

This Springer imprint is published by the registered company Springer Nature Switzerland AG
The registered company address is: Gewerbestrasse 11, 6330 Cham, Switzerland

The ASA Press

ASA Press, which represents a collaboration between the Acoustical Society of America and Springer Nature, is dedicated to encouraging the publication of important new books as well as the distribution of classic titles in acoustics. These titles, published under a dual ASA Press/Springer imprint, are intended to reflect the full range of research in acoustics. ASA Press titles can include all types of books that Springer publishes, and may appear in any appropriate Springer book series.

ASA PRESS

The Acoustical Society of America

On 27 December 1928 a group of scientists and engineers met at Bell Telephone Laboratories in New York City to discuss organizing a society dedicated to the field of acoustics. Plans developed rapidly, and the Acoustical Society of America (ASA) held its first meeting on 10–11 May 1929 with a charter membership of about 450. Today, ASA has a worldwide membership of about 7000.

The scope of this new society incorporated a broad range of technical areas that continues to be reflected in ASA's present-day endeavors. Today, ASA serves the interests of its members and the acoustics community in all branches of acoustics, both theoretical and applied. To achieve this goal, ASA has established Technical Committees charged with keeping abreast of the developments and needs of membership in specialized fields, as well as identifying new ones as they develop.

The Technical Committees include acoustical oceanography, animal bioacoustics, architectural acoustics, biomedical acoustics, engineering acoustics, musical acoustics, noise, physical acoustics, psychological and physiological acoustics, signal processing in acoustics, speech communication, structural acoustics and vibration, and underwater acoustics. This diversity is one of the Society's unique and strongest assets since it so strongly fosters and encourages cross-disciplinary learning, collaboration, and interactions.

ASA publications and meetings incorporate the diversity of these Technical Committees. In particular, publications play a major role in the Society. *The Journal of the Acoustical Society of America* (JASA) includes contributed papers and patent reviews. *JASA Express Letters* (JASA-EL) and *Proceedings of Meetings on Acoustics* (POMA) are online, open-access publications, offering rapid publication. *Acoustics Today*, published quarterly, is a popular open-access magazine. Other key features of ASA's publishing program include books, reprints of classic acoustics texts, and videos. ASA's biannual meetings offer opportunities for attendees to share information, with strong support throughout the career continuum, from students to retirees. Meetings incorporate many opportunities for professional and social interactions, and attendees find the personal contacts a rewarding experience. These experiences result in building a robust network of fellow scientists and engineers, many of whom become lifelong friends and colleagues.

From the Society's inception, members recognized the importance of developing acoustical standards with a focus on terminology, measurement procedures, and criteria for determining the effects of noise and vibration. The ASA Standards Program serves as the Secretariat for four American National Standards Institute Committees and provides administrative support for several international standards committees.

Throughout its history to present day, ASA's strength resides in attracting the interest and commitment of scholars devoted to promoting the knowledge and practical applications of acoustics. The unselfish activity of these individuals in the development of the Society is largely responsible for ASA's growth and present stature.

In memory of Dr. Minoru Nagata, friend and mentor.
(1925–2018)

Foreword

What for me sets Yasu Toyota aside from other acousticians is that he does not listen to music only as an acoustician but also with musical ears. The greatest difficulty is to find the right balance between transparency and reverberation. Neither of these can really work without the other and in my view he hears that perfectly.

Tel-Aviv, November 2018 *Daniel Barenboim*

I met Yasu Toyota when we began to work on the Walt Disney Concert Hall project together many years ago. He was more collaborative than I had experienced with acousticians, and he was open and interested in exploring new ideas and innovative solutions. He was very respectful of the architectural considerations necessary to make user-friendly spaces. When we started our work, we agreed that Hans Scharoun's Berlin Philharmonie had the qualities that we wanted in the Walt Disney Concert Hall. It became a starting point for our conversations, especially in terms of the feeling of the room and the feeling of connectivity between the performers and the audience. This hall created a communal experience unlike any other space that either of us had experienced. The audience connected with the performers; the performers felt the adulation and performed better; the audience responded with more love. It was a truly a virtuous cycle, and Yasu and I have spent the last twenty years studying and developing new and different ways to achieve this connectivity in our halls. Our work together has resulted in innovative solutions, and we have created some very successful rooms for music.

I rely on Yasu, and I trust him. He has been my great collaborator all of these years, and it has been my honor and privilege to work with this great and talented man.

Los Angeles, January 2019 *Frank Gehry*

My name is Valery Gergiev and I'm a friend of Yasu Toyota. If I had been born 300 years ago, I would have been a friend of Antonio Stradivari or whichever of the Guarneri family you prefer, many would say del Gesù. Of course, there were many smaller luthiers who followed: I love Guadagnini instruments, other members of the Guarneri family, Bergonzi, Amati, and da Salò. But Yasu Toyota, I put him in the league which I previously introduced, of Stradivarius and Guarneri.

In the 20th century, in the last thirty years all of a sudden, a success story was emerging belonging to a group of acousticians from Japan making fantastic halls. Even in Europe, America, and Russia, people were saying "Oh, there is a fantastic hall in Tokyo, a fantastic hall in Kyoto, and in other Japanese cities.' It is difficult to explain how Japanese professionals learned so much about acoustics while in Europe, they could enjoy the Amsterdam Concertgebouw and Vienna Musikvereinssaal.

I am very honored to be friends with Yasu Toyota. We have had several unbelievable, unforgettable experiences together building several halls in St. Petersburg and a hall in Moscow, but even his halls in Japan, Hamburg, Paris, and the United States happen to be a part of my life story. When I am brought to a hall with Yasu's name attached, the typical characteristics of the sound are recognizable immediately. We are all lucky that the greatest acousticians of all time did not live 350 years ago. Some of them are living now.

St. Petersburg, June 2019 *Valery Gergiev (transcribed)*

デジタライズしてしまった現代の音響理論がまだ到達できていない演奏空間を、交響するひとつの楽器に組みあげる永田音響理論。それは、究極の完成型としての、京都コンサートホールである。
Nagata Acoustics has achieved a method of creating performing arts spaces which resonate like a musical instrument, a feat which is still not possible through modern, digitalized acoustical technology. The ultimate expression of this theory is our collaboration on Kyoto Concert Hall. (*translation by book authors*)

Oita, April 2019 *Arata Isozaki*

Yasuhisa Toyota has a unique combination of talents as he is not only a master of the acoustical science but also a profound music lover with great knowledge of the various choirs of the orchestra.

I have had only the most wonderful experiences with him in Israel where he not only created the acoustics of our new chamber music hall but also completely revived the acoustic and therefore the sound in the Charles Bronfamn Auditorium.

Mr. Toyota has designed so many halls all over the world that, as we say, he needs no introduction. I wish him the best on all his future projects.

Los Angeles, December 2019 *Zubin Mehta*

Pleasures Declared and Shared

Among the greatest pleasures, the most amazing hallucinatory experiences that this discipline made for mad perfectionists—architecture—can offer, there are the encounters with the great, the indispensable, specialists who come with, change and strengthen the architectural work. These people are artists of precision, inventors of details that turn out to be essential within the complementary dimensions made up of structures, landscapes, scenographies, sounds, music... Among all such sorcerers, I've been particularly fascinated by one who works on the depth and subtlety of sound, a certain Yasuhisa Toyota. For a long while I dreamed of taking him with me on a great architectural and musical adventure, but there was a hitch: he was always booked up exclusively by one of my great architect friends... Finally, in 2002, I succeeded in getting him to embark with me on the Copenhagen concert hall... a hall that looks like a 'meteorite' that's fallen from the sky into a blue cube on a brand new polder, where Danish Radio was having its new buildings erected on a slab of concrete, without realizing what was about to happen later... much later. To cut a long story short, the brief involved an internal structure, tucked in behind blue screens, with a concert hall and three other auditoriums. That's where I saw Yasu the sorcerer at work, with his miniature of the hall, an acoustic model in which, on every one of the 1,850 seats, a quietly concentrating audience member was sitting. The whole thing was on a scale of 10cm to a metre, though every now and then there would be a notable exception to this, in the area of the stage: Yasu would suddenly be sitting there, on stage, in the flesh! Japanese sized, he'd managed to put himself in the audience's place to actually listen to real musical pieces, look at the balconies at eye level, and assess the slope of the handrails, striking them to assess the way they reverberated... Those sessions have stayed with me and with my team. Those images have become cult images, so many proofs of the said sorcerer's commitment, his ability to look ahead, to hear sound before it's actually made. This alertness to a preconceived reality has made his reputation and that of Nagata and Motoo Komoda. Each of the four halls is an instrument to be listened to and that has been produced in the same spirit.

With Yasu we studied five projects; the second of these, one we worked on in 2006, was an island music hub, in the middle of the river in Seoul, that had a concert hall and an opera auditorium... a golden mass emerging from the rocks and trees, with the whole thing reflected in the water. It was the winning project, but... once the Mayor of Seoul became the President of the Republic, he forgot all about his Opera House... causing me extreme frustration, as I would so love to have heard Yasu's acoustics resonate there and see him sitting once more in the middle of the new rocky model...

Then it was the Philharmonie de Paris, a winning project, a built project. Nagata was already under contract to an architect friend of mine... So I designed the concert hall with Harold Marshall as an auditorium, conjuring up ethereal layered expanses of music and light and leaving listeners-spectators suspended, giving them the impression of being surrounded by, and immersed in, the music and the light. But the client wanted to have his own consultant acoustician, in the presence of Eckhard Kahle, whom I'd got to know with Russ (Russell Johnson) in the 1990s, when I was building the concert hall in Lucerne. This situation meant I could have Yasuhisa Toyota, whom my client greatly admired, as my personal consultant on the development of the hall. My design for having the music envelop the audience was challenged by a few grumpy bureaucrats, so I asked Yasu what he thought of it, and he told me it didn't serve any purpose but that it could be done, and... he went back to his model, to test it – he's forever testing in a bid to improve! The point of the story is that, as far as acoustics go, the complementary nature of the Marshall-Toyota duo was what really enabled the hall to see the light of day and to be recognized as being unique.

Then it was the turn of Germany and the Kronberg project, a modest musical ensemble integrated into a small town, a simple jewel in a room made of wood, with a roof over it... Eternal regrets.

Finally, there was the competition for the Munich concert hall, a precious composition that was misunderstood... Lost illusions.

But thank you, Mr. Yasuhisa Toyota, for being with me in the days of glory and in the days of oblivion...

We will carry on, in complete complicity, and once again try to force them (the clients, the music-lovers) to see and hear because, Yasu, as you've told me clearly enough, we can also see with our ears.

Paris, March 2020 *Jean Nouvel*

Concert halls are instruments. Whatever we all do, however excellently we perform, we are nothing without the space into which the music travels. I really believe that the general public has no idea what an astonishing difference great acoustic can make to a performance, particularly in emotional terms. Yasu has long been the Stradivarius of concert hall design, and in his spaces we no longer have any excuse, we simply HAVE to perform at our highest level. It is not only like a mega instrument but also a vessel which transports us on our journey, enabling music in a profound sense. My dear Yasu, we are, as we sometimes say in English, pathetically grateful for the chance to play as we would wish, not an everyday experience for musicians! So we all wish you more power to your elbow, your ears, and the unusual sensitivity of yourself and your magnificent team.

London, September 2019 *Sir Simon Rattle*

During my forty years of performing around the world I've learned to assign the experience of acoustics to one of the four categories below:

1. Musicians on stage perceive the acoustical properties of the hall as an obstacle, something between the music and the audience. I often think of a membrane as a metaphor. Intuitively we feel that somehow we have to get through it to get to the other side, to pierce it to reach out to the listeners.
2. The music has presence in the space, but there is no focus. The layering and balancing of complex scores is difficult or impossible. I happen to believe that true expression in music is impossible without clarity, balance and focus.
3. The experience on stage is neutral. The strangest category. The hall doesn't distort the musical thought, but it doesn't add anything either. We don't feel inspired or supported by the acoustics, but we can do a professional job, rarely anything more than that.
4. The best category: we experience the hall as a high quality instrument which the orchestra plays. The acoustics not only reproduce, but support and inspire. Over time, we learn to trust the hall. Instead of fighting it, we achieve a perfect symbiosis: the expression itself is been amplified in the space. The music, the orchestra and the audience become one. This kind of magic (I've no better word for it) happens often in spaces designed by Nagata Acoustics.

The design and building process of the Walt Disney Concert Hall in Los Angeles led by Frank Gehry and Yasuhisa Toyota has been one of the most exciting and joyous projects in my entire life. It certainly changed the fate of the LA Philharmonic and classical music generally in the city. Now, one-and-a-half decade later, it can be concluded that the hall also changed the fate of the entire Downtown of LA.

In the superb acoustics of the Walt Disney Concert Hall, music can go beyond a mere aesthetical experience into the realm of a powerful psychophysical/emotional phenomenon. Warmth and transparency. Body and soul. Yin and Yang. Yasuhisa Toyota and Nagata Acoustics have created a masterpiece.

Helsinki, May 2019 *Esa-Pekka Salonen*

Introduction

I learned acoustics and acoustic design at the Kyushu Institute of Design (KID, now a department of Kyushu University). After I graduated in 1977, I joined Nagata Acoustics. KID was a unique college where I could learn two different aspects of acoustics: acoustics as engineering and music as art. I was an enthusiastic classical music lover when I was a high school student, playing saxophone in junior high band and oboe in high school orchestra. My goal was to work close to classical music, possibly as a recording engineer for classical music or as an acoustician who designs the acoustics of concert halls.

I decided to join Nagata Acoustics, which had been established only a few years prior by Dr. Minoru Nagata. Dr. Nagata had been working as an acoustics researcher at NHK (Nippon Hoso Kyokai, the Japanese National Broadcasting Company similar to BBC in the UK), in a technical laboratory under the chief acoustic researcher Mr. Yasuo Makita. Makita had been in charge of the acoustic design of Tokyo Bunka Kaikan (1963), the most important concert hall in Tokyo before the opening of Suntory Hall in 1986. After retiring from NHK, Makita became a professor in the acoustic design department at KID.

My first few years at Nagata Acoustics were quite busy with noise measurements and noise control projects, for road traffic noise, train and subway noise and vibration, aircraft noise, and so on. Noise pollution was growing constantly, together with Japan's economic growth. This was a good opportunity for me to learn the fundamentals of sound engineering, since noise control uses physical thought processes and mathematical calculations, which are also important for the acoustic design of concert halls.

In the 1980s, there was a boom in building multipurpose halls in Japan. Nagata Acoustics was in charge of acoustic design of many halls, probably more than one hundred publicly funded cultural centers. This prepared us for the boom in concert halls which accompanied the economic bubble in the second half of the 1980s.

Suntory Hall in Tokyo opened in 1986. Many concert halls dedicated to classical music, using natural acoustics, followed Suntory Hall one after another for the next decade, in all parts of Japan. Fortunately, the other concert halls built in Tokyo after Suntory Hall were in many different styles, such as shoebox, fan shape, and of many different sizes. We were very lucky to be in Tokyo as acoustical consultants, since we could experience many different concerts by many different performers—not only from Tokyo but also from foreign countries. Listening to those many different performers in different concert halls helped greatly in developing our acoustic knowledge and skill.

I worked closely with Dr. Minoru Nagata for these important two decades, 1980s and 1990s, before moving to Los Angeles to open a new office of Nagata Acoustics. Dr. Nagata was my boss in the company and a mentor for me in my life. He was not only an excellent acoustic engineer with a rigorous scientific background but he was also a wonderful music lover. He always suggested that I go out of the office where we do desk work every day, out into the field and listen to concerts in concert halls. He valued hearing actual concerts and communicating with musicians. What I learned through these experiences is a treasure. This book is an homage to Dr. Minoru Nagata who passed away in the summer of 2018 while this book was being written.

After moving to Los Angeles in 2000, the first important project was Walt Disney Concert Hall. We had started working on Disney Hall already in 1989, which was even before the start of the design of Sapporo Concert Hall in 1993. The two concert halls, Disney and Sapporo, are related to each other, a relation that benefited both projects. When Sapporo started design work in 1993, the Disney project had already completed the schematic design phase as well as a 1:10 scale model acoustic test. But, Sapporo opened first, in 1997. When Disney started construction in 1999, Sapporo had already been open for 2 years, and Disney would open in 2003. Sapporo learned many lessons from Disney, and Disney learned many lessons from Sapporo.

After Disney, we have been very lucky and happy with many high-profile concert hall projects around the world: Mariinsky Theatre Concert Hall (St. Petersburg, Russia 2006), Danish Radio Concert Hall (Copenhagen, Denmark 2009), New World Center (Miami, USA 2011), Helsinki Music Centre (Helsinki, Finland 2011), Shanghai Symphony Hall (Shanghai, China 2014), Philharmonie de Paris (Paris, France 2015), Elbphilharmonie (Hamburg, Germany 2017), and Pierre Boulez Saal (Berlin, Germany 2017). The list continues, with more than 30 halls in total. Many of the concert halls we worked on belong to the so-called vineyard typology, where the stage is surrounded by the audience, and the audience is divided into several groups. These groups are configured at different levels, so that small wall surfaces among the audience groups can provide acoustically effective reflections. The most important aspect of the vineyard-style concert hall is its visual, architectural intimacy, coupled with the acoustical intimacy, created between the stage and each audience member and also among the audience members: they can see and feel each other's faces over the stage. This is why many modern concert halls take the vineyard-style hall shape and seating configuration.

In this book, we list 32 concert hall projects of our office, with physical data, drawings, photos, and essays for each. Following the projects descriptions are some topics on acoustics we learned by working on the projects, which help us consider design directions for concert hall projects in the future.

Los Angeles, CA, USA Yasuhisa Toyota

Contents

Part I Project Anthology

1 Suntory Hall ... 3
2 Kyoto Concert Hall ... 9
3 Sapporo Concert Hall "Kitara" ... 17
4 Richard B. Fisher Center for the Performing Arts ... 23
5 Walt Disney Concert Hall ... 33
6 Muza Kawasaki Symphony Hall ... 45
7 Mariinsky Concert Hall ... 53
8 Danish Radio Concert House ... 61
9 New World Symphony ... 71
10 Helsinki Music Centre Concert Hall ... 79
11 Kauffman Center for the Performing Arts: Muriel Kauffman Theatre ... 87
12 Kauffman Center for the Performing Arts: Helzberg Hall ... 95
13 Soka Performing Arts Center ... 101
14 Isabella Stewart Gardner Museum ... 111
15 USC Brain and Creativity Institute ... 119
16 Bing Concert Hall ... 125
17 Charles Bronfman Auditorium ... 131
18 Auditorium Giovanni Arvedi ... 141
19 Shanghai Symphony Hall ... 147
20 NOSPR Katowice ... 157
21 Auditorium, Fondation Louis Vuitton ... 167
22 Radio France Concert Hall ... 177
23 Philharmonie de Paris ... 185
24 Musco Center for the Arts ... 197
25 Lotte Concert Hall ... 205
26 Elbphilharmonie Hamburg ... 215
27 Pierre Boulez Saal ... 231
28 La Seine Musicale ... 241

29 Repino Hall ... 251

30 Jinji Lake Concert Hall ... 259

31 Zaryadye Concert Hall ... 267

32 The Conrad Prebys Performing Arts Center ... 277

Part II Topics on Acoustics and Acoustical Design

1 Vienna Musikvereinssaal as the Starting Point of Discussion ... 285

2 Berlin Philharmonie and the Birth of the Vineyard-Style Concept ... 289

3 Surround Seating Paradigm ... 291

4 Foundational Projects of Nagata Acoustics ... 293

5 Reverberation Time, Other Metrics, and Underlying Goals ... 295

6 Computer Simulation ... 297

7 Scale Model Testing ... 299

8 The Importance of Stage Acoustics ... 305

9 Ceiling Height, Ensemble Reflectors, and Soffits ... 307

10 Acclimation to a New Hall ... 311

11 Orchestral Balance ... 313

12 Orchestra Layout ... 315

13 Excessive Sound Exposure on Stage ... 319

14 Stage Risers ... 321

15 Stage Structure ... 325

16 Variable vs. Fixed Acoustics ... 329

17 Sound System Design in Concert Halls ... 331

18 Semi-staged Productions in Concert Halls ... 333

19 Concert Hall Organ Design ... 335

20 A New Direction of Concert Hall Design ... 339

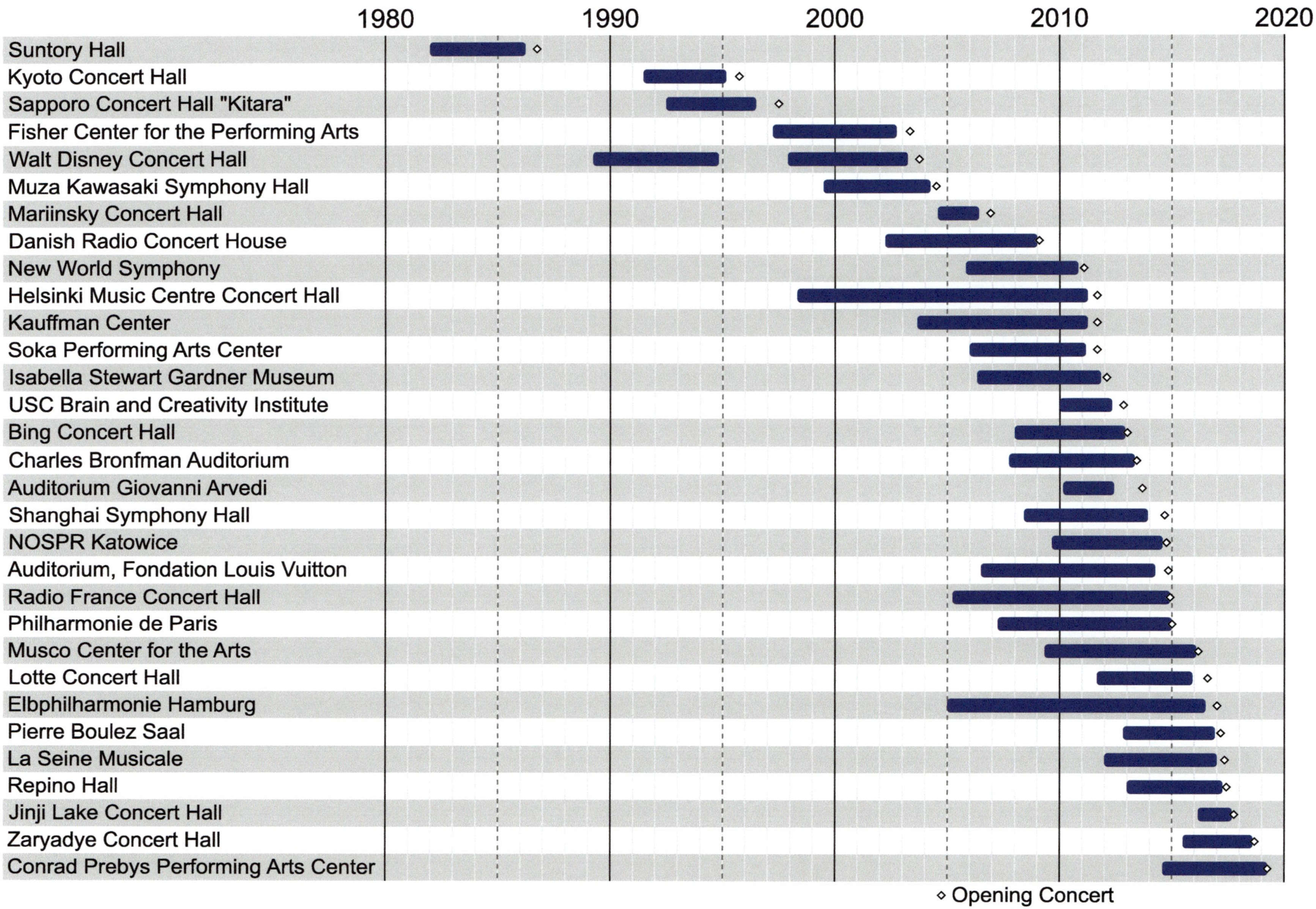
1980
1990
2000
2010
2020
Suntory Hall
Kyoto Concert Hall
Sapporo Concert Hall "Kitara"
Fisher Center for the Performing Arts
Walt Disney Concert Hall
Muza Kawasaki Symphony Hall
Mariinsky Concert Hall
Danish Radio Concert House
New World Symphony
Helsinki Music Centre Concert Hall
Kauffman Center
Soka Performing Arts Center
Isabella Stewart Gardner Museum
USC Brain and Creativity Institute
Bing Concert Hall
Charles Bronfman Auditorium
Auditorium Giovanni Arvedi
Shanghai Symphony Hall
NOSPR Katowice
Auditorium, Fondation Louis Vuitton
Radio France Concert Hall
Philharmonie de Paris
Musco Center for the Arts
Lotte Concert Hall
Elbphilharmonie Hamburg
Pierre Boulez Saal
La Seine Musicale
Repino Hall
Jinji Lake Concert Hall
Zaryadye Concert Hall
Conrad Prebys Performing Arts Center
◇ Opening Concert

Part I

Project Anthology

1. Suntory Hall

View from center balcony. Inspired by the Berlin Philharmonie, the stage is placed at the center of the audience seating

Y. Toyota et al., *Concert Halls by Nagata Acoustics*, https://doi.org/10.1007/978-3-030-42450-3_1

Suntory Hall is the brainchild of the late Suntory president and owner, Keizō Saji. Known for his pioneering attitude and following the Suntory slogan "Yatte Minahare," translated roughly as "just try it" without overthinking, Saji decided to build the first single purpose concert hall in Tokyo. To this end Saji created the Suntory Music Foundation in 1969 to nurture classical music in Japan and retained Yasui Architects to begin design work in 1982. Maestro Herbert von Karajan, a personal friend of Saji, acted as an advisor for the project.

The hall is nestled in a commercial development which was constructed as a part of the revitalization of the Akasaka-Roppongi district in central Tokyo. A sunken courtyard, aptly named Herbert von Karajan Platz, leads to the hall lobby complete with marble walls and a 30-sided chandelier. To the left, the small, "Blue Rose" hall has a system of floor lifts that can reconfigure the space for chamber concerts and banquets alike. Continuing ahead past the café leads to the main hall.

Free from the restrictions of public projects, there was considerable latitude to experiment with the form of the hall. The vineyard-style typology was decided from the start, strongly influenced by von Karajan and the Berlin Philharmonie. The new hall would feature terraced seating areas surrounding the stage and the orchestra would sit on curved stage risers made of resonant wood.

Some vestiges of the historical seating layout remain, such as the expansive main floor seating with a gentle rake seating 870. A single small wall divides the seating areas and provides space for circulation. A minimal number of seats were placed under the balcony near the control room. The balcony, side, and chorus seating areas form eleven distinct terraces which hold the remainder of the 2,006 viewer-listeners. The balcony fronts are angled downwards to provide the audience with early reflections that would come from narrow sidewalls in a shoebox hall. Balcony fronts and terrace walls are covered in smooth sheets of ivory colored marble which complement the clean, crisp lines of the walls. Subtle reflections off the polished stone create the impression that the stage glows in the spotlight, focusing our attention to the center of the space.

At the time of design, studies showed that lateral reflections were particularly important in achieving good acoustics. For the seating areas flanking the stage, there are no sidewalls to provide early reflections of any kind. The issue was the source of much consternation as the acoustical success and larger viability of the vineyard layout was at stake. The design solution was to accordion fold the walls to send incoming reflections from the ceiling sideways. The wood finished walls are largely featureless save for this pleated profile in plan.

In another inspiration from the Berlin Philharmonie, nine plexiglass ensemble reflectors are suspended above stage. While one of the local orchestras chooses to perform with the panels raised all the way to the main ceiling, the general consensus is that the stage acoustics are difficult unless the reflectors are lowered to an elevation approximately 12 m above stage. Since the panels are clear and have minimal thickness, they do not impede the view of the ceiling, thereby preserving the impression of a singular, unified volume. A tented ceiling profile is created by gypsum board segments which comes to its apex directly above the front edge of the stage, another influence from Berlin Philharmonie.

Suntory Hall was the first case where we used a 1:10 scale model to investigate the acoustical properties. Before computer modeling was available, reflective paper, laser pointers, and measuring tape were used to map reflection delay times throughout the hall. Upon completion of the hall, the differences in acoustical character between various seating areas became a case study that has become the foundation of our design process.

Today, Suntory Hall is one of the most popular venues in Tokyo. While there is no resident orchestra, the hall hosts more than 400 concerts each year. There is some perception that the hall has matured as it aged, but in reality the local orchestras have improved in quality, giving the impression that the hall has changed. Since the exceptional clarity of the hall tends to leave the ensemble somewhat exposed, the orchestras must improve the quality and balance of their sound.

After 30 years, the hall was refurbished in 2016. Care was taken not to change any of the interior materials in order to preserve the visual and acoustical impression that Tokyo audiences have come to love. Notably, the stage risers were extended and the thickness of the stage floor decking was decreased to improve resonance, changes which have been well received by musicians. An updated, line array sound system was installed in the main hall for lectures and other educational programming.

Erik Bergal

View from side terrace. The ceiling is composed of overlapping plates of gypsum which meet at their highest point just in front of the stage edge. This point is much higher than the elevation of the suspended, plexiglass ensemble reflectors

View from upstage terrace. While the main floor seating is relatively flat, reminiscent of historical halls, the terraced seating is arranged in blocks with a steep rake

Suntory Hall	
Location	Tokyo, Japan
Owner	Suntory Limited
Architect	Yasui Architects
Acoustical consultant	Nagata Acoustics
	Minoru Nagata, Yasuhisa Toyota, Akira Ono
Design start	1982
Construction start	Q3 1983
Construction end	Q2 1986
Opening date	October 12, 1986
Building size	12,000 m^2
Main hall	
Seating capacity	2,006
Room volume	21,000 m^3
Surface area	6,700 m^2
Volume/seat	10.5 m^3/seat
Volume/surface area	3.1 m
Finish material	
Ceiling	Gypsum board
Ensemble reflectors	Plexiglass
Walls	Wood veneer on gypsum board on concrete
Audience floor	Wood board on concrete
Stage floor	Pine
Seat manufacturer	Kotobuki Seating
Organ builder	Rieger Orgelbau
Model scale	1:10

Table 1.1 Suntory Hall—acoustical metrics at 500 Hz

RT unoccupied	2.6 s
RT occupied	2.1 s
EDT	2.5 s
C_{80}	−1.4 dB
D_{50}	27%

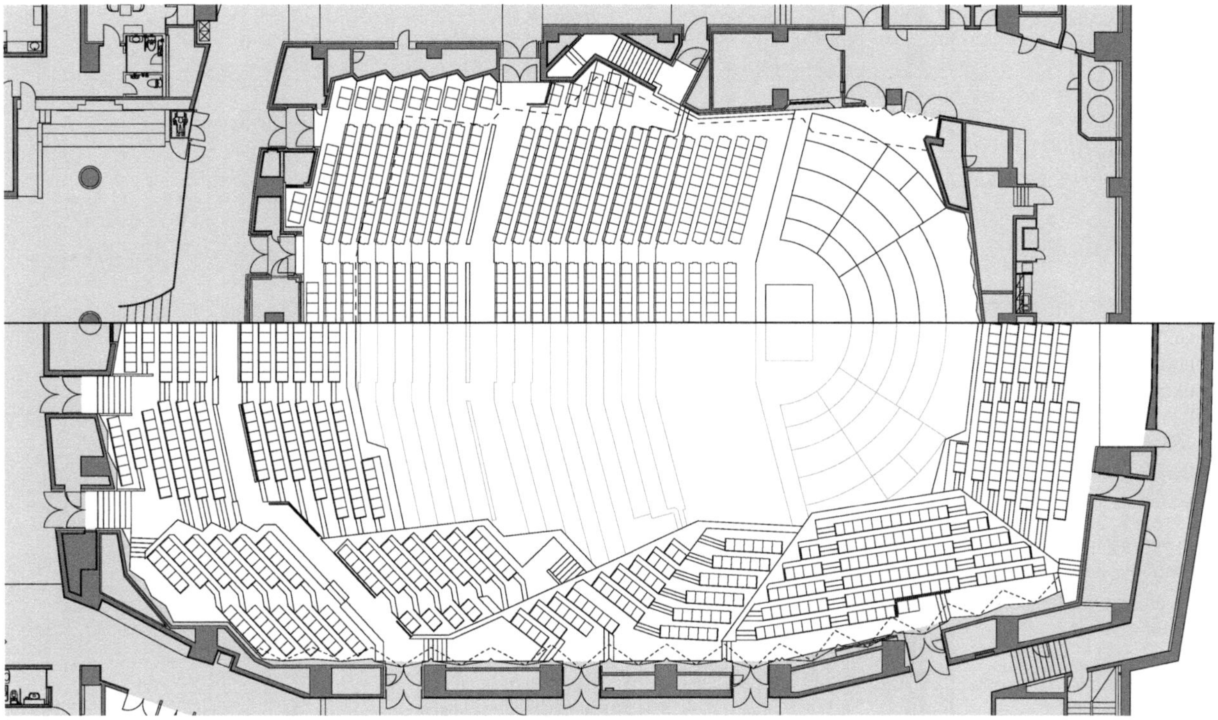

Fig. 1.1 Stage and terrace levels plan
Courtesy of Yasui Architects

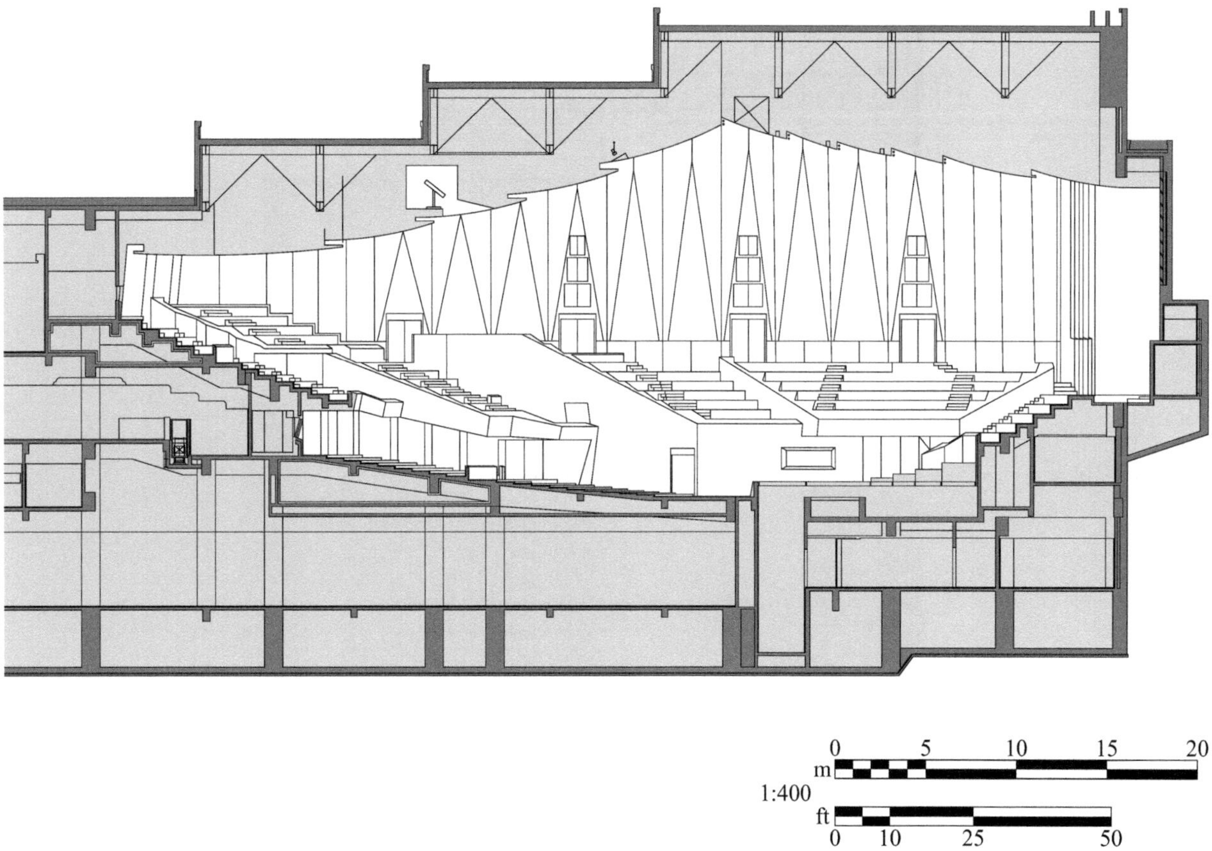

Fig. 1.2 Longitudinal section
Courtesy of Yasui Architects

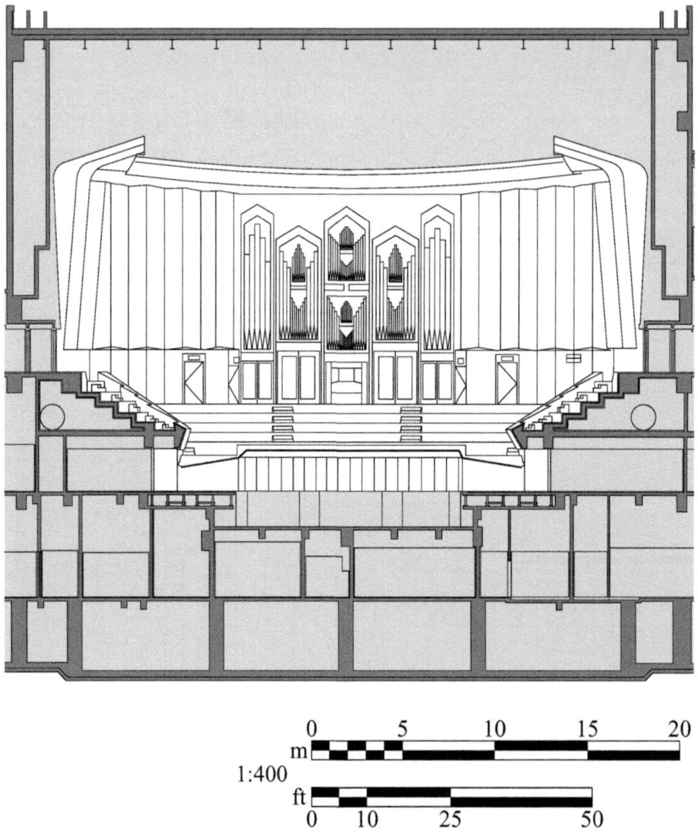

Fig. 1.3 Cross section
Courtesy of Yasui Architects

2. Kyoto Concert Hall

View from balcony. The axial ceiling panel contains an array of lights amid pyramidal acoustical diffusion. The rest of the ceiling and walls are left intentionally clean and stark in contrast

Y. Toyota et al., *Concert Halls by Nagata Acoustics*, https://doi.org/10.1007/978-3-030-42450-3_2

The Kyoto Symphony Orchestra holds a unique position as an official part of the city government, making its musicians truly public servants. The construction of a new concert hall was planned for the orchestra as part of the celebrations surrounding the 1,200th anniversary of establishing the city—then known as Heian-kyō—as the capital of Japan in 794. The new hall opened on October 12, 1995 with a 1 year delay. The venue is easily reached by subway in the north of Kyoto at the edge of a large botanical garden. Upon entering the building, a large spiral ramp leads up to the concert hall lobby.

The city of Kyoto was interested in a traditional, shoebox style concert hall and wrote their design competition accordingly. At the time, Suntory Hall in Tokyo was still under construction meaning that there were no nearby vineyard style halls to use as references. Arata Isozaki was already acknowledged as a talented architect and chose to break from the project brief as much as possible since he found the limitations constricting. Audience seating is arranged on three levels familiar from the historical shoebox layout: the main floor, a parterre, and a balcony broken up into loges. Seating areas were added around the stage. A rake was added to the main audience floor and balcony loges were rotated to improve sightlines. While the plan mostly follows a shoebox scheme, the visual impression is distinctly contemporary compared to the shoeboxes of antiquity. Seats are upholstered with iridescent blue velvet that shift from lavender, through teal, to mint depending on the viewing angle.

The most striking architectural feature of the hall is the highly diffuse ceiling texture on a band running down the center of the hall. The field of pyramids jutting down serve double duty as acoustic diffusion and disguise for the theatrical components. When viewed from stage, loudspeakers and swiveling theatrical lights can be found scattered among the chaos. On either side of this strip are stark, white surfaces that prevent the ceiling from becoming too visually cluttered. Movable panels at the corner with the walls can reveal additional theatrical lighting when necessary. A handful of large, decorative columns give the impression of supporting the ceiling.

Retractable lighting rigs. At the corner of the walls and ceiling, retractable lighting can be used for specialized performances. When not in use, the white walls are left clean. Spherical lights embedded in the central ceiling panel are used for everyday performances

Late in design, the location of the organ was offset to the stage left to further break from the usual symmetry of traditional halls. Isozaki sat in the acoustical scale model and sketched the design for the boxes that make up the organ. Aside from the visual impression, this change afforded space to add a gallery for off-stage musicians.

Kyoto Concert Hall was a turning point in the design of stage construction. In an effort to optimize the resonance achieved by the stage structure itself, several mockups were evaluated by the acoustic team and musicians. Fourteen podiums were constructed from a variety of wood types. Each podium was evaluated both while a cellist and a bassist sat on the podium and played. The results favored the Japanese cypress, Hinoki, that had been used for traditional Japanese construction for centuries. The acoustical benefits of this species was confirmed in a similar test years later in Los Angeles as part of the design of Walt Disney Concert Hall design.

The acoustics in Kyoto Concert Hall are well balanced and have a more subdued impression than other halls. Because of their sister city relationship, the inaugural concert was performed by Orchestre de Paris on October 12, 1995, although the Kyoto Symphony Orchestra did have a ceremonial concert a few days earlier. Following the inauguration, the Kyoto Symphony Orchestra began the first concert series in their new hall. As one of those first regular concerts, Mahler's Symphony No. 8 was chosen due to its commanding presence but was logistically challenging. The short rehearsal time and difficulty in securing enough professional chorus members led to a slightly muddled concert that did not allow the acoustics of the hall to shine. As always, as the orchestra has spent more time in the space, the performance quality has improved dramatically.

Erik Bergal

Balcony loges. The blocks of balcony seats are broken into groups and angled towards the stage. This improves sightlines and provides visual interest to the otherwise rectangular plan

Rehearsal seen from main floor. The early rehearsals are especially exciting and tense moments in the birth of a new hall as everyone experiences the acoustics for the first time. From centerline, the asymmetry of the organ position is highlighted. The boxes on the upper left can be used for the occasional off-stage ensemble

Kyoto Concert Hall	
Location	Kyoto, Japan
Owner	City of Kyoto
User	Kyoto Symphony Orchestra
Architect	Arata Isozaki & Associates
Acoustical consultant	Nagata Acoustics
	Yasuhisa Toyota, Keiji Oguchi
Theater consultant	Jukoh Sato, Celebration of the lights
Construction cost	JPY 16.7 billion
Design start	July 1991
Construction start	October 1992
Construction end	March 1995
Opening date	October 12, 1995
Main hall	
Seating capacity	1,833
Room volume	20,000 m^3
Surface area	6,300 m^2
Volume/seat	10.9 m^3/seat
Volume/surface area	3.2 m
Finish material	
Ceiling	Concrete, fiberglass reinforced concrete
Wall	Wood veneer on gypsum board on concrete, fiberglass reinforced concrete
Stage floor	Hinoki
Seat manufacturer	Kotobuki Seating
Organ builder	Klais Orgelbau
Model scale	1:10

Table 2.1 Kyoto Concert Hall—acoustical metrics at 500 Hz

RT unoccupied	2.2 s
RT occupied	2.0 s

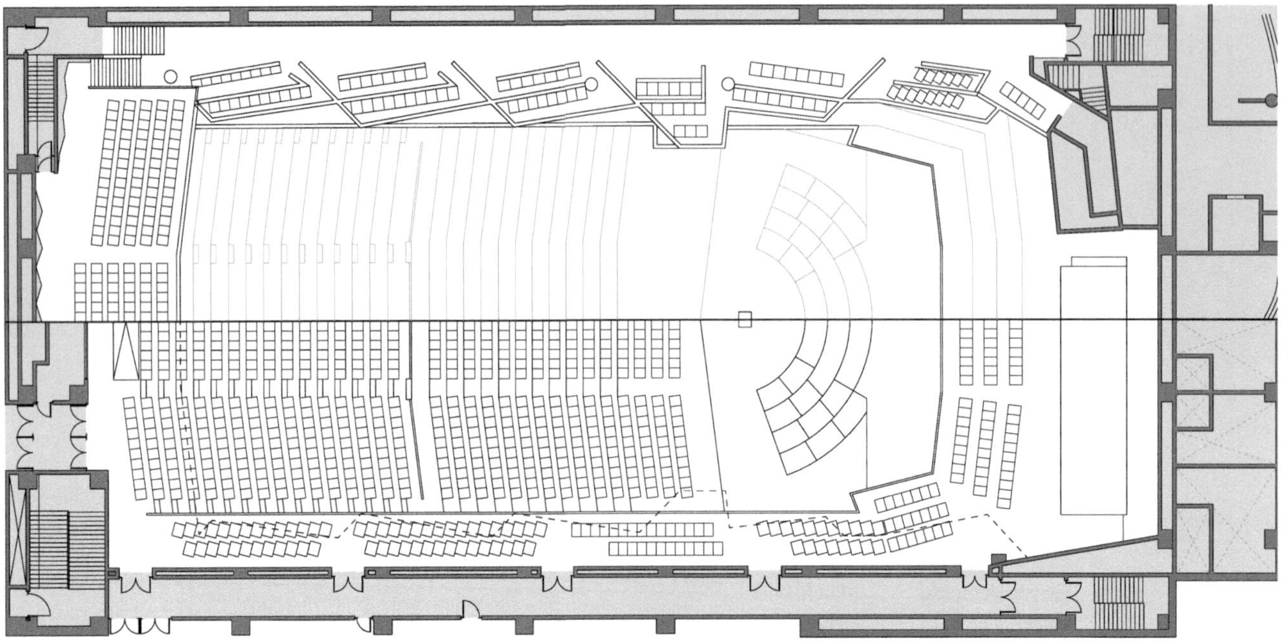

Fig. 2.1 Stage and balcony levels plan
Courtesy of Arata Isozaki & Associates

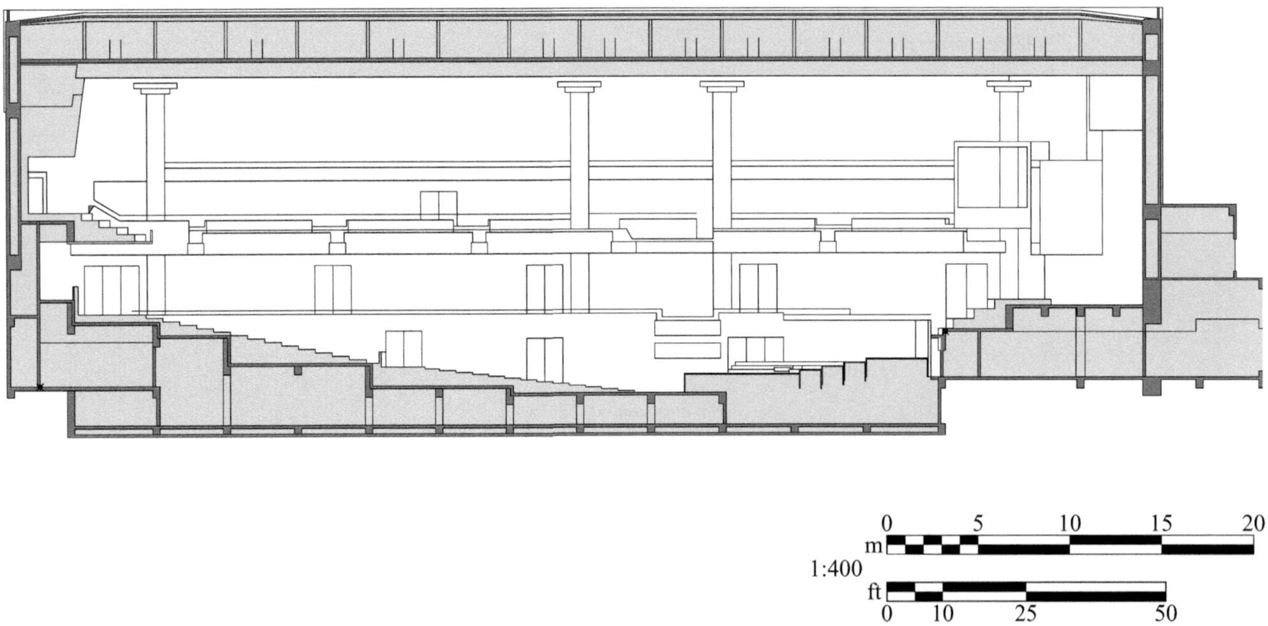

Fig. 2.2 Section
Courtesy of Arata Isozaki & Associates

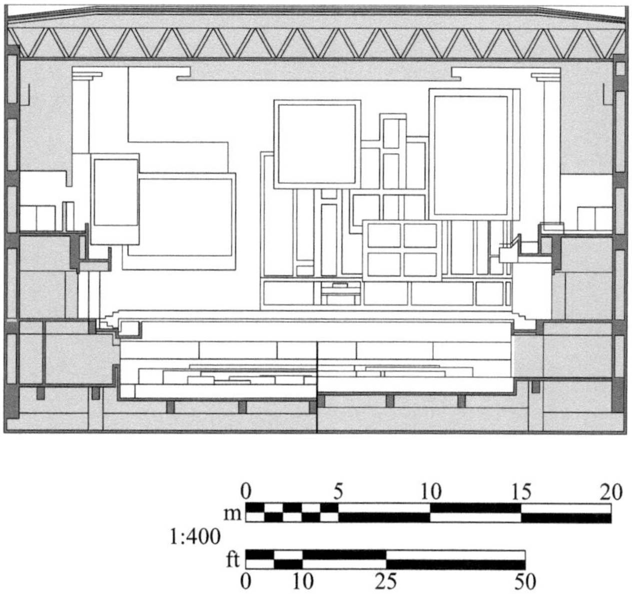

Fig. 2.3 Cross section
Courtesy of Arata Isozaki & Associates

3. Sapporo Concert Hall "Kitara"

View from rear seats. Seating terraces give a variety of views towards the stage and fellow audience viewer-listeners. The tent-like ceiling has a suspended ensemble reflector at the middle of the tallest area

Y. Toyota et al., *Concert Halls by Nagata Acoustics*, https://doi.org/10.1007/978-3-030-42450-3_3

Sapporo Concert Hall is commonly referred to by its moniker "Kitara," a portmanteau of "kita" or north (Sapporo is the largest city on the island of Hokkaido, the northern-most of Japan's four largest islands), and "kuru" or "welcome" or "please come." The name is an invitation to come north, as well as a type of ancient Greek lyre, the "kithara." The hall is the home of the Sapporo Symphony and the Pacific Music Festival.

The Sapporo Symphony was founded in 1961, and was funded jointly by the city of Sapporo, the prefecture of Hokkaido, and the regional newspaper Hokkaido Shimbun. There had been long years of discussions about how and whether to build a dedicated concert hall for the orchestra, which finally solidified in 1990 with the establishment of the Pacific Music Festival by Leonard Bernstein, in Sapporo. Bernstein sought an Asian city in which to hold the Pacific Music Festival, as a vehicle of peace through musical cooperation, and the dry, mild summertime climate of Sapporo presented the ideal location. However, it was clear that the existing facilities were serviceable for the rehearsals by the centerpiece orchestra composed of young musicians from around the world, especially Pacific Rim countries, but did not meet the standards expected for performances by major international ensembles. The Pacific Music Festival was instrumental in encouraging the city to start the new project for the Sapporo Concert Hall.

In 1992, the city organized an architectural competition for the project, which would eventually include the centerpiece 2,008-seat hall, a small chamber music hall, and some office and rehearsal spaces. Six large architectural offices in Japan were invited to participate in the competition, resulting in the choice of Hokkaido Engineering Consultant Company (now known as Docon) as the designer. Nagata Acoustics was chosen as the acoustical consultant for the project, on the basis of a long-standing and fruitful relationship with the city.

The concert hall is designed as a typical vineyard-style concert hall, with the stage near the center of the room, and viewer-listeners surrounding it on all sides. This was one of the first large halls to be designed using our computer simulation technique. The hall is characterized by a broad and comfortable floor plan, eschewing balconies and overhangs. It has a

View from the side terrace. The audience seats are steeply raked

tent-like ceiling shape and a large ensemble reflector above the stage and front audience, which together with the absence of overhangs are reminiscent of the Berlin Philharmonie. The division of the audience into multiple small blocks also gives a good impression of intimacy, without the sensation of being in a large, anonymizing space.

The broad footprint of the hall presented a particular challenge in the management of very early reflections. The room shape is quite wide, when both the overall room shape and each of the individual seating blocks are considered. Rather than relying on side walls for the very early reflections, as is traditionally expected, very early reflections are created by surfaces behind the audience. Often, reflections created by surfaces behind individual viewer-listeners are assumed to be detrimental or dangerous. But, if the possibility of using surfaces on all sides of the viewer-listener for the provision of very early reflections was allowed, the flexibility in designing the hall is greatly increased. This was the first design where we made this decision, which was ultimately very successful and effective. 1:10 scale model tests were also used to confirm the effectiveness.

The effectiveness of these small walls among the audience having been verified, the idea was incorporated into other projects. Most notably, Walt Disney Concert Hall (which had experienced a prolonged work pause during the majority of the design and construction process of Sapporo Concert Hall) was adjusted to incorporate the small walls. This cross-pollination between the projects is noteworthy since Walt Disney Concert Hall had basically been designed before the project of Sapporo Concert Hall had been launched by the city, and the Sapporo architects had visited Gehry's studio in Los Angeles and were introduced to the design. Even though few similarities are immediately visible between the two halls, an important acoustical lesson which was learned during the design and scale model test of Walt Disney Concert Hall was applied to this hall: the soffits at the side of the stage for acoustical feedback. Between those two elements, a true cross-pollination of acoustical ideas can be said to have occurred.

Daniel Beckmann

View along centerline. The hall is symmetrical. Soffits can be seen along the entire length of the side walls, behind all audience seats

Sapporo Concert Hall "Kitara"	
Location	Sapporo, Japan
Owner	City of Sapporo
User	Sapporo Symphony Orchestra
	Pacific Music Festival
Architect	Hokkaido Engineering Consultant Co., Ltd. (Docon)
Acoustical consultant	Nagata Acoustics
	Yasuhisa Toyota, Ayumi Ozawa
Construction cost	JPY 16.5 billion
Design start	1992
Construction start	1994
Construction end	1996
Opening date	July 4, 1997
Building size	20,746 m^2
Seating capacity	2,008
Room volume	28,800 m^3
Surface area	8,200 m^2
Volume/seat	14.3 m^3/seat
Volume/surface area	3.5 m
Finish material	
Ceiling	Precast concrete
Walls	Gypsum board
Audience floor	Wood board on concrete
Stage Floor	Hinoki
Seat manufacturer	Kotobuki Seating
Organ builder	Alfred Kern et Fils Manufacture d'Orgues
Model scale	1:10

Table 3.1 Sapporo Concert Hall "Kitara"—acoustical metrics at 500 Hz

RT unoccupied	2.2 s
RT occupied	2.0 s
EDT	2.0 s
C_{80}	0.8 dB
D_{50}	38%
G	3 dB

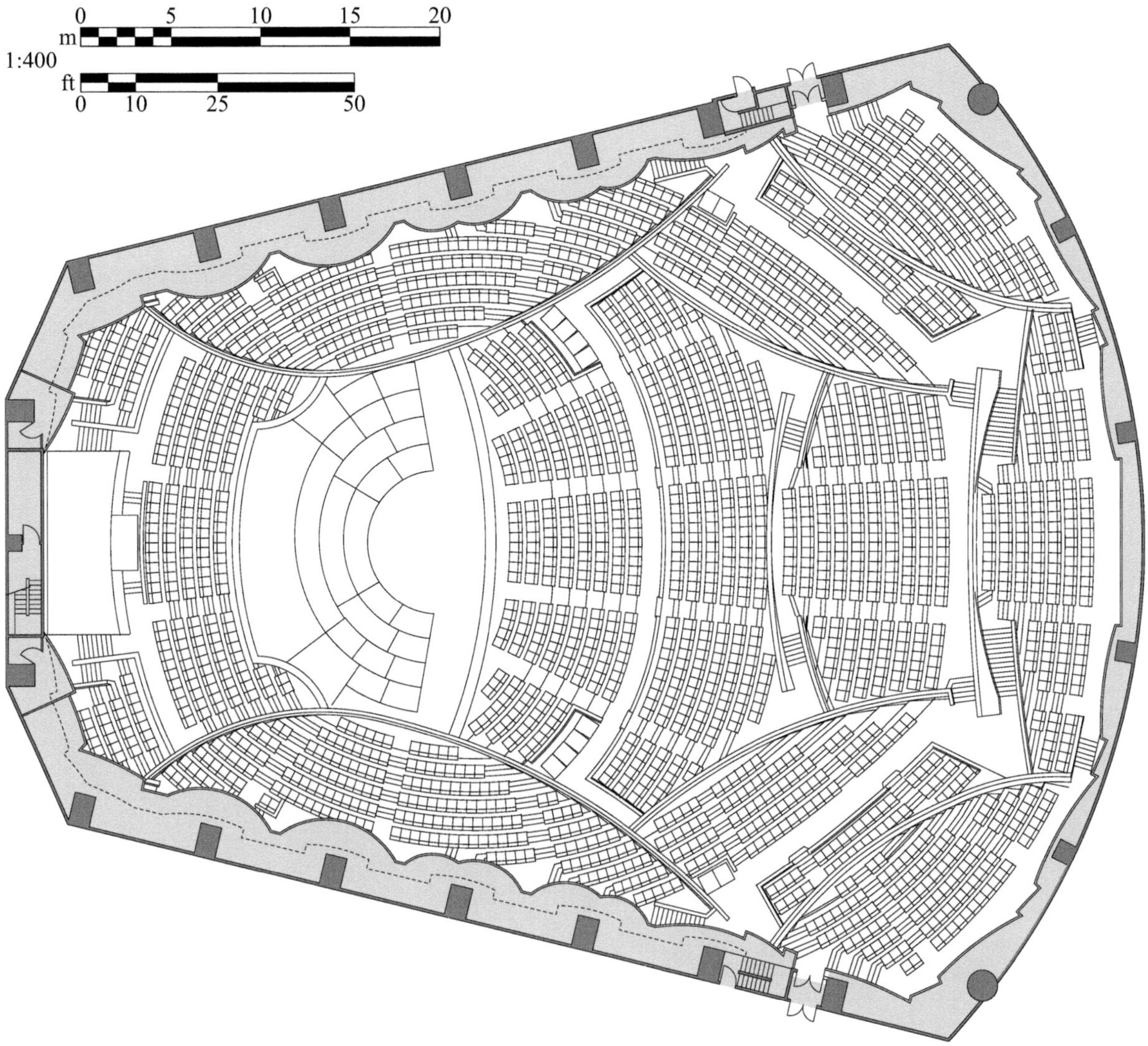

Fig. 3.1 Plan
Courtesy of Hokkaido Engineering Consultant Co., Ltd.

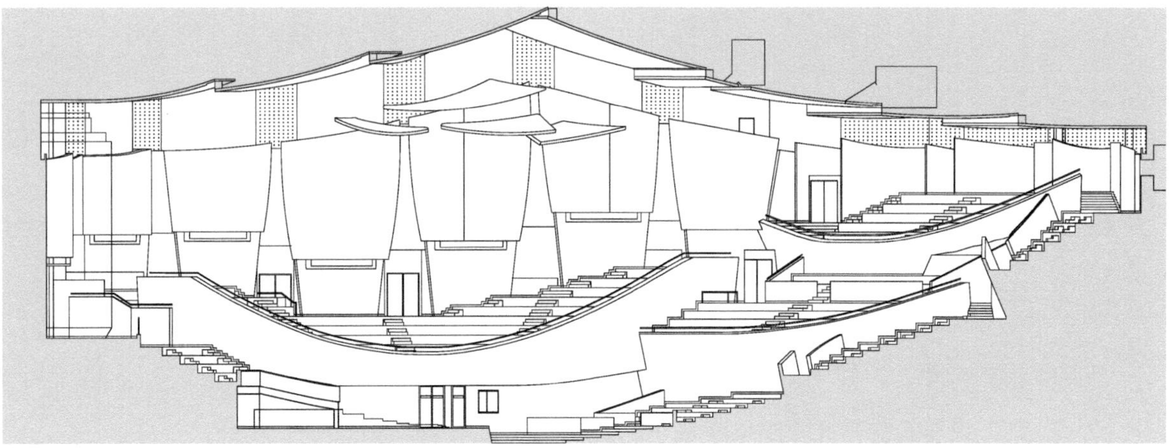

Fig. 3.2 Section
Courtesy of Hokkaido Engineering Consultant Co., Ltd.

4. Richard B. Fisher Center for the Performing Arts

View from main floor. For orchestral performances, a movable shell encloses the stage. This provides reflecting surfaces for the orchestra and unifies the auditorium and stage in a single, continuous volume

Y. Toyota et al., *Concert Halls by Nagata Acoustics*, https://doi.org/10.1007/978-3-030-42450-3_4

Bard College is a small college located in Annandale-on-Hudson, 150 km upstream from New York City. The school focuses on liberal arts and includes a music conservatory. Bard College is host to a music festival every summer starting in 1990. Each summer, the festival is dedicated to focus on a single composer with musical programming, lectures, and film to explore both their music and historical context in great depth. The festival now has a home in the Fisher Center, a considerable upgrade to the tent that originally housed the festival.

The vision for the new building came from Dr. Leon Botstein. Botstein has been president of Bard College since 1975 and music director of the American Symphony Orchestra since 1991. Under the leadership of Botstein, Bard College has grown to include several buildings by internationally renowned architects.

The Fisher Center marks the first completed collaboration between Nagata Acoustics and architect Frank Gehry, opening in April 2003, just a few months ahead of Walt Disney Concert Hall in Los Angeles. The two projects are further connected by their facades, both clad in undulating stainless-steel panels. Here, the panels effectively hide the stage tower from disrupting the pastoral setting. In addition to the main theater, the building includes a black box theater and several smaller dance studios.

Exterior. The arts center is situated in the middle of a grassy field, clad in warped surfaces unmistakably by Gehry

The hall seats 900 viewer-listeners distributed over a main floor and two tiered balconies. While small, this is enough to seat more than a third of the student body. The architectural design of the interior is defined by a contrast between unfinished concrete and amber wood. Monolithic concrete walls are embellished with wooden ribbons which add visual interest as well as acoustic diffusion. Cylindrical segments make up the lofty ceiling profile and hide the technical lighting. Minimal lighting rigs hang from the underside of the balconies to supplement the ceiling lighting.

Shell stored backstage. When not in use, the shell is stored in an annex off of the stage. The individual segments of the shell telescope downwards to aid in rolling them away

While the space needs to accommodate theatrical performances, orchestral performance acoustics could not suffer. The solution is a large orchestra shell that closes off the stage tower during concert performances. Due to the construction budget, a fully mechanical system was unfeasible; however, sufficient mass was required to achieve good acoustics. Ease of use is sacrificed for acoustic quality as the process of assembling or dismantling the shell is time and labor intensive, requiring nearly 40 person-hours of work. To be stowed, the walls of the shell are first broken into 13 towers. Next, the top half of each tower is lowered mechanically, then the entire tower is wheeled to a backstage storage alcove. In order to provide appropriate airflow for the musicians, the shell walls have vertical slits where air enters the stage enclosure.

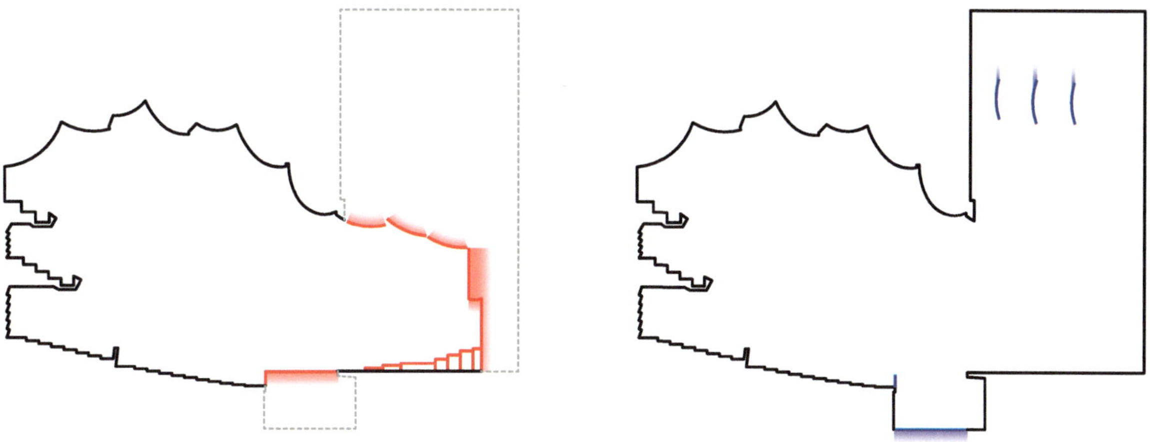

Fig. 4.1 Scope of spatial transformations

Several key design elements, such as the hexagonal footprint, are inherited from Tokyo Bunka Kaikan, a larger auditorium in Ueno Park, Tokyo. One of the most important aspects is the movable pit floor rises to stage level extending the downstage area. This allows as many musicians as possible to move in front of the proscenium. The increased ceiling height and volume shared with the audience is particularly beneficial for the strings sections.

For the opening, Botstein chose Mahler's Symphony No. 3, a huge ensemble for such a small hall. However, under Botstein's direction, the performance was successful. This was a particular relief since several critics and musicians from Los Angeles attended the opening concert with curious anticipation of the upcoming Disney Hall opening.

Erik Bergal

View from stage. Stepped, descending balconies reference the loges found in historical opera houses, but in a distinctly modern style

View from side terrace. Seven convex, prismatic clouds form the ceiling in order to disperse acoustic reflections. Wood ribbons adorn the exposed concrete walls to add texture for acoustic diffusion

Richard B. Fisher Center for the Performing Arts	
Location	Bard College, Annandale-on-Hudson New York, USA
Owner	Bard College
Architect	Frank O. Gehry (Gehry Partners, LLP)
Acoustical consultant	Nagata Acoustics (room acoustics) *Yasuhisa Toyota* Robert F. Mahoney & Associates (isolation and noise control)
Theater consultant	Theatre Projects Consultants
Construction cost	USD 60 million
Design start	April 1997
Construction start	August 1999
Construction end	Q4 2002
Opening date	April 25, 2003
Building size	10,000 m^2
	Sosnoff Theater
Seating capacity	900
Room volume	9,040 m^3
Surface area	3,740 m^2
Volume/seat	10.0 m^3/seat
Volume/surface area	2.4 m
Finish materials	
Ceiling	Plywood
Floor, walls	Concrete
Seat manufacturer	Figueras Seating Solutions

Table 4.1 Richard B. Fisher Center for the Performing Arts—acoustical metrics at 500 Hz

	Concert configuration	Opera configuration
RT unoccupied	1.9 s	1.3 s
RT occupied	1.7 s	1.1 s
EDT		1.1 s
C_{80}		3.6 dB
D_{50}		51%

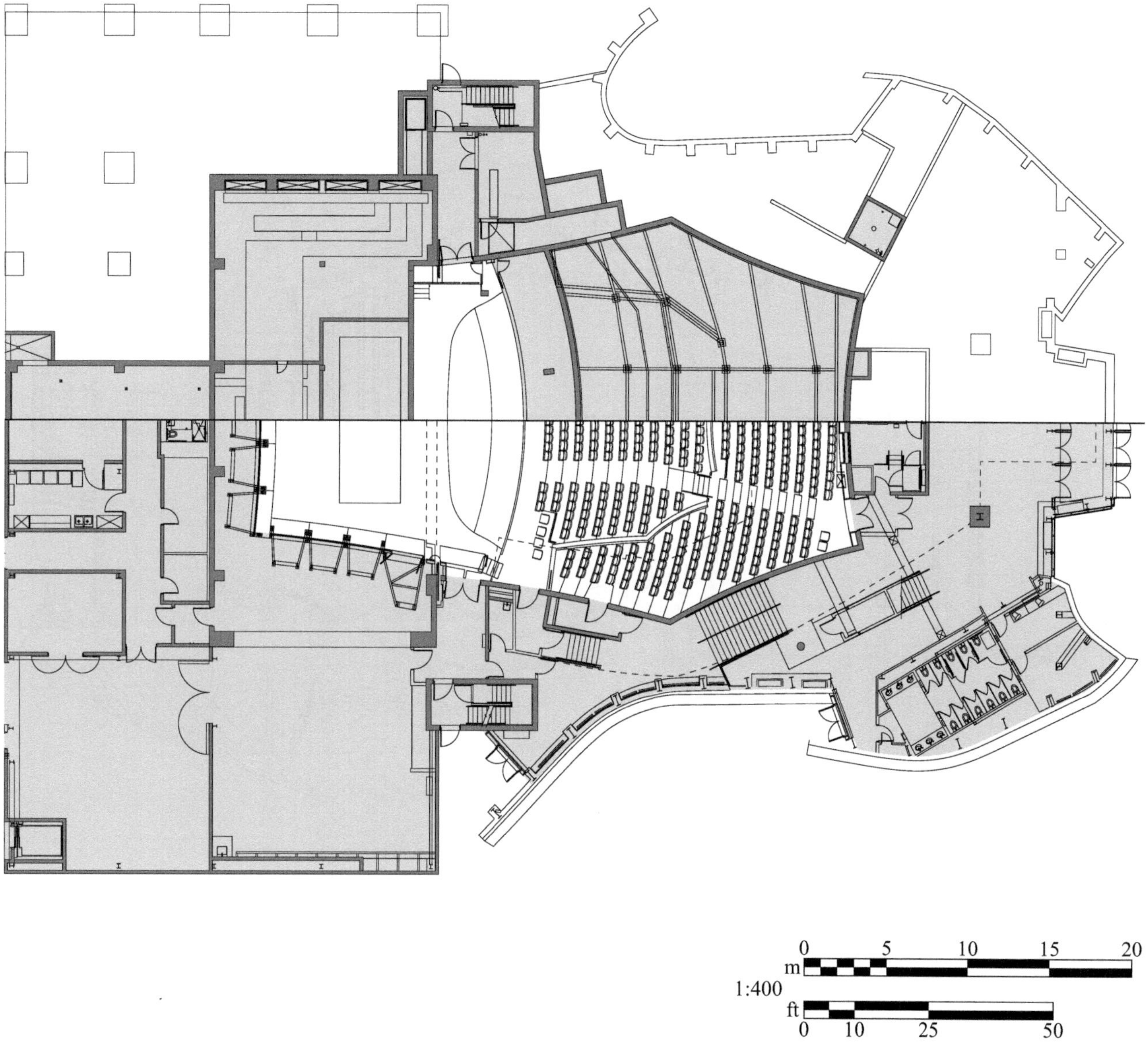

Fig. 4.2 Pit and stage levels plan
Courtesy of Gehry Partners, LLP

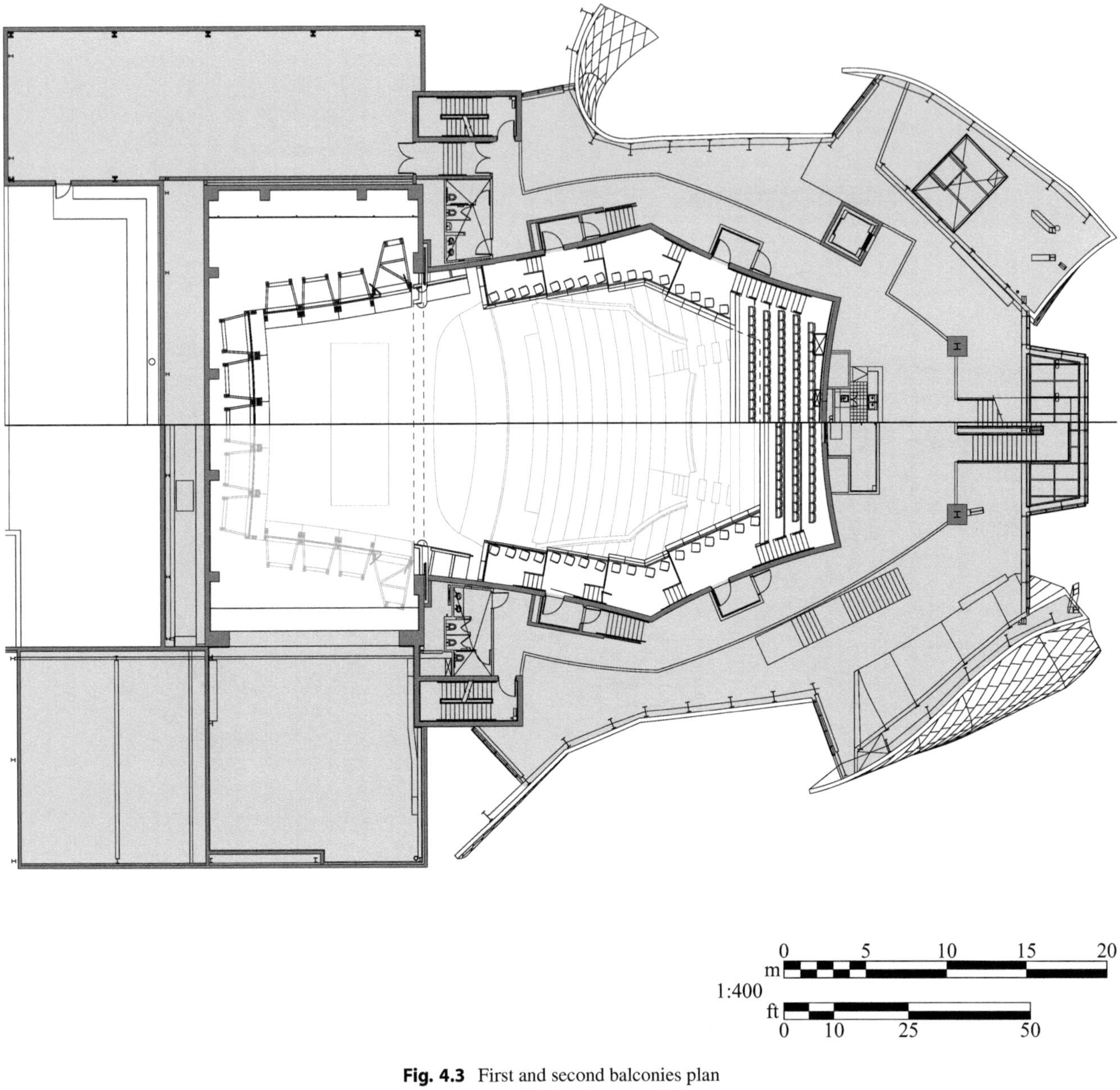

Fig. 4.3 First and second balconies plan
Courtesy of Gehry Partners, LLP

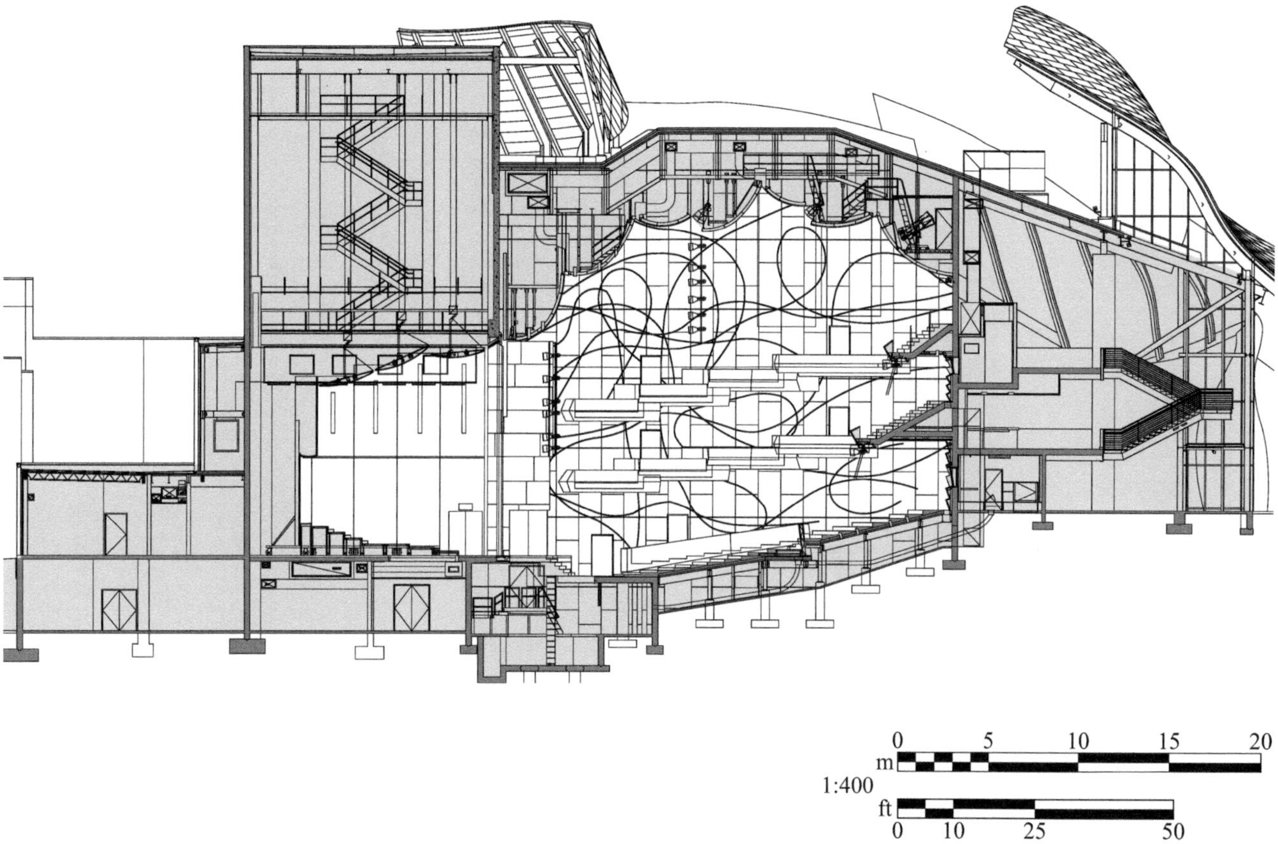

Fig. 4.4 Section
Courtesy of Gehry Partners, LLP

5. Walt Disney Concert Hall

View from balcony. The audience ascends steeply from stage. Even at the higher levels, one feels very close to the stage, an impressive feat for a hall over 2000 seats. Curved, wood, organ pipes encase the unique instrument. The ceiling curves away from its lowest point above stage

Y. Toyota et al., *Concert Halls by Nagata Acoustics*, https://doi.org/10.1007/978-3-030-42450-3_5

Following a generous donation from the widowed Lillian Disney in 1987, planning started to create a new performing arts center in honor of the late Walt Disney. Originally, the project included a 2500-seat concert hall, 1000-seat chamber hall, and rehearsal and administrative space for the Los Angeles Philharmonic and Los Angeles Master Chorale. Part of the urban renewal of Bunker Hill, the concert hall would expand The Music Center of Los Angeles County, which includes the Ahmanson Theatre, Mark Taper Forum, and Dorothy Chandler Pavilion, the latter of which was the home of the Los Angeles Philharmonic since its completion in 1964.

The project was spearheaded by Ernest Fleischmann, the Executive Director of the Los Angeles Philharmonic. Whereas Dorothy Chandler Pavilion is a multipurpose proscenium theater accommodating opera, ballet, and musicals, Fleischmann's vision was for a concert hall dedicated to classical music without compromise. The project brief emphasized the relationship between the building, the orchestra, and the public as well as acoustic intimacy:

> It must be a unique building but it also must be a building of humanism, a quality based on the purity of functionalism and the principle of expressing scale as a relationship between man and space, as an ideal of freedom and comfort. The building must be warm and intimate, not cold or institutional in nature. The design must incorporate a feeling of openness, inviting the public to participate.

Four shortlisted architects submitted detailed designs: Gottfried Böhm, Hans Hollein, James Stirling and Michael Wilford, and Frank Gehry. Simultaneously, the planning committee began investigating potential acousticians and visited Tokyo Bunka Kaikan and Suntory Hall.

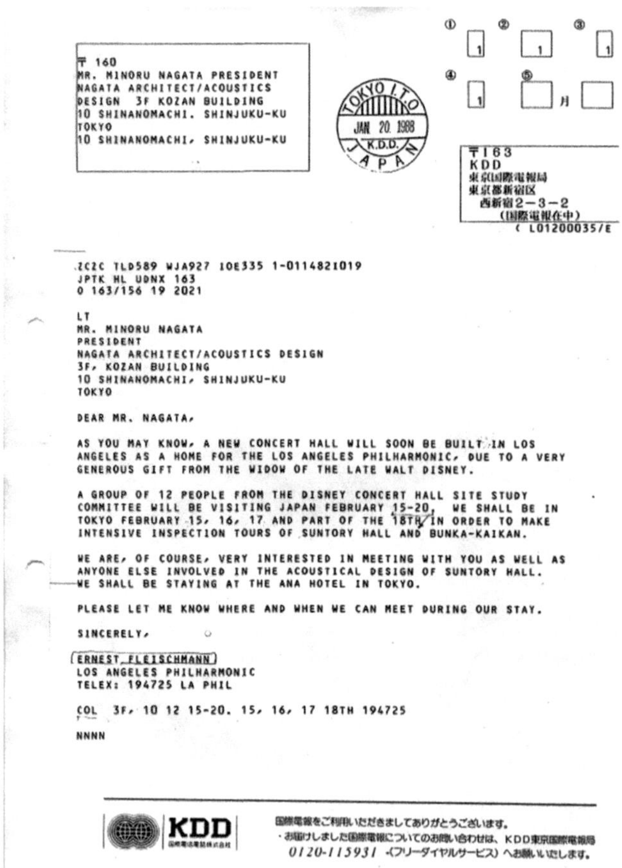

〒 160
MR. MINORU NAGATA PRESIDENT
NAGATA ARCHITECT/ACOUSTICS
DESIGN 3F KOZAN BUILDING
10 SHINANOMACHI. SHINJUKU-KU
TOKYO
10 SHINANOMACHI, SHINJUKU-KU

TOKYO I.T.O. JAN 20 1988 K.D.D. JAPAN

〒163
KDD
東京国際電報局
東京都新宿区
西新宿2−3−2
(国際電報在中)
(L01200035/E

ZCZC TLD589 WJA927 10E335 1-0114821019
JPTK HL UDNX 163
0 163/156 19 2021

LT
MR. MINORU NAGATA
PRESIDENT
NAGATA ARCHITECT/ACOUSTICS DESIGN
3F, KOZAN BUILDING
10 SHINANOMACHI, SHINJUKU-KU
TOKYO

DEAR MR. NAGATA,

AS YOU MAY KNOW, A NEW CONCERT HALL WILL SOON BE BUILT IN LOS ANGELES AS A HOME FOR THE LOS ANGELES PHILHARMONIC, DUE TO A VERY GENEROUS GIFT FROM THE WIDOW OF THE LATE WALT DISNEY.

A GROUP OF 12 PEOPLE FROM THE DISNEY CONCERT HALL SITE STUDY COMMITTEE WILL BE VISITING JAPAN FEBRUARY 15-20. WE SHALL BE IN TOKYO FEBRUARY 15, 16, 17 AND PART OF THE 18TH IN ORDER TO MAKE INTENSIVE INSPECTION TOURS OF SUNTORY HALL AND BUNKA-KAIKAN.

WE ARE, OF COURSE, VERY INTERESTED IN MEETING WITH YOU AS WELL AS ANYONE ELSE INVOLVED IN THE ACOUSTICAL DESIGN OF SUNTORY HALL. WE SHALL BE STAYING AT THE ANA HOTEL IN TOKYO.

PLEASE LET ME KNOW WHERE AND WHEN WE CAN MEET DURING OUR STAY.

SINCERELY,

ERNEST FLEISCHMANN
LOS ANGELES PHILHARMONIC
TELEX: 194725 LA PHIL

COL 3F, 10 12 15-20. 15, 16, 17 18TH 194725

NNNN

KDD
国際電報をご利用いただきましてありがとうございます。
・お届けしました国際電報についてのお問い合わせは、KDD東京国際電報局 0120-115931 (フリーダイヤルサービス) へお願いいたします。

Initial contact from Los Angeles Philharmonic

Gehry was announced as the project architect early in December 1988, with Nagata Acoustics officially selected in April 1989.

Approaching the hall we are first struck by the now iconic facade made up of smoothly contorted extrusions sheathed in stainless steel. Passing through the glass doors, we enter a lobby distributed over several levels which both assist in ticket control and in breaking up the crowds to give the impression of a small concert experience. Architectural details preview the interior finishing, such as the carpet pattern matching the audience seat upholstery and lighting and air outlets which allude to oversized organ pipes.

To replace the chamber hall which was cut from the building program, a large alcove was added off the main lobby entrance. This area is usually populated with rows of movable seats where concert goers are invited to free lectures introducing the musical and historical context for the concert they are about to enjoy. Since these lectures and other events often use a mix of acoustic and amplified music and speech, the acoustic conditions were studied in depth. Perforated panels hide acoustic diffusion which prevent problematic focusing from the complex, concave shape.

The hall itself is conceptually thought to be composed of an outer "box" and inner "boat." In plan, the box is rectangular, with its white concrete walls sloping out 6° on the sides and rear and by 17° on the upstage face. As we walk through the sound lock into the circulation space, this tilt directs our gaze upwards where we notice that we are already within the hall volume. A skylight in each of the four, lofted corners of this interstitial space let indirect light into the hall, both adding acoustic volume and creating the impression of an open expanse in which the "boat" is nestled. Passing through an open entryway, we enter the wood clad inner structure.

While the Los Angeles Philharmonic was clear that the orchestra "will not be separated from the audience on a proscenium stage … contributing to the sense of intimacy," the seating layout took several months to settle. Gehry's competition design was far too wide even for a surround style hall so Nagata Acoustics recommended to reduce the width, although this led to the task of relocating the lost seats. Thirty small models of various seating configurations were created for discussion.

Standing at the stage edge looking back towards the rear of the hall, the steep rake of the audience is apparent: there is a wall of seats extending up. Impressive for a hall of 2265 seats, the farthest seat is less than 36 m from the stage edge looking down at an angle of 30°. The effect is to create a giant bowl with the stage at its base. The first balcony wraps around the orchestra level seating creating side walls which emulate the width of a shoebox hall. A second balcony is backed by large windows which let in natural light during early concerts.

View from upstage terrace. From the upstage seats, the verticality of the seating arrangement is apparent. Small walls and narrow side walls provide acoustic reflections for the main floor

On either side of the stage, doubly curved walls extend the length of the hall. A terrace and balcony are inset into these bulkheads. Similarly, the upstage seating consists of six rows of chorus benches below slim balconies flanking the organ. Sitting between the upstage audience balconies, an explosion of curved, wooden pipes hides the rows of metal pipes that make up the 72 stops of the singular organ, itself a collaboration between Gehry and Rosales Organ Builders located in Los Angeles. Despite their unconventional form, all the pipes are functional and indeed, tuning slides can be seen from the balconies. In order to provide the best acoustic environment, the organ is fully situated within the acoustic volume of the "box" without the extruded form that often forms the organ box. The manual is located at the base of the organ adjacent to the last rows of the chorus seating area, but can also be played from a movable manual on stage via electronic umbilical.

The ceiling segments are made up of layers strips of Douglas fir backfilled with shotcrete to increase the surface density, in some places reaching 170 kg/m^2. Despite the massive construction, the ceiling shape is light and dynamic, with billowing, doubly curved surfaces giving the impression of the underside of a giant fish. In shape, the ceiling is made up of three convex curves in transverse section: a central underbelly and two fins. The center portion is further broken into another three convex curves to ensure the area above stage is of somewhat uniform height but never parallel to the floor. Starting with the lowest point 15.5 m above the stage, the ceiling is split into steps in order to follow the steep profile of the seats. Theatrical lighting and air supply are hidden in the vertical jumps between steps. The two ventral fins slope down to meet the walls, softening the junctions with the walls.

Despite a fully developed design and completed 1:10 scale model test, work came to a halt in late October 1994 due to lack of funding leaving only an expensive parking structure in the middle of LA. During this time, construction of Sapporo Kitara was completed. Due to the similarities between the halls, Sapporo can be seen as the sibling of Walt Disney Concert Hall and a one-to-one test. By the time work restarted in 1997, several design points were incorporated from Sapporo. First, two reflecting walls were added to the front audience seating with angles designed to reflect sound to the audience seated directly in front of them. Similar angles were introduced on either side of the stage to improve stage acoustic conditions. Sapporo also proved that as long as there is sufficient clarity, a rich reverberance is welcome, so acoustically absorptive carpet in the audience balconies was removed from the design of Disney.

Rehearsals began in the hall on June 30, 2003 under the command of Esa-Pekka Salonen. Walt Disney Concert Hall opened on October 24 later that year, 15 years after the initial donation.

Several years after the opening, variable acoustical curtains were added to the four skylight shafts for use in amplified performances. These curtains had been proposed in the design process but were quickly cut due to Fleischmann's unflinching vision that the hall be an immutable space for acoustic performances. In one correspondence with the project managers in 1992, he quipped that "there is no need to provide for costly adjustments to accommodate anything other than symphony and similar concerts, and recitals." Further discussion of variable acoustics was summarily dropped.

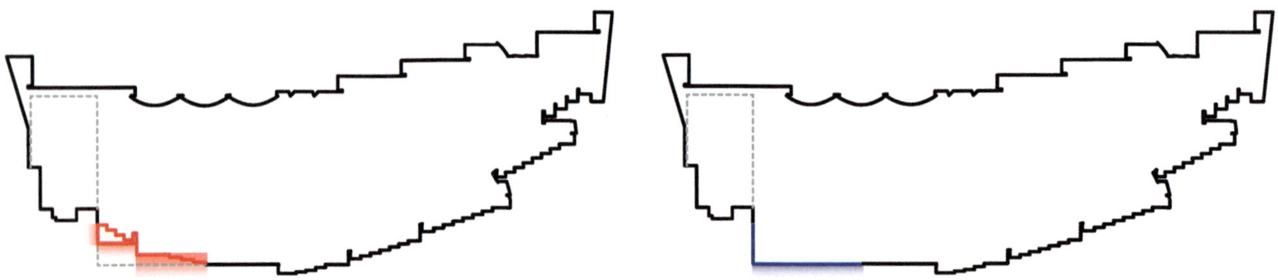

Fig. 5.1 Scope of spatial transformations

The stage is comprised of twelve mechanical risers for flexible orchestra arrangements. The stage depth of 14.4 m is just barely enough for the Los Angeles Philharmonic. For performances which require more elaborate staging, the first four rows of chorus seating can be lowered to stage level increasing the centerline stage depth to 18.3 m. This condition is used for semi-staged opera or the occasional pop or crossover concert. The Philharmonic is keen to play with experimental performances including intricate projection mapping onto the organ, and semi-staged opera sets designed by architects such as Jean Nouvel, Zaha Hadid, and Gehry himself.

Erik Bergal

Exterior. The curved, stainless steel surfaces of the facade have become an iconic landmark of downtown Los Angeles

View from side terrace before a concert

View from second balcony

Preconcert area. The lobby features a preconcert area which can accommodate lectures and banquets. Skylights make this space bright and lively before afternoon concerts. Acoustic shaping behind perforated panels prevents focusing from the concave wall surfaces

	Walt Disney Concert Hall
Location	Los Angeles
	California, USA
Owner	Los Angeles County
User	Los Angeles Philharmonic
Architect	Gehry Partners, LLP
Acoustical consultant	Nagata Acoustics (room acoustics)
	Minoru Nagata, Yasuhisa Toyota, Suzuyo Yokose
	Charles M. Salter Associates (building acoustics)
Theater consultant	Theatre Projects Consultants
Construction cost	USD 274 million
Design start	April 1989
Construction start	December 8, 1999
Construction end	Q2 2003
Opening date	October 24, 2003
Building size	27,200 m^2
Seating capacity	2,265
Room volume	30,600 m^3
Surface area	8,360 m^2
Volume/seat	13.5 m^3/seat
Volume/surface area	3.7 m
Finish material	
Ceiling	Douglas fir backfilled with shotcrete
Walls	Douglas fir backfilled with shotcrete
Audience floor	Oak
Stage floor	Alaskan Yellow Cedar
Seat manufacturer	Poltrona Frau
Organ builder	Rosales Organ Builders
Model scale	1:10

Table 5.1 Walt Disney Concert Hall—acoustical metrics at 500 Hz

RT unoccupied	2.2 s
RT occupied	2.0 s
EDT	1.9 s
C_{80}	1.2 dB
D_{50}	45%
G	13 dB

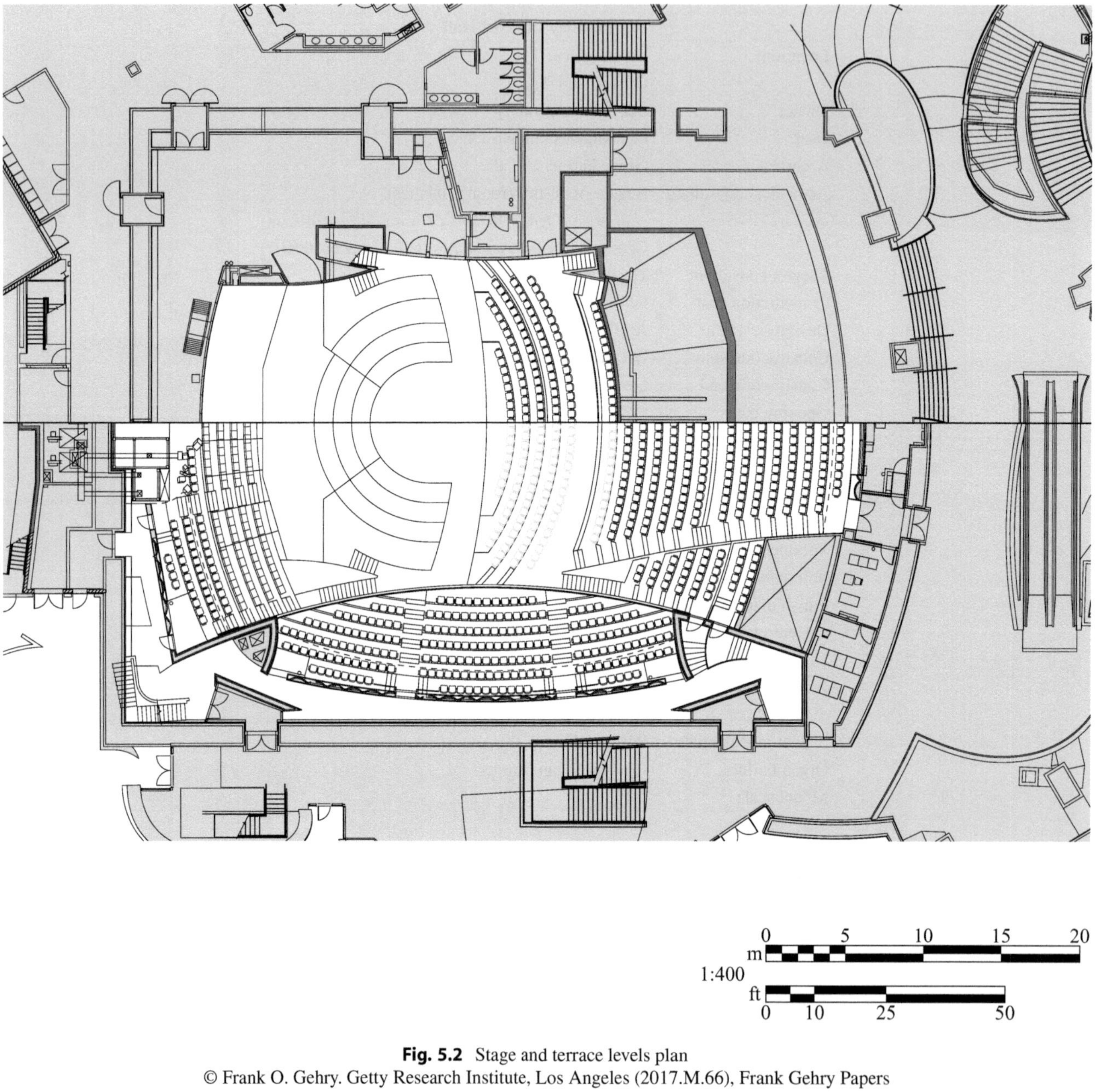

Fig. 5.2 Stage and terrace levels plan

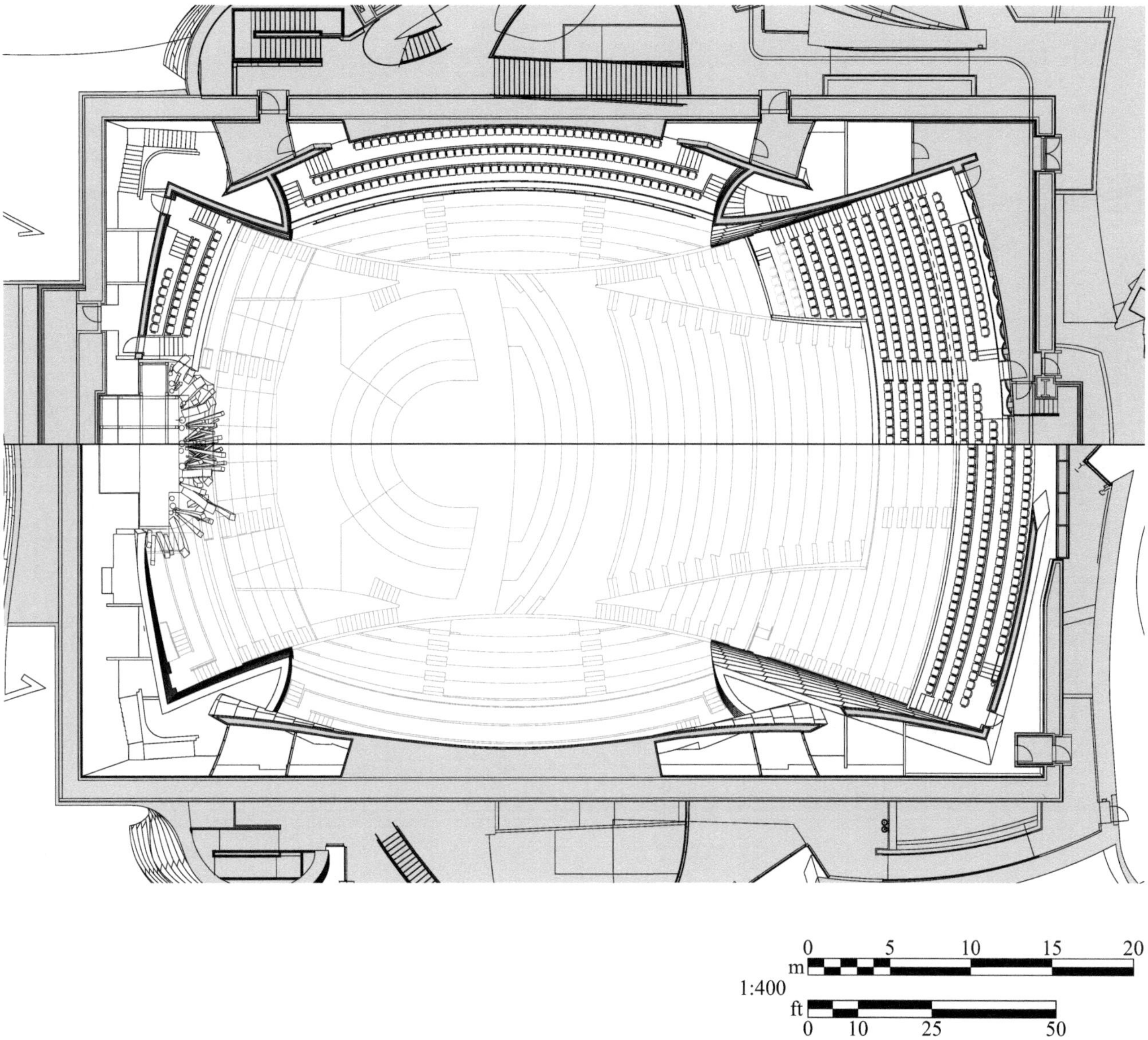

Fig. 5.3 First and second balconies plan
© Frank O. Gehry. Getty Research Institute, Los Angeles (2017.M.66), Frank Gehry Papers

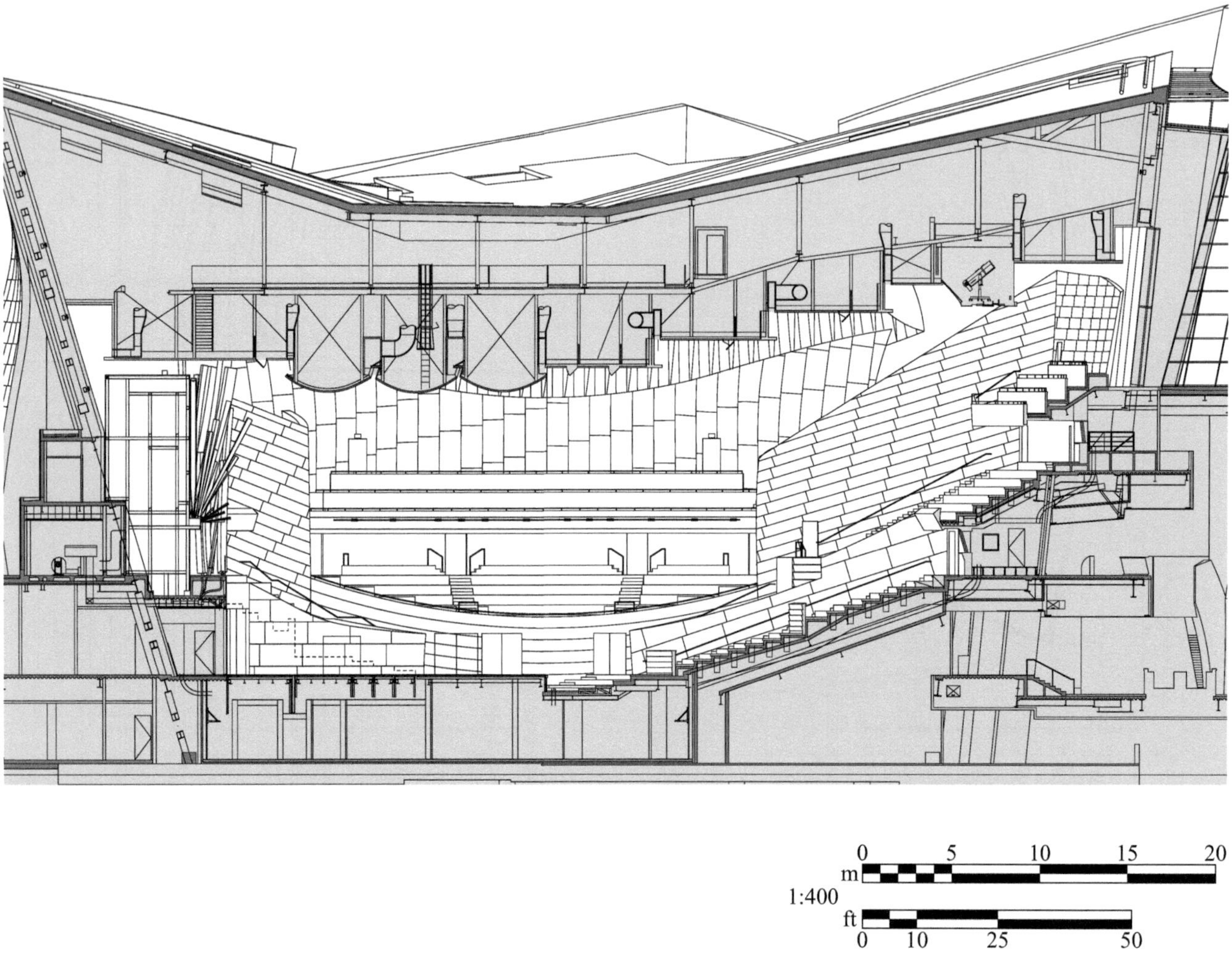

Fig. 5.4 Longitudinal section

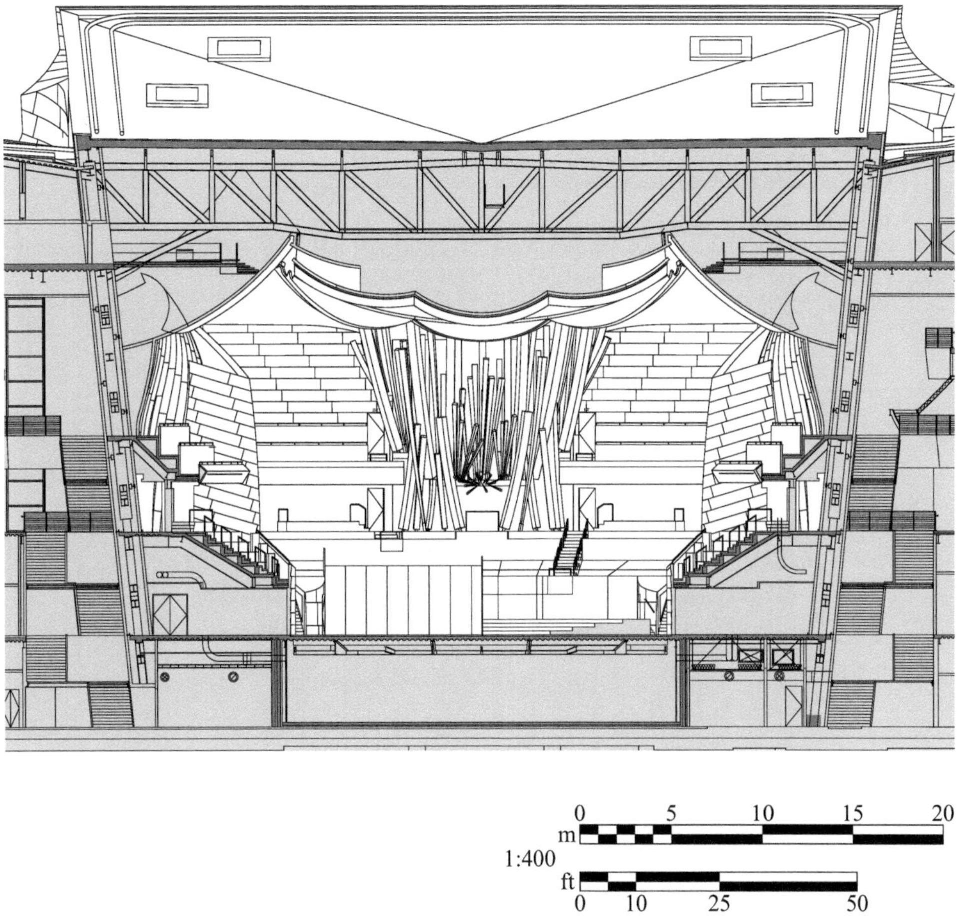

Fig. 5.5 Cross section

6. Muza Kawasaki Symphony Hall

View from rear seats. The white balcony fronts and terrace walls show the spiral in the design

Y. Toyota et al., *Concert Halls by Nagata Acoustics*, https://doi.org/10.1007/978-3-030-42450-3_6

Kawasaki is a major city in Japan, with more than one million inhabitants. However, it is often overshadowed by its two larger neighbors, the capital Tokyo to the north and the major port and industry city of Yokohama to the south. In the early 1990s, Kawasaki began a program to improve the city's image and to formulate a more positive identity within the greater Tokyo Metropolitan Area. One field where the city recognized support that could differentiate the city is music. From 1994 to 1998, the city investigated building a concert hall with more than two thousand seats.

A major aspect of the investigation was the plan for programming the hall after completion. Frequently in Japan, the considerations of programming did not begin until after the completion of the facility, often with lackluster results and the rare shuttering of a facility. The leaders of Kawasaki knew that to support their vision of drawing attention and visitors to the city, a strong concert program would be necessary. The city resolved to work with an existing orchestra in Tokyo, and reach an agreement whereby the orchestra would "reside" at the new facility, both for rehearsals and the majority of performances. At the time, there were more than ten professional symphony orchestras in Tokyo, most of which did not concentrate their performances and rehearsals at a single venue. The Tokyo Symphony Orchestra would become the resident orchestra in the hall, making most of their performances in the hall and using it as their rehearsal space.

As a second aspect of the programming plan, visiting orchestras and musicians were seen as crucial components of the plan to enhance Kawasaki's visibility. The city had the foresight to set aside the funds necessary to engage major international touring orchestras, and after the opening, MUZA Kawasaki would become a destination for, among others, the Berlin Philharmonic, Vienna Philharmonic, London Symphony Orchestra, and the Boston Symphony Orchestra.

Key to the visibility and convenience of the hall is its location directly outside of the main Kawasaki rail station, on the very heavily traveled Tokyo–Yokohama line. The design of this very large project, which includes a shopping mall and office tower, naturally involved many firms, with MHS Planner, Architects and Engineers finally responsible for the design of the concert hall itself.

The initial design idea called for several overhanging balconies, not unlike a horse-shoe opera house, though eventually the architects were convinced to employ a vineyard-style layout that avoids overhangs. The architect's motto for designing the hall was "Beyond Berlin," referring to the Berlin Philharmonie. The guiding idea for the seating layout was to use a spiral, and indeed several "arms" of a spiral can be seen in the walls which define the audience blocks. Similar to the Berlin Philharmonie, many seats are on slanted floors, which give an unusual sense of dynamism to the space. For diffusion of reflections, the white walls within the audience are inscribed with stripes which also have dynamically alternating angles.

Seven years after the opening in the hall in 2004, the hall was badly damaged in the 2011 Great East Japan earthquake. The hall closed for roughly 2 years in order to entirely rebuild the ceiling.

Daniel Beckmann

View along centerline. Though the view towards the stage is symmetrical, the hall itself is not. Audience seats are on slanted floors, with each seat at a slightly different elevation from its neighbor. The seats behind the stage clearly show this design

View from upstage side balcony. Almost all audience seats are visible in the hall as it narrows towards the back. Seating areas are arranged three-dimensionally, with no seats having a sensation of being under an overhang. Sloping white bands visually connect the audience areas, and bring the focus to the stage

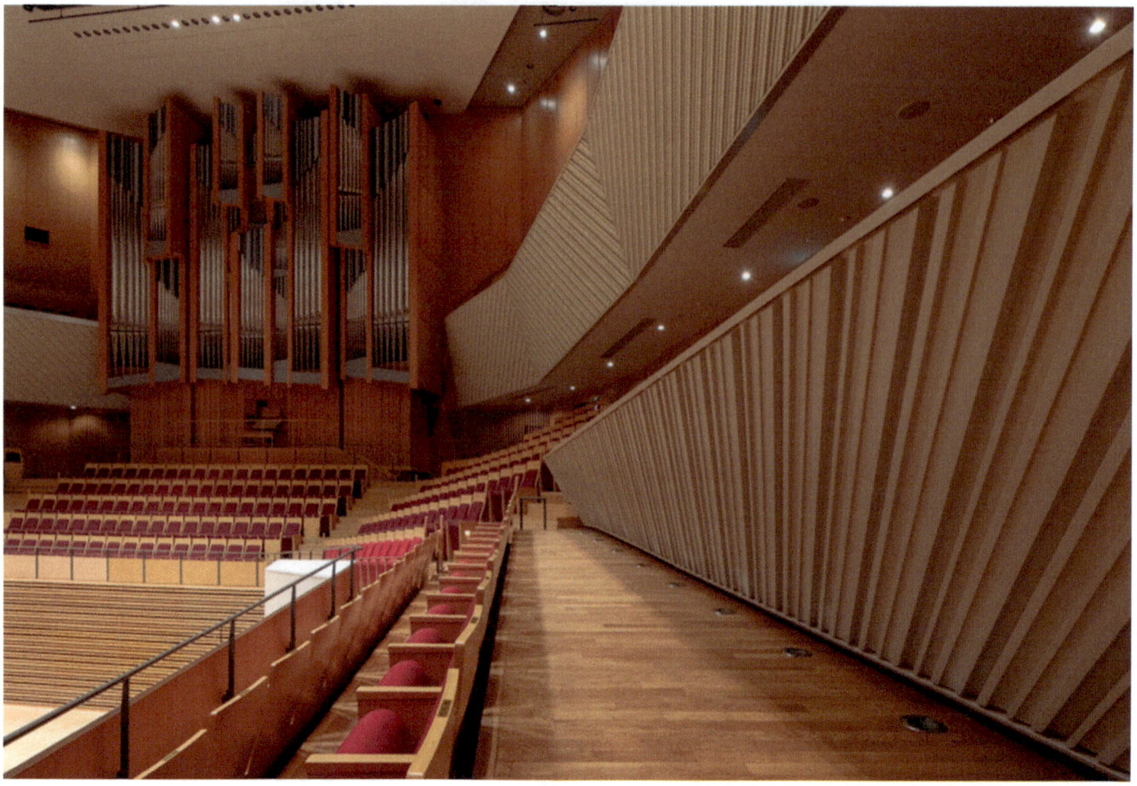

Terrace wall detail. The articulation, composed of stripes of varying depths, on the wall at the right is slanted, and alternates its angle between different segments of the wall

Public approach from train station. The concert hall is at the top of the building on the left, which also houses a shopping mall. Also part of the development is the connected office tower

Muza Kawasaki Symphony Hall	
Location	Kawasaki, Kanagawa, Japan
Owner	Kawasaki City
User	Tokyo Symphony Orchestra
Architect	MHS Planners, Architects and Engineers
Acoustical consultant	Nagata Acoustics
	Yasuhisa Toyota, Akira Ono, Chiaki Ishiwata
Theater consultant	ACT Planning
Design start	1999
Construction start	2001
Construction end	2004
Opening date	July 1, 2004
Building size	35,100 m^2
Seating capacity	1,997
Room volume	27,300 m^3
Surface area	8,400 m^2
Volume/seat	13.6 m^3/seat
Volume/surface area	3.3 m
Finish material	
Ceiling + Canopy	Fiber reinforced gypsum board
Wall	Fiber reinforced gypsum board
Audience floor	Wood flooring
Stage floor	Hinoki
Seat manufacturer	Kotobuki Seating
Organ builder	Orgelbau Kuhn AG
Model scale	1:10

Table 6.1 Muza Kawasaki Symphony Hall—acoustical metrics at 500 Hz

RT unoccupied	2.3 s
RT occupied	2.0 s
EDT	1.7 s
C_{80}	2.0 dB
D_{50}	45.6%
G	5.5 dB

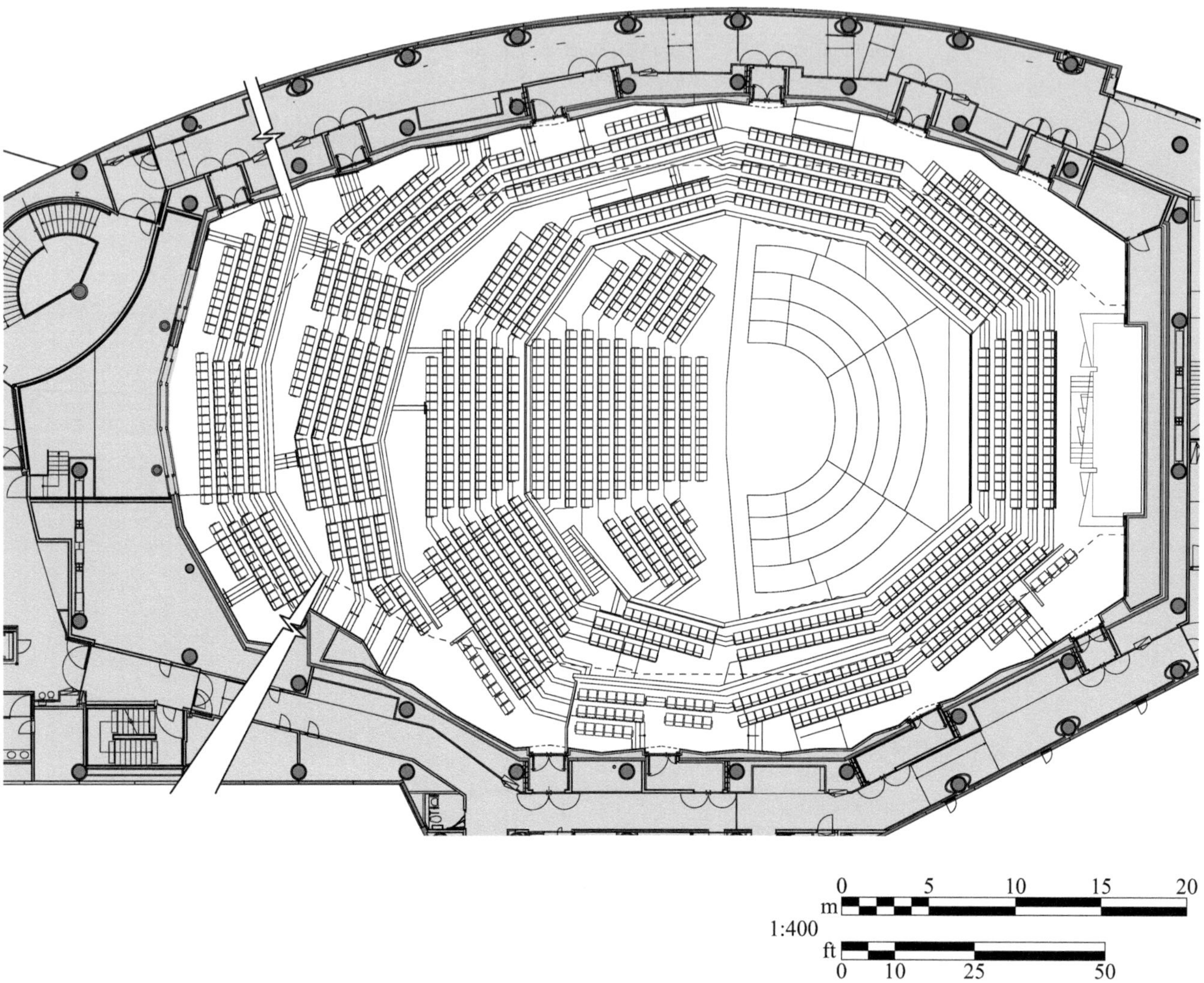

Fig. 6.1 Main level plan
Courtesy of MHS Planners, Architects and Engineers

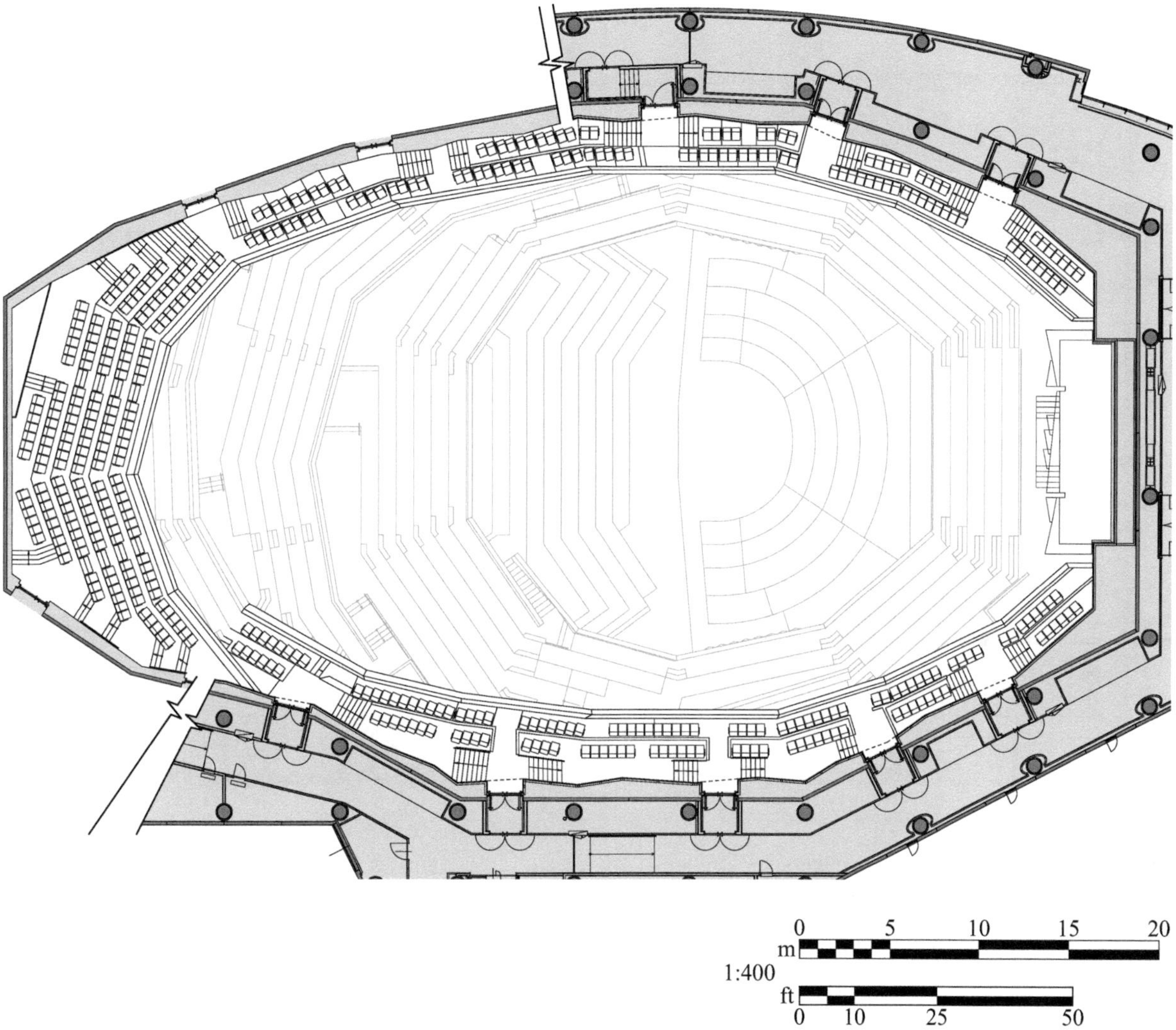

Fig. 6.2 Balcony plan
Courtesy of MHS Planners, Architects and Engineers

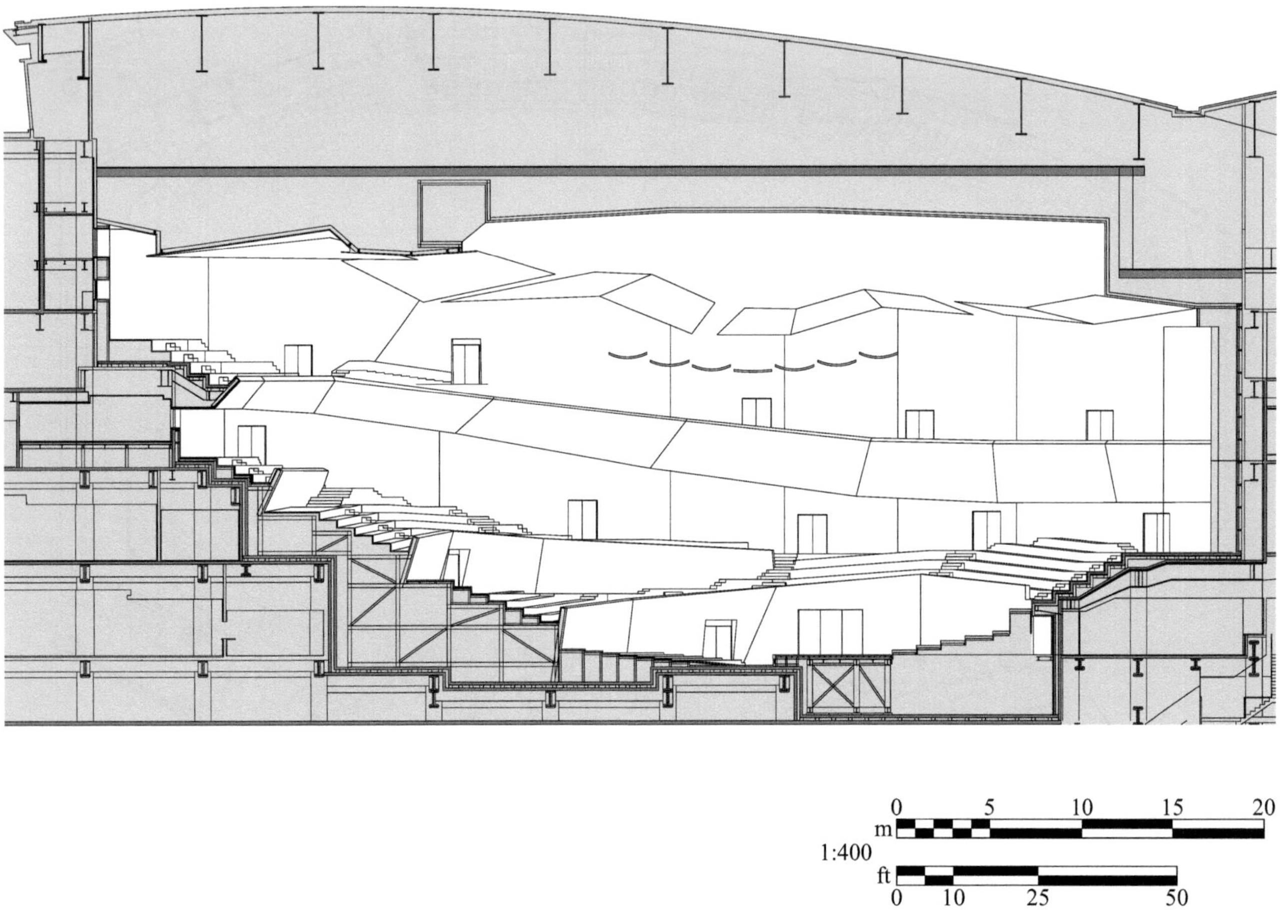

Fig. 6.3 Section
Courtesy of MHS Planners, Architects and Engineers

7. Mariinsky Concert Hall

View from main audience area. The narrow hall width was dictated by the existing building's footprint. A few unique design elements were introduced, such as the two balcony level seating behind the stage and the both sides

Y. Toyota et al., *Concert Halls by Nagata Acoustics*, https://doi.org/10.1007/978-3-030-42450-3_7

History of Mariinsky Theatre

St. Petersburg, Russia, is the home of the historic Mariinsky Theatre, an opera and ballet company founded in 1783. During the Soviet era, it was known as the Kirov Theatre, and became known around the world with many successful tours, most famously by the ballet company. The historic Theatre building, with its distinctive pale, pastel green facade, was built in 1860 as an imperial Theatre, and is registered as a UNESCO World Heritage site as an integral member of the historic St. Petersburg Baroque and Neoclassical ensemble.

The Mariinsky Theatre consists of an opera company, a ballet company, and several orchestras which support those companies and give many concerts themselves. The main facilities are the historic Mariinsky Theatre, the Mariinsky Concert Hall which opened in 2006, and the newest Mariinsky II opera house which opened in 2013.

Several blocks away from the historic theater, which is linked by a footbridge to the new Mariinsky II, is a scenery warehouse and offices built in 1900, designed by architect Viktor Schoreter. A great many stage sets were constructed in this workshop, many of which are still used for current productions.

From Workshop Space to the New Concert Hall

In September 2003, most of the opera sets, props, and costumes were lost in a devastating fire at the warehouse. It was extraordinarily difficult at that time to present the performances which had already been scheduled.

The building itself was irreparably damaged, except for the facade of the building including the historic street-facing portion. Artistic Director Valery Gergiev devised a plan to give a silver lining to this tragedy, and put it into action with uncommon speed. The silver lining was to use the miraculously surviving portion of the building as the framework for a new concert hall. It was fortunate that the original warehouse building was well suited to be used for a medium-sized shoebox concert hall.

With this, a new concert hall was added to the Mariinsky's facilities and a space dedicated to orchestra performance was born.

Project Team and Schedule

Paris-based architects Fabre/Speller and Philippe Pumain were responsible for the architectural design of the renovation.

The design and construction schedule were very fast. Eleven months after the fire, construction began in August 2004. Work at the site progressed at a rapid pace, even during the extremely cold winter months. Two years later, the opening concert took place in November 2006.

Innovative Evolution of the Shoebox Style

While the hall's design was started from shoebox style due to the rectangular prism building shape, we made two significant modifications. First, the rake of the main floor seats was designed considerably steeper. Second, two levels of balconies were added behind the stage and the sides of the stage. Both changes to the typical shoebox style had the effect of greatly enhancing the sense of intimacy and presence in the concert hall. Incorporating some design elements from arena-style halls resulted in a completely new type of concert hall with a markedly different character than existing famous shoebox style concert hall such as the Vienna Musikvereinssaal.

Two levels of balconies at the sides provide reflections back to the stage, as well as the central audience areas, by using the 90° combination between the wall and the ceiling formed by the cantilevered balconies. Two forward-inclined terrace walls divide the main audience area, and provide very early reflections to the seats directly in front of them.

View from upstage balcony. The room acoustical design elements can be seen, such as the corner part under the balcony seats, the side walls shaped with gently convex horizontal alternate strips and the perforated surfaces for sound absorption adjustment and the balcony front surfaces shaped with random wooden blocks

Ceiling texture. The fine grooves with depth and spacing random were designed, aiming scattering sounds at high range

The ceiling is shaped with gently convex surfaces, in order to prevent flutter echoes with the stage floor below, and fine grooves with random spacings and widths are incised into the surface to scatter high-frequency sounds. The upper side walls are articulated with alternating, gently convex horizontal strips. The walls surrounding the stage and the main floor audience area, as well as the balcony fronts, are shaped using random wood blocks, in order to obtain diffusion effects at middle and high frequencies.

Wood was used as the interior surface for the walls, ceiling, and floors, and the major efforts were to achieve the acoustically necessary mass for reflecting low-frequency sounds. The ceiling in particular is approximately 20 cm thick.

On portions of the upper walls, perforated boards obscuring sound-absorbing fiberglass were installed. The fiberglass was intended to be potentially removed after the completion of construction, in order to adjust the necessary amount of sound absorption while listening to live music in the hall. In the end, most of the fiberglass intended for adjustment was removed, with only the audience seats remaining as the sound-absorbing areas in the hall

Stage Risers and Orchestra Pit

The stage is equipped with the operable risers as are found in many other concert halls, to allow efficient preparations of the stage for the busy schedule of the hall. Audience seats behind the stage can be used as chorus seats, which merge almost seamlessly with the orchestra on risers.

A unique feature of this mid-sized concert hall is the orchestra pit at the front of the stage. The pit is formed by sinking the front three rows of seats. While a standard feature of opera houses and theaters, such a pit is very seldom found in concert halls.

Fig. 7.1 Scope of spatial transformations

Staged opera configuration. As one of the major features of this hall, there can be a unique orchestra pit configuration, by sinking the forefront seats. The pit can be used for stage performances like opera or ballet, by arranging the orchestra in the sunken area

The idea of the pit came from Gergiev himself. In the 10 years since the opening of the concert hall, many opera productions have been staged in the concert hall, a unique use of this type of space which is not easily seen elsewhere. By necessity, these productions are quite unique, much different than those in traditional opera houses which offer typical stage equipment to accommodate stage sets.

Opening Concert and Its Preparations

Before the opening, many different ensembles were scheduled to rehearse in the hall in order to familiarize themselves and the artistic management of the hall with the acoustics of the new space. All those present found a high quality acoustic, capable of supporting the collected ensembles such as solo string and wind instruments, male and female vocals, and ensembles of varying sizes up to full symphonic orchestra. The hall supports equally well the fortissimo of full orchestra and the pianissimo of solo strings and harp. The hall gives good volume in all corners, and there is an impression of fullness with rich reverberation. Ensemble sound is clear, and each instrument can be heard distinctly. Musicians on the stage commented that they can hear themselves very well, as well as other voices on stage. Gergiev praised the acoustics of the hall, saying "it is as if listening while inside of a Stradivarius."

On November 29, 2006, the hall officially opened, the program starting with Rimsky-Korsakov's Capriccio Espagnol. Works by Tchaikovsky, Shostakovich, and Prokofiev were performed by stars of the Mariinsky Theatre and famous soloists rounded out the 3 hour concert.

The concert hall opened to the public in April 2007, with a celebratory performance of a cycle of Mahler's symphonies, to mark the 100th anniversary of the great Austrian composer and conductor's first concert in St. Petersburg.

Since then, the Mariinsky Theatre has presented orchestra concerts, chamber music, solo recitals, and many other performances, along with guest performers and ensembles. The number of events increases every year, with one particular highlight being the "White Nights Festival" in June.

Wonderful Sound of the Pipe Organ

A pipe organ was installed roughly 2 years after the opening, manufactured by Daniel Kern and his Manufacture d'Orgues of Strasbourg, France. Since the instrument's inauguration on October 1, 2009, one or two organ recitals are held every month.

Pipe organ. In 2009, the wonderful sound pipe organ was newly installed, incorporated in the front wall, replacing the glass window

The idea of adding an organ came after opening the hall; it was never an element of the original design. A large window on the upstage wall was originally intended to allow some natural light into the hall (first passing through a multipurpose rehearsal space behind the stage). In 2008, Gergiev decided to have an organ installed and the pipes were installed in the wall behind the stage, replacing the window.

Churches and cathedrals generally have very high ceilings, and the large room air volume brings a very rich sound. In concert halls, though, a lack of reverberance is often felt, and the organ sound itself tends to be transmitted directly to the audience. This is only disadvantageous if the sound of the organ is not beautiful, as much of the charm is lost. However, in the Mariinsky Concert Hall, the highest quality instrument has been installed. That sound is unbelievably beautiful, it feels like you want to immerse forever in this space of stereoscopic and deep music.

Motoo Komoda

	Mariinsky Concert Hall
Location	Saint Petersburg, Russia
Owner	Mariinsky Theatre
Design start	August 2004
Construction start	May 2005
Construction end	June 2006
Opening date	November 29, 2006
	Main hall
User	Mariinsky Orchestra
	Mariinsky Ballet
	Mariinsky Opera
Architect	Fabre/Speller Architectes
	Philippe Pumain Architecte
Acoustical consultant	Nagata Acoustics
	Yasuhisa Toyota, Keiji Oguchi, Motoo Komoda
Seating capacity	1,100
Room volume	12,000 m^3
Surface area	4,000 m^2
Volume/seat	10.9 m^3/seat
Volume/surface area	3.0 m
Finish material	
Ceiling + Walls	Finland birch
Audience floor	Wood
Stage floor	Alaskan Yellow Cedar
Seat manufacturer	Quinette Gallay
Organ builder	Daniel Kern Manufacture d'Orgues

Table 7.1 Mariinsky Concert Hall—acoustical metrics at 500 Hz

RT unoccupied	2.2 s
RT occupied	1.9 s
EDT	2.1 s
C_{80}	−0.3 dB
D_{50}	35%

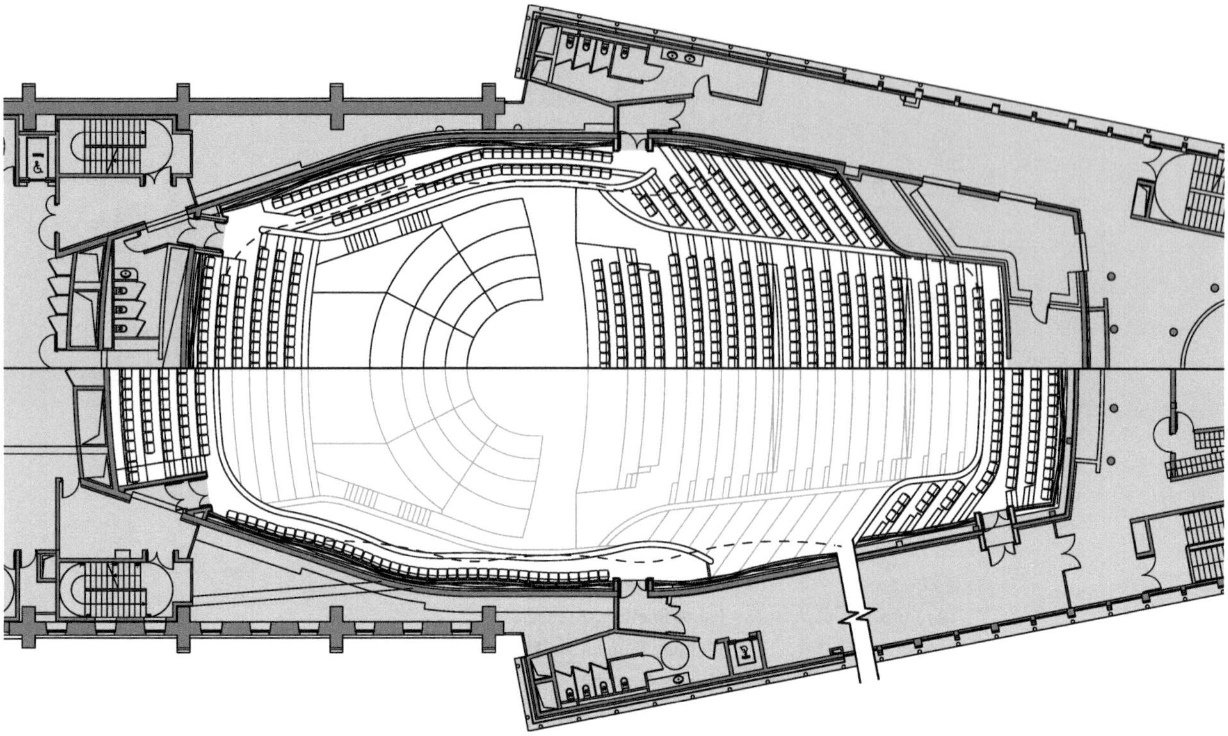

Fig. 7.2 Main and balcony levels plan
Courtesy of Fabre/Speller Architects

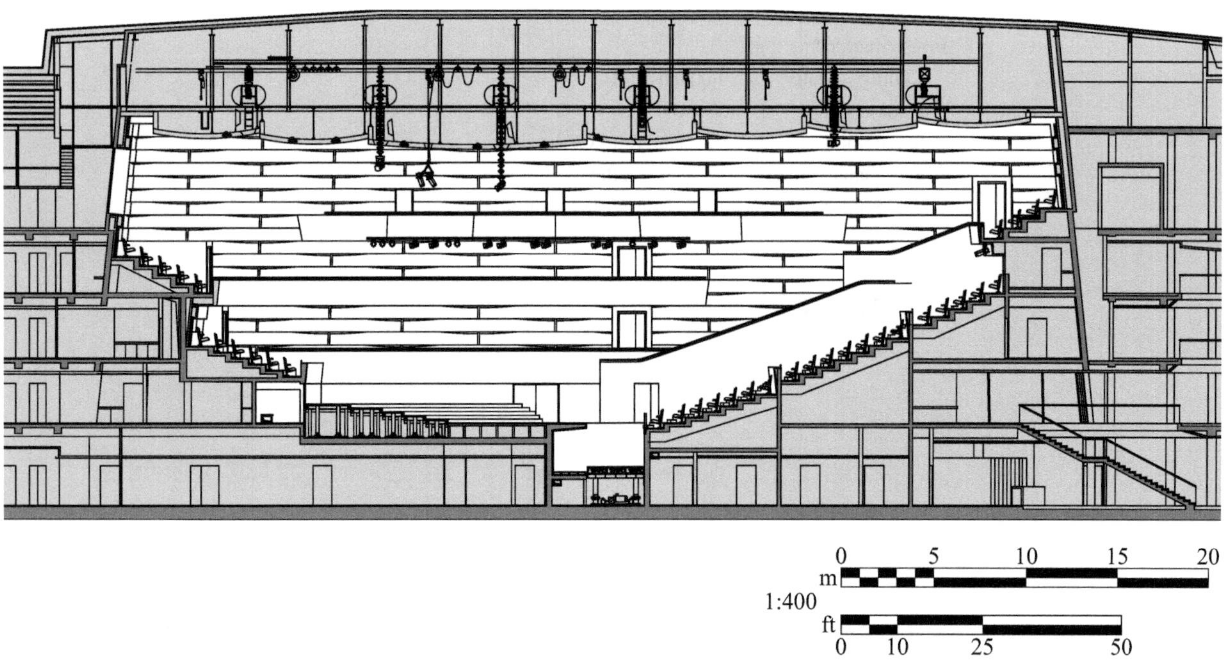

Fig. 7.3 Section
Courtesy of Fabre/Speller Architects

8. Danish Radio Concert House

View from the main audience area. The vineyard style with drastic asymmetry, focused on the visual and acoustical intimacy that the audience can gain by looking at each other's faces
©2020 Jean Nouvel / Artists Rights Society (ARS), New York / ADAGP, Paris

Y. Toyota et al., *Concert Halls by Nagata Acoustics*, https://doi.org/10.1007/978-3-030-42450-3_8

In the early 2000s, Danish Broadcasting Corporation (DR) decided to consolidate its headquarters which were scattered in and around Copenhagen. Four buildings would make up the new DR campus. The first three structures completed contain a variety of offices and TV studios, while the last houses a large concert hall and its attached facilities.

The competition for selecting designers of this concert hall building was held from the end of 2001 to the beginning of 2002 at which point the design team of Ateliers Jean Nouvel based in Paris and Nagata Acoustics was selected.

The Architecture

Jean Nouvel conceived a unique facade design which creates a clear separation between the concert experience from the mundanity of everyday. The hall form resembles a gigantic cocoon covered in large scales and entirely surrounded by a rectangular, semi-transparent blue mesh. While these screens may seem a little strange in the daytime, as the sun sets and evening concerts begin, various images are projected on the surfaces from powerful projectors. The scenery is fantastical and beautiful, bringing some much needed illumination to the long nights of Northern European winters. The color also reminds us of blue screen technology frequently used in film and TV productions.

Concert hall building exterior at night. The concert building is entirely surrounded by a rectangular, semi-transparent blue mesh used as a projection screen at night
©Philippe Ruault ©2020 Jean Nouvel / Artists Rights Society (ARS), New York / ADAGP, Paris

Koncertsalen

The 1800-seat concert hall which occupies the majority of the new building was planned mainly for classical music concerts. It became the new home for the Danish National Symphony Orchestra in addition to hosting many musicians on international tours. This concert hall functions as part of the broadcast corporation's facility and was often referred to as Studio 1 (although currently known as Koncertsalen), alluding to the use for live broadcasts and program recordings.

A stage width of 22 m and a depth of 15 m equipped with a motorized stage riser system was secured so as to accommodate full-size orchestras. Discussions with the musicians and other DR staff iteratively honed the design of the orchestra layout. Only after orchestral rehearsals on a mock-up in the otherwise completed hall, the riser system finalized and installed.

Room Shape

Regarding the shape, the owner, DR, decided on the vineyard style over a shoebox hall early in the programming and planning stage before the design competition began. The client focused on the visual and acoustical intimacy that the audience can gain by looking at each other's faces. Building on the reference to Berlin Philharmonie, Nouvel added further and more drastic asymmetry. The layout of the hall places audience seats all around the stage in 15 scattered terraces, the smallest of which being the royal box with only 8 seats.

Wood is used abundantly in various forms on almost all surfaces of floor, wall, and ceiling. The upper part of the side walls is composed of a flowing curved surface, enhanced by the painted finish of the surface by French artist Alain Bony. The elegant surfaces are illuminated by indirect lighting which, in conjunction with the red color scheme throughout the hall, create a feeling of warmth and intimacy. The horizontal texture and wave finish are contrasted by the shining, vertical pipes of the organ installed behind the stage.

View from the audience area behind the stage. The layout of the hall places audience seats all around the stage in 15 scattered terraces

View of fully occupied hall. In the fully occupied hall, the audience densely populates the entire field of view, creating an intimate and shared concert experience

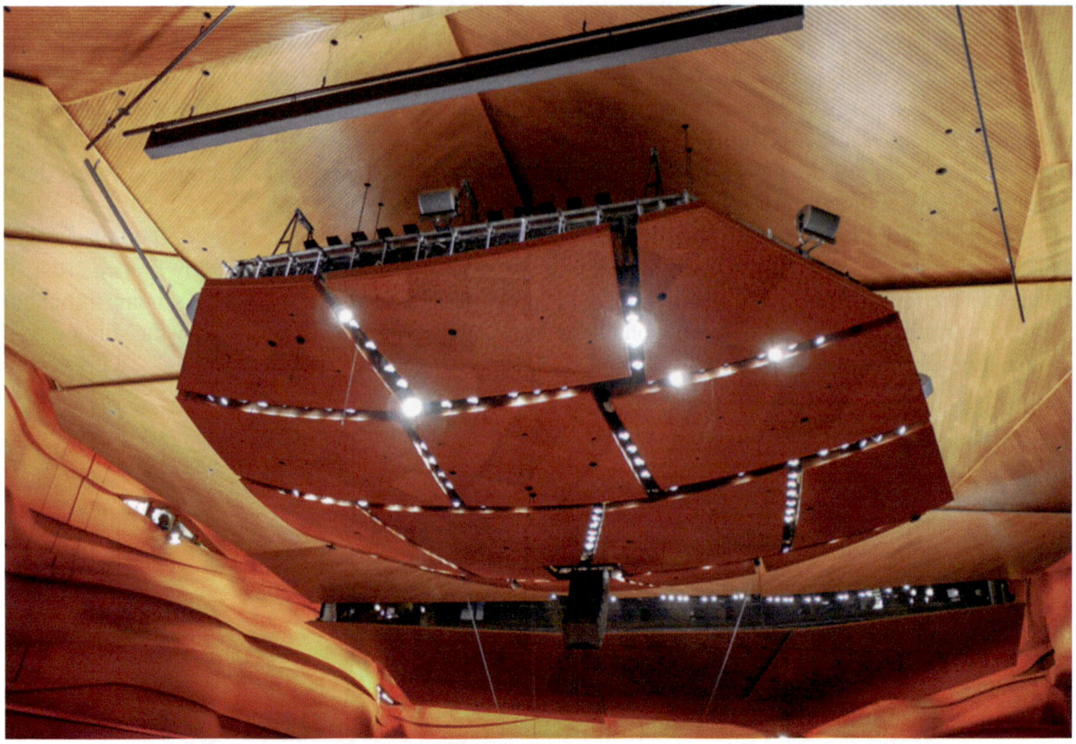

The acoustical ensemble reflector. A gentle convex shape acoustical ensemble reflector is suspended under the ceiling. Although the reflector is divided into several parts, the minimum dimension of each subsection was kept within 3–4 m in consideration of reflection in the low frequency range

The highest part of the ceiling is about 23 m from the stage floor surface, and a gentle convex shape acoustical ensemble reflector is suspended under the ceiling. Although the reflector is divided into several parts, the minimum dimension of each subsection was kept within 3–4 m in consideration of reflection in the low frequency range.

At the back of each audience seat block, a wall for obtaining very early reflected sounds is placed and tilted inward. In addition, on the upper part of the side walls, the soffit ceiling surface with a small depth almost entirely encircles the hall, providing effective early acoustic reflections.

Interior Material

The ceiling consists of the heaviest and the most rigid material in the hall, laminated gypsum boards with wood finish, to fully reflect low frequencies. Scaly panels are arranged to overlap, and fine random grooves were designed on the surface of each panel for high-frequency scattering.

Regarding the audience walls, wood was used as the surface material on the base of multiple layers of gypsum board. The lower terrace walls have horizontal grooves which look like scratches and the upper walls are wavy, both aiming at the scattering of reflected sound.

Shallow texture on ceiling panels. Scaly panels are arranged to overlap, and fine random grooves were designed on the surface of each panel for high-frequency scattering
©2020 Jean Nouvel / Artists Rights Society (ARS), New York / ADAGP, Paris

Surface texture on terrace walls. The lower terrace walls have horizontal grooves which look like scratches and the upper walls are wavy, both aiming at scattering reflected sound
©2020 Jean Nouvel / Artists Rights Society (ARS), New York / ADAGP, Paris

Sound Absorption Curtains

Although the main use of this concert hall is classical music, the hall is made to permit various entertainment such as jazz (Copenhagen has been famous for jazz for a long time), pops, conferences, award ceremonies, and dinner shows. To this end, a system of motorized curtains was introduced to cover the entire upper part of the walls to reduce the reverberation.

Acoustical Scale Model Test

Investigations of whether or not harmful echoes occur were performed on a physical 1:10 scale model. After thoroughly checking the complicated shapes of ceiling and walls, the angles of several walls were slightly changed. In symmetrical halls, it is sufficient to check listening positions on only one side, whereas in this asymmetric hall, such a shortcut is not possible. In short, the scale model testing took much longer than usual.

The acoustical model was also used for architectural interior and lighting design and received many visitors. As could be expected when the client is a broadcast corporation, the acoustical testing was briefly live streamed with a small camera and microphone in the model as a promotional TV program.

The Three Studios

Several floors below the main hall are three large studios. Studio 2 can welcome a maximum of 540 viewer-listeners and is designed to be able to cope with various concert configurations—in addition to recording as a studio—centering on classical music. Sporting an interior inspired by Hollywood's large studios, there are movable wall panels printed with portraits of musicians and artists, such as the local Danish composer Carl Nielsen, which serve to adjust the sound field. A key feature is the motorized stage riser system of approximately the same size as the concert hall, meaning Studio 2 can serve as a backup rehearsal space for a full orchestra.

Studio 3, also called rhythmic studios, fits 170 people and accommodates mainly pop music events and concerts. The interior is designed with a grand piano as a motif. The opening and closing panels that remind us of the lid of the piano are distributed over the ceiling and the walls, so that the absorptive power can be adjusted.

Studio 4 is called the chorus studio and can fit 180 chorus members on its movable podiums. The ceiling and walls are all finished in red mostly treated with rotary-type, sound reflective–absorptive panels in the shape of triangular prisms. Rotating the prisms adjusts the distribution and quantity of absorption so that the reverberant characteristics can correspond to a wide variety of events.

In addition to these large size studios, many medium and private rehearsal rooms are located on each floor.

Opening

On January 17, 2009, the concert hall was inaugurated by Queen Margrethe II of Denmark with other members of the royal family in attendance. Among the projects of Nagata Acoustics, this was the first facility completed in Western Europe.

The opening gala concert had a relatively brief program where ten works by Danish and French composers were performed. In contrast, the subsequent regular performance series featured Mahler's Symphony No. 2 "*Resurrection*," Beethoven's Symphony No. 9, and Stravinsky's three big ballets one after another.

In concerts, it is particularly impressive that the sound of each instrument is clearly audible; the nuances of solo vocals and the resonance of the chorus were very moving. This beautiful building became a new landmark of Copenhagen and quickly began to gather the attention of classical music enthusiasts.

Motoo Komoda

DR Koncerthuset	
Location	Copenhagen, Denmark
Owner	Danmarks Radio (Danish Broadcasting Corporation)
User	Danish National Symphony Orchestra
Architect	Jean NOUVEL—Ateliers Jean Nouvel
Acoustical consultant	Nagata Acoustics
	Yasuhisa Toyota, Motoo Komoda, Ayako Hakozaki
Theater consultant	Scenography: Jacques LE MARQUET, dUCKS scéno (Michel COVA)
Design start	April 2002
Construction start	June 2003
Construction end	January 2009
Opening date	January 17, 2009
Building size	25,000 m^2
Koncertsalen	
Seating capacity	1,800
Room volume	28,000 m^3
Surface area	7,900 m^2
Volume/seat	15.6 m^3/seat
Volume/surface area	3.5 m
Finish materials	
Ceiling and canopy	Microshaped veneer over gypsum board
Balcony and stage walls	Microshaped multiplex over gypsum board
Wave wall	Layered gypsum board
Rear wall	Perforated gypsum board + 50 mm Mineral wool + Air space
Audience floor	Wood parquet over layered gypsum board
Stage floor	Port Orford Cedar
Seat manufacturer	Figueras Seating Solutions
Organ builder	J.L. van den Heuvel Orgelbouw B.V.
Model scale	1:10

Table 8.1 Danish Radio Concert Hall—acoustical metrics at 500 Hz

RT unoccupied	2.0 s
RT occupied	1.8 s
EDT	1.9 s
C_{80}	1.4 dB
D_{50}	45%
G	3.5 dB

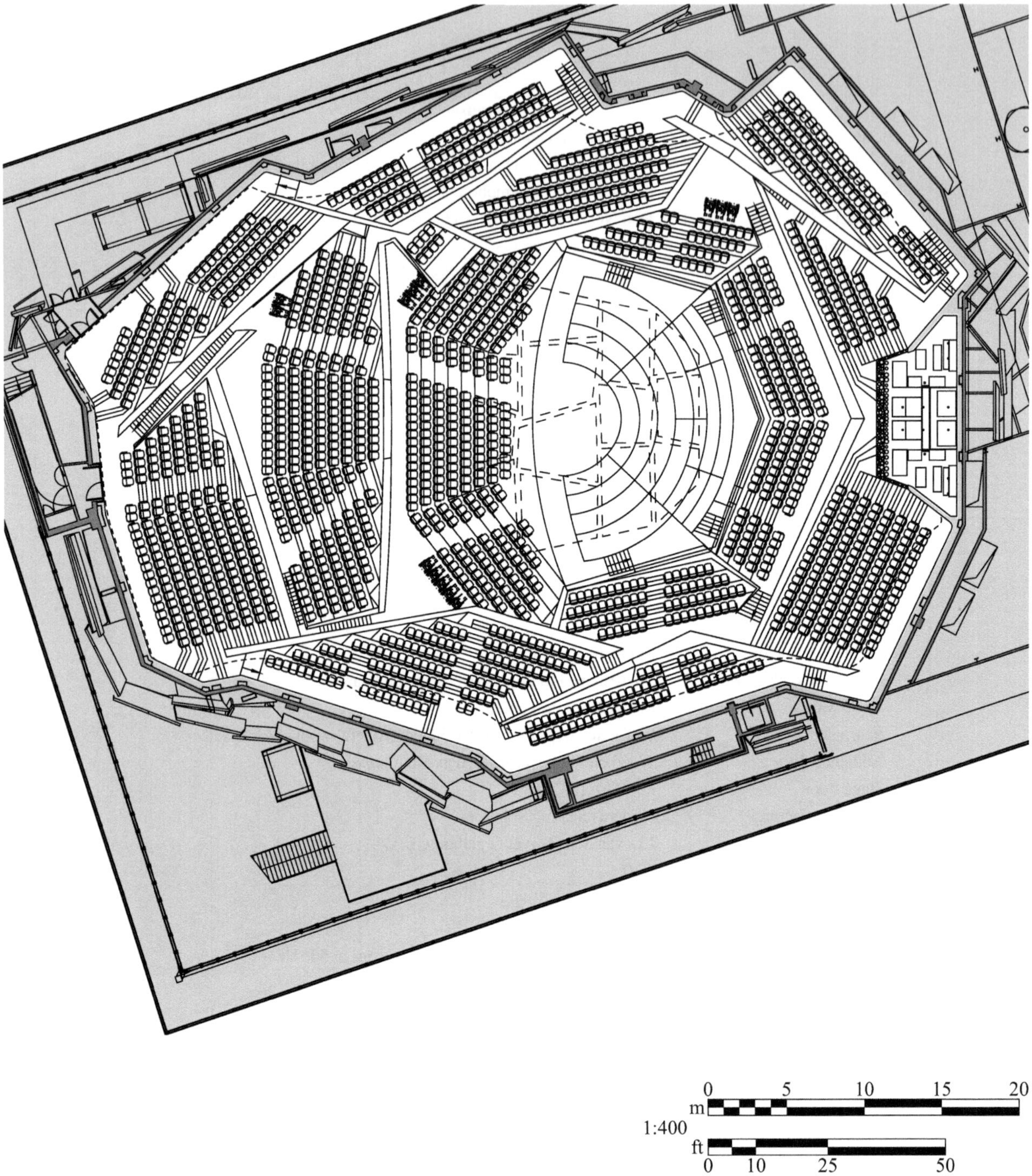

Fig. 8.1 Plan

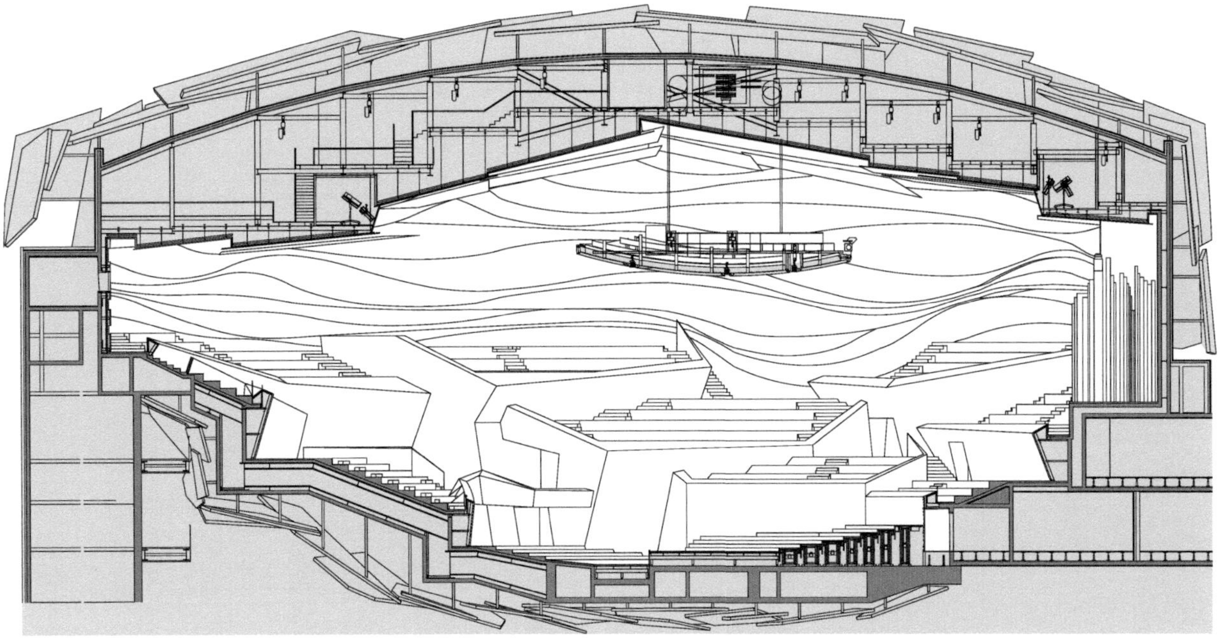

Fig. 8.2 Longitudinal section

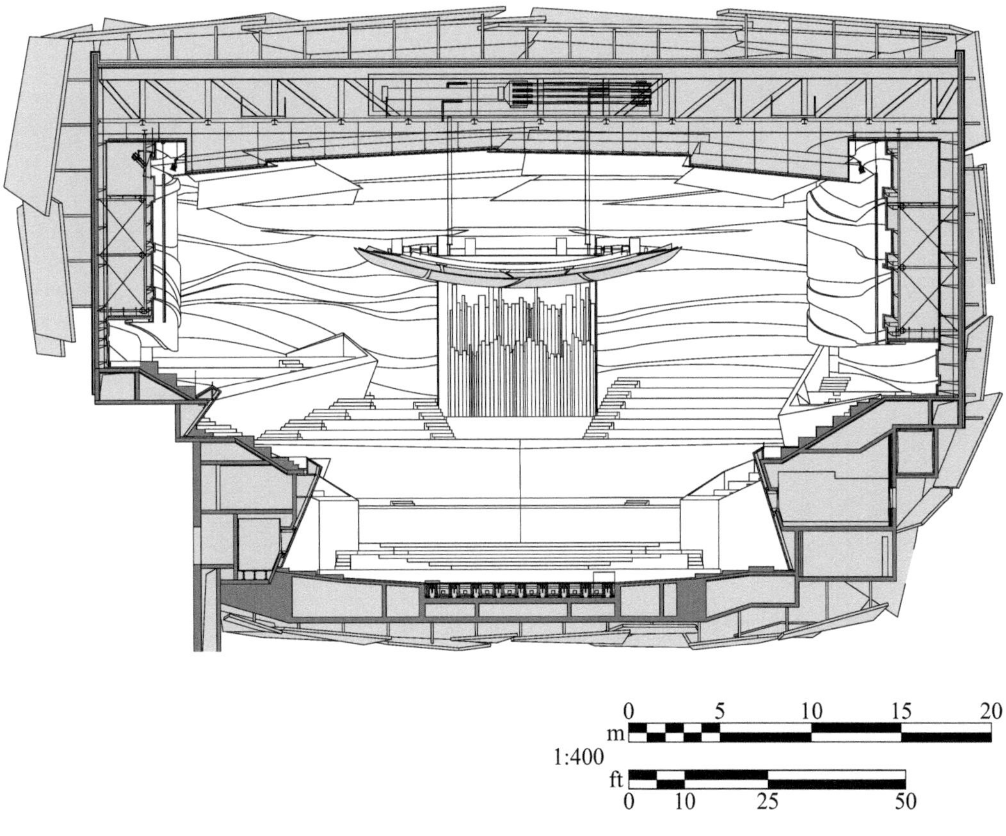

Fig. 8.3 Cross section

9. New World Symphony

View from rear corner. The room is flooded with natural light from the skylight above, and the window to the street behind the stage. Upstage and stage-right seats are bench seats

Y. Toyota et al., *Concert Halls by Nagata Acoustics*, https://doi.org/10.1007/978-3-030-42450-3_9

Miami Beach has been the home of the New World Symphony—"America's Orchestral Academy"—since its founding in 1987 by conductor Michael Tilson Thomas. Located one block away from the historic art-deco pedestrian mall of Lincoln Road, the 9,500 m^2 (100,000 ft^2) New World Center houses a wide range of facilities: a 756-seat concert hall, a multi-purpose performance room seating up to 170, twenty-four individual practice rooms, four chamber ensemble rooms, three percussion studios, and multiple offices and audio/visual facilities. The center was designed by architect Frank Gehry, who has known Tilson Thomas since childhood.

Concert Hall

The concert hall is designed to serve as a rehearsal space for the orchestra, the main focus of the Academy. The orchestra is placed in the center of the room, with audience on all sides of the stage. Several different room configurations are possible in order to support the wide variety of programming offered by the New World Symphony. In addition to the central stage area, four small additional "performance platforms" are located in the audience area. These allow smaller ensembles or antiphonal portions of a larger ensemble to be flexibly arranged without requiring a time-consuming change-over of the main stage. Three of the performance platforms are at the corners of the stage, between audience seating blocks, and one platform is cantilevered above the main audience area, adjacent to the control room. The audience area also provides flexibility in the hall configuration, since the first eight rows of audience seats, or 235 seats, are on retractable risers and can provide a completely flat floor condition. All stage risers, including the front "apron" where soloists usually stand, can also be set at this lowest floor level, allowing a flexible open area of 455 m^2 (4,900 ft^2) for various uses such as banquets or experimental performances.

Sails as projection surfaces. Sails at the upper half of the room can be used as projection surfaces, often with site-specific video art made to accompany the music

The sculptural sails that form the acoustic reflectors in the upper half of the room are all painted smooth white, to better act as screen surfaces for the fourteen different video projectors. The sails are sprayed with a thin layer of powdered marble (one component of a system normally used to create acoustically absorptive surfaces that appear to be hard and reflective) in order to absorb a small amount of high-frequency sound. The slight absorption mimics the effect of the small-scale irregularities which would normally be applied to the majority of primary reflecting surfaces in the room, but would otherwise hinder the use of the sails as projection screens. Sophisticated computer software stitches together the output of the fourteen projectors and maps them to the curved surfaces of the sails to create one continuous canvas, wrapping almost the entire audience.

The combination of immersive moving imagery with music performed live, in a setting of excellent natural acoustics, is a unique experience. Video artists are commissioned to create work which is performed alongside musical works, reinforcing each other rather than "movie music" which simply serves as an accompaniment. Amplified performances are supported by a sophisticated sound system and by vertically retractable banners at the outside walls of the perimeter walkway, which connects all the viewer-listener seats.

While the extensive use of projectors would imply an absence of natural light, in fact, when the projectors are not used, there is an abundance of natural light in the concert hall. A large skylight above the control room provides ample indirect light from the strong Miami sun. Indirect views to the street are provided through a window wall behind the stage, below a sculptural extension of the interior acoustical sails through the facade, giving an idea of the contents of the large white cube to the people on the busy street. Behind the main audience area, a window opens a view of the hall into the large atrium.

Building Organization

The exterior of the building is relatively straightforward and planar, keeping the highly sculptural forms for which the architect is well known exposed to the interior of the building. The building is divided into three parts: the concert hall on the right (when faced from the park), the "bar" that contains administrative offices and many smaller practice rooms on the left, and between them a six-story, glass-fronted atrium filled with the sculptural forms of the larger rehearsal rooms. The large,

Atrium. The light-filled atrium houses several rehearsal and studio facilities in free-standing sculptural forms

white, exterior wall of the concert hall volume acts as $650\,m^2$ ($7{,}000\,ft^2$) projection screen for the lawn viewing area of the $10{,}000\,m^2$ (2.5 acre) "Soundscape Park" adjacent to the building, where the public is invited to enjoy free "Wallcasts" of live concerts in the hall and other programs.

Daniel Beckmann

View from upstage performance platform. The room is flooded with natural light from the large skylight. One of the four performance platforms is cantilevered above the audience, adjacent to the control room

Seat wagons stored. One-third of the audience seats can be retracted, as shown in the photo. An acoustically transparent cover to the stored seats is not shown

New World Symphony	
Location	Miami Beach
	Florida, USA
Owner	New World Symphony
User	New World Symphony
Architect	Gehry Partners, LLP
Acoustical consultant	Nagata Acoustics (room acoustics)
	Yasuhisa Toyota, Keiji Oguchi, Motoo Komoda, Kayo Kimotsuki Kallas, Daniel Beckmann
	Robert F. Mahoney & Associates (isolation and noise control)
	Fred Vogler (sound system)
Theater consultant	Theatre Projects Consultants
Construction cost	USD 160 million
Design start	November 2005
Construction start	January 2008
Construction end	November 2010
Opening date	January 25, 2011
Building size	9,350 m^2
Performance Hall	
Seating capacity	756
Room volume	14,200 m^3
Surface area	5,800 m^2
Volume/seat	18.3 m^3/seat
Volume/surface area	2.43 m
Finish material	
Ceiling	Shotcrete behind veneer plaster, BASWA Phon base
Walls	Shotcrete behind veneer plaster, BASWA Phon base
Audience floor	White Oak
Stage floor	Alaskan Yellow Cedar
Seat manufacturer	Poltrona Frau
Model scale	1:24

Table 9.1 New World Symphony—acoustical metrics at 500 Hz

RT unoccupied	2.2 s
RT occupied	1.9 s

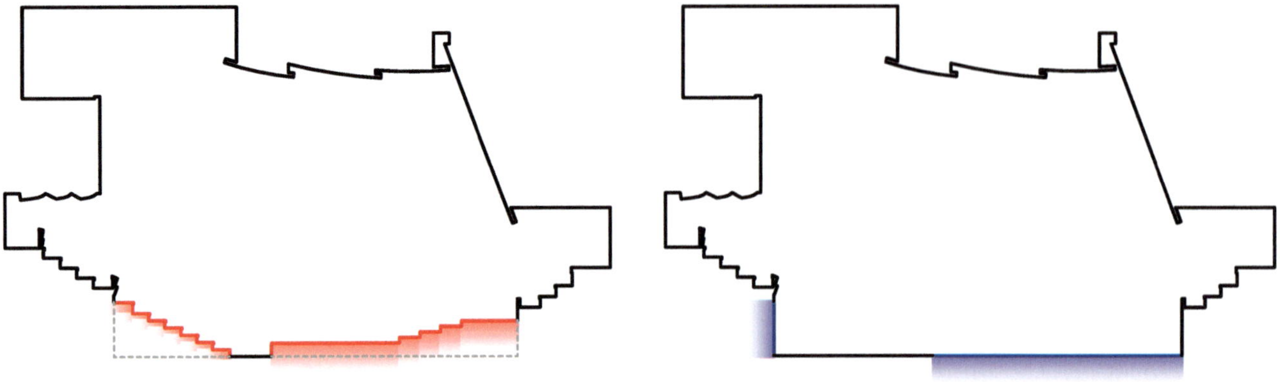

Fig. 9.1 Scope of spatial transformations

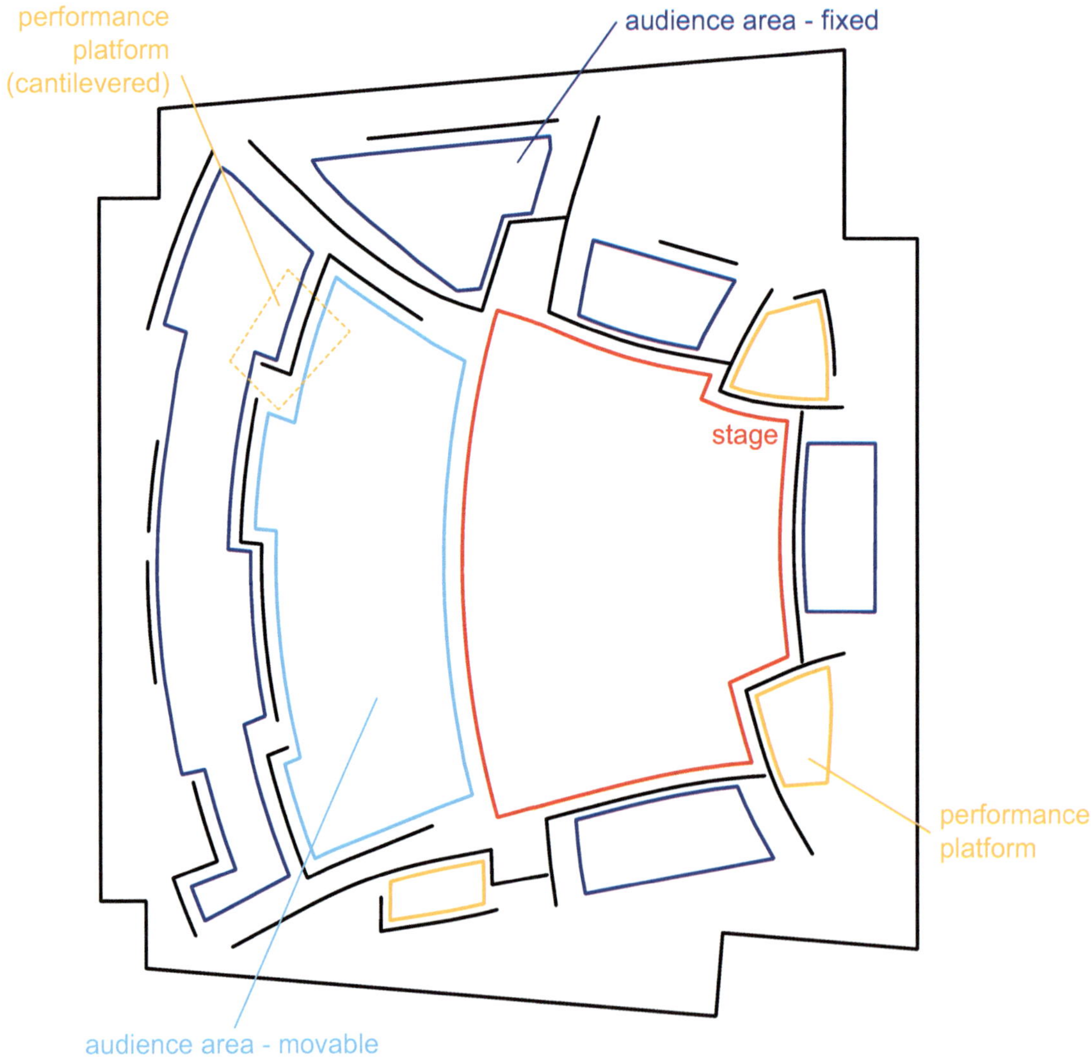

Fig. 9.2 Stage and audience configuration

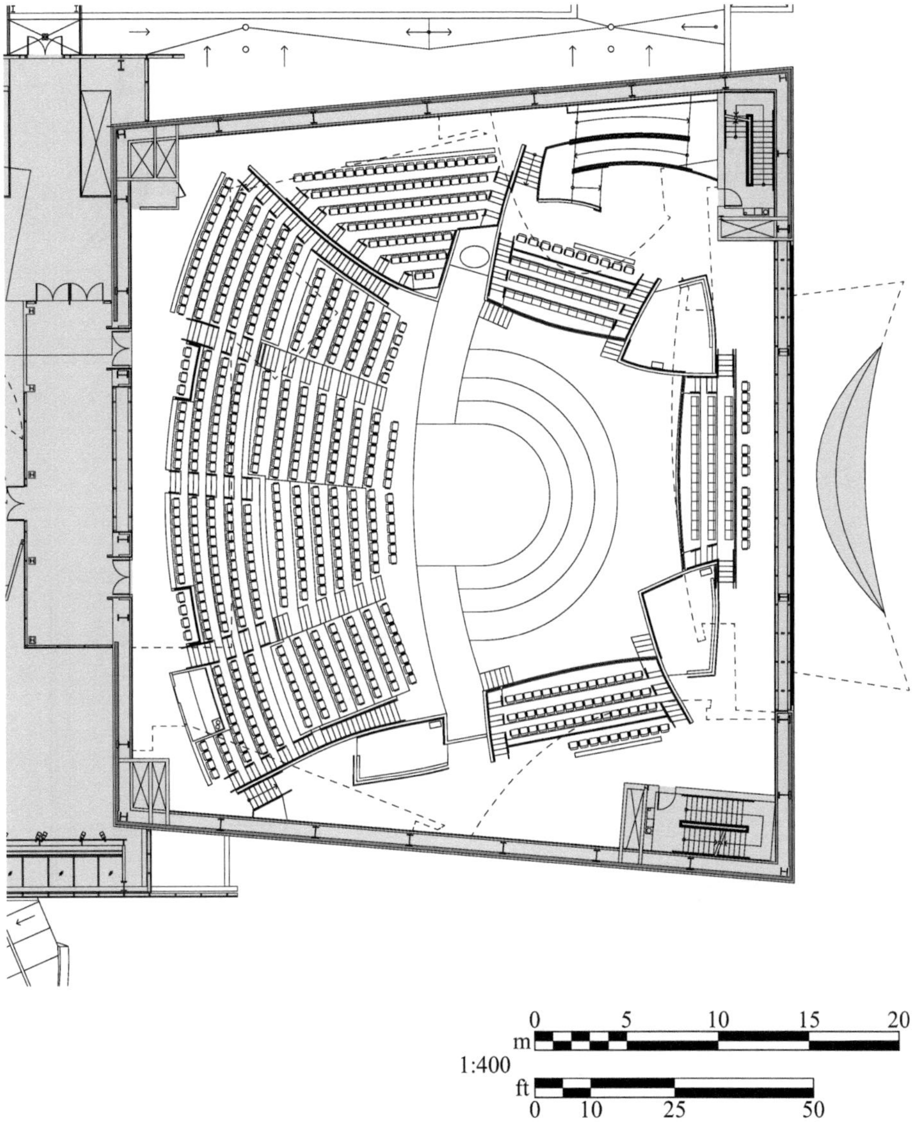

Fig. 9.3 Plan
Courtesy of Gehry Partners, LLP

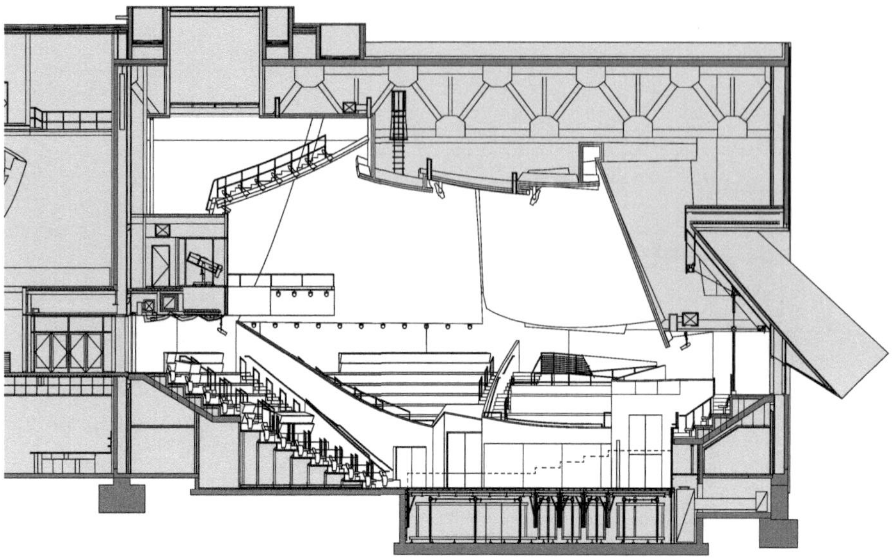

Fig. 9.4 Longitudinal section
Courtesy of Gehry Partners, LLP

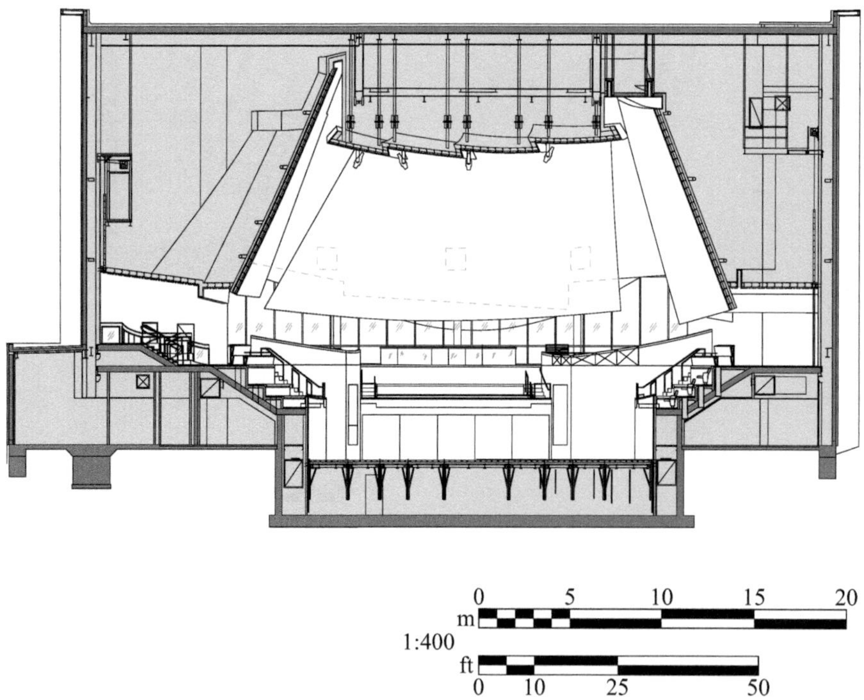

Fig. 9.5 Cross section
Courtesy of Gehry Partners, LLP

10. Helsinki Music Centre Concert Hall

View from side of central seating tier. Terraced seating offers unobstructed sightlines to the stage and divides audience in intimate blocks. The wooden walls, balcony soffits, and fascias feature surface texture designed to scatter sound reflections

Y. Toyota et al., *Concert Halls by Nagata Acoustics*, https://doi.org/10.1007/978-3-030-42450-3_10

In May 1999, a consortium formed by the Finnish national government, the city of Helsinki, and Finnish Broadcasting Company launched a competition to select a design architect and concept for a new music center in Helsinki, which would include a large concert hall dedicated to orchestral programs, with shared residency by Helsinki Philharmonic Orchestra and Finnish Radio Symphony Orchestra. Invited the previous year, Nagata Acoustics had been appointed as an advisor to the competition committee and jury and as the acoustical design consultant for the project.

Helsingin Musiikkitalo (Helsinki Music Centre) finally opened to the public on August 31, 2011.

Twenty Years in the Making

The project for a new music center and a new concert hall finds its roots in 1992 and an initiative of Sibelius Academy. Joined by Helsinki Philharmonic Orchestra and Finnish Radio Symphony Orchestra, the planning process formally started in 1994. The new facility was set to become a companion to Finlandia Hall, designed by Alvar Alto and opened in 1971; originally designed as a mixed-used conference center, its acoustics had never been truly satisfactory for unamplified musical performances and orchestral programs.

The competition committee received over two hundred entries for the first stage of the selection process. No clear design direction had been instructed to the participants, and design proposals varied in typologies and concepts. Therefore, when the committee established specifically that the new hall should feature an arena-style and vineyard-type layout, they decided to open the second stage of competition to all the designers who had submitted to the first stage. In 2000, LPR Architects from Turku (Finland) were designated as the lead designer for the new music center.

The chosen site, designated in 1998, offered a prime location for the new hall, both for its centrality in the city and its relation with other major cultural facilities. Situated along Mannerheimintie, the city's main street, the former railway yard sits between Finlandia Hall to the north and Kiasma Museum (designed by New York based architect Steven Holl) to the south. Less than a kilometer north along the same street is the Finnish National Opera. Directly across the street stands the national Parliament House. However, the former railyard and its abandoned warehouses had become an informal gathering place for local residents, and the project plans drew strong oppositions. It was not until a dramatic fire destroyed much of the remaining structures in 2006 that the plot could finally be cleared to make way for the new construction.

Design work on the building and the hall effectively started in 2005, and the project was officially approved and confirmed by national and local authorities in 2008. The foundation stone was laid on-site on October 22, 2008 and construction was completed in just two and a half years. The first rehearsal took place in the hall on May 5, 2011, four months before the hall was inaugurated.

The Design of the Hall

The hall is fully inscribed in a 52 m long, 39 m wide rectangular footprint. A ring of glazed walls at the level of the main lobby splits the space horizontally at mid-height. The two halves are accessed each from their own dedicated foyer level and are only connected by nine solid cores housing entrance sound locks, vertical circulations, and the volume reserved for the pipe organ.

The lower half of the hall is occupied by the large stage fitted with mechanical risers arranged in a semi-circle around the conductor and first rows of strings. The main seating tier rises quickly with a steep rake, segmented into ten blocks arranged asymmetrically. Lit stairs wander between the seating banks up from the stage level. To the rear, they meet the level of the main foyer and a circulation which follows the glazed perimeter walls all around the hall. This circulation serves additional seating terraces to the sides and back of the stage, once again arranged asymmetrically.

The upper half of the hall is populated by a ring of seating divided in eleven blocks. Their varied arrangements of rows and rakes follow the same language as the terraced seating of the lower section. With their guardrails reduced to metal wires, the ring appears as a collection of cantilevered platforms rather than a uniform balcony, however, they are constrained on the top by a common circulation and foyer level and on the bottom by the same soffit elevation.

The main ceiling is shaped as wide alternating slopes crossing along the center line of the hall and peaking at a height of approximately 20 m above the level of the stage. They are separated by gaps which serve as storing positions for theatrical light trusses. Above the stage, the effective ceiling height is lowered to 15 m by virtue of a suspended circular reflector, which can be accessed from above for maintenance by a motorized lift. Alternating shallow troughs are cut into the surface of the ceiling and canopy to integrate stage lights.

During daytime, the peripheral window fills the space with natural light permeating through the glazed main facade and offers direct views to the hall from the surrounding public space. During performances, if so desired, curtains can be deployed between the two layers of glass, for privacy or black-out.

View from upstage seating. Seating is divided into compact blocks arranged all around the stage asymmetrically. The glazed walls separating the lower seating tiers from the balcony ring lets daylight in from the facade and creates visual connection with the surrounding foyer

Although the walls delimiting the audience appear vertical, they effectively consist of a first acoustically transparent face (comprising of wooden slats backed by a light fabric and a thin wire mesh) which conceals a second solid concrete surface with segments tilted forward to direct early sound reflections. Walls of the lower tier as well as balcony blocks feature a dark wooden finish with randomly scattered wooden grooves designed to scatter high frequency sound reflections. Peripheral cores and upper walls, as well as the main ceiling consist of cast-in-place concrete finished with textured fiberglass wallcovering painted dark gray.

To prevent harsh direct reflections from the extensive glazed surfaces of the hall, careful design considerations were given to the shape and position of the windows. Sound reflections through glazed surfaces were designed to be only second-order reflections, with their paths meeting at least one surface with surface irregularities, typically the balcony soffit. Additionally, the geometry follows a sawtooth plan shape to prevent sharp reflections back to the stage.

Audience chairs feature a truly unique design developed by the architects. They exhibit a lean and compact triangular profile, with the folded seat bottom and connected armrests flush with the main front surface. The back panel is recessed with a similar triangular shape to allow for more legroom. Sides and back are cladded with dark-gray metal panels, while seat and seatback are upholstered with thin cushions covered in dark fabric sprinkled with thin light-gray lines motif. Armrests and edges of the triangle are bound in black fabric. Overall, they add to the resolutely modern and angular design of the hall and its sharp, dark lines.

The design team reserved an ample volume behind the stage and choir seating for later installation of a pipe organ. The space remained empty for the first years of operation, concealed behind an acoustically transparent metallic curtain. In May of 2018, the design and construction of the instrument was commissioned to experienced Austrian firm Rieger Orgelbau. It is due for completion at the end of 2022.

Different Strokes

While the hall is featured prominently in the overall design of the building, it is only the proverbial tip of the iceberg. The Music Centre also houses six smaller halls as well as the world-famous Sibelius Academy which trained a host of renowned composers, conductors, and instrumentalists.

The six halls are buried on the west side of the plot, between the main concert hall and the street. They were each designed to address a specific program and type of use, directly reflected in their respective names:

- Rehearsal Hall Paavo replicates the stage of the main hall for full orchestra rehearsals, but is also frequently used to host small concerts and public or corporate events for up to 240 attendees.
- Organo can seat up to 140 people and be used for performances as well as classes and rehearsals. It features three different pipe organs with distinct styles.
- Camerata hosts concerts for small ensembles in natural acoustics, but is also suited for lectures. Its fixed raked seating can accommodate 240 people.
- Sonore was designed primarily for vocal performances and staged performances. A motorized platform can be adapted to create an orchestra pit, extend the stage platform for larger ensembles, or install additional seats. Its seating capacity ranges from 200 to 280 people.
- Black-Box offers retractable telescopic seating and was designed for performances with amplified music or speech. Its capacity ranges from 220 with raked seating to 400 on the flat floor. It is complete with a wire-mesh grid technical ceiling and a small side stage for increased versatility.
- Auditorio is a small, 80-seat lecture, and conference hall.

Ensuring Quietness in a Dense Building and Urban Environment

Due to the direct adjacency of a busy street and a tram line to the building, the position of the respective rooms in the building as well as their structures was specified to prevent air-borne and structure-borne noise and vibrations from being transferred to the halls and between the halls.

As the most sensitive space, the concert hall was moved as far away from the street and tram tracks as possible on the site, and the structure of the hall is floated on resilient elastomer pads. The peripheral window has two distinct layers of glass, spaced apart approximately one meter. The other halls were located underground along the street and built with box-in-box structures, with the inner box comprising a cast-in-place, floating slab supporting precast concrete walls, and shotcrete ceiling. The precast panels assembled to erect the walls were formed to create the interior acoustical room shape, which is visually concealed by acoustically transparent perforated panels, with distinct patterns across the respective rooms.

Opening Night

The grand opening concert on August 31, 2011 featured both resident ensembles. It started with a truly unique rendition of Jean Sibelius's *Finlandia* conducted by Jukka-Pekka Saraste and performed by instrumentalists from both resident ensembles. They were joined by a choir of Sibelius Academy students and personnel, and amateur singers from the greater Helsinki. The singers stood in the aisles among the audience and made the final and beloved *Finlandia Hymn* a true moment of communion between performers and listeners.

Marc Quiquerez

Building facade. The fully glazed facade and large windows between foyer and concert hall allow for daylight in the hall and direct views from outside and public spaces

	Helsinki Music Centre
Location	Helsinki, Finland
Owner	State of Finland
	City of Helsinki
	Finnish Broadcasting Company
Architect	Arkkitehtitoimisto Laiho-Pulkkinen-Raunio
Construction cost	EUR 188 Million
Design start	May 1998
Construction start	October 2008
Construction end	April 2011
Opening date	August 31, 2011
Building size	38,600 m^2
	Concert Hall
User	Helsinki Philharmonic Orchestra
	Finnish Radio Symphony Orchestra
	Sibelius Academy
Acoustical consultant	Nagata Acoustics
	Yasuhisa Toyota, Keiji Oguchi, Motoo Komoda, Marc Quiquerez
Seating capacity	1,704
Room volume	24,000 m^3
Surface area	6,900 m^2
Volume/seat	14.1 m^3/seat
Volume/surface area	3.5 m
Finish material	
Ceiling + Canopy	Painted textured fiberglass wallcovering on concrete
Wall	Wood work on concrete, Wood grille, Glazing
Audience floor	Wood on concrete
Stage floor	Finnish Pine
Seat manufacturer	Riihimäen Metallikaluste Oy
Organ builder	Rieger Orgelbau
Model scale	1:10

Table 10.1 Helsinki Music Centre Concert Hall—acoustical metrics at 500 Hz

	Reflective condition	Curtains deployed
RT unoccupied	2.5 s	2.1 s
RT occupied	2.0 s	1.8 s

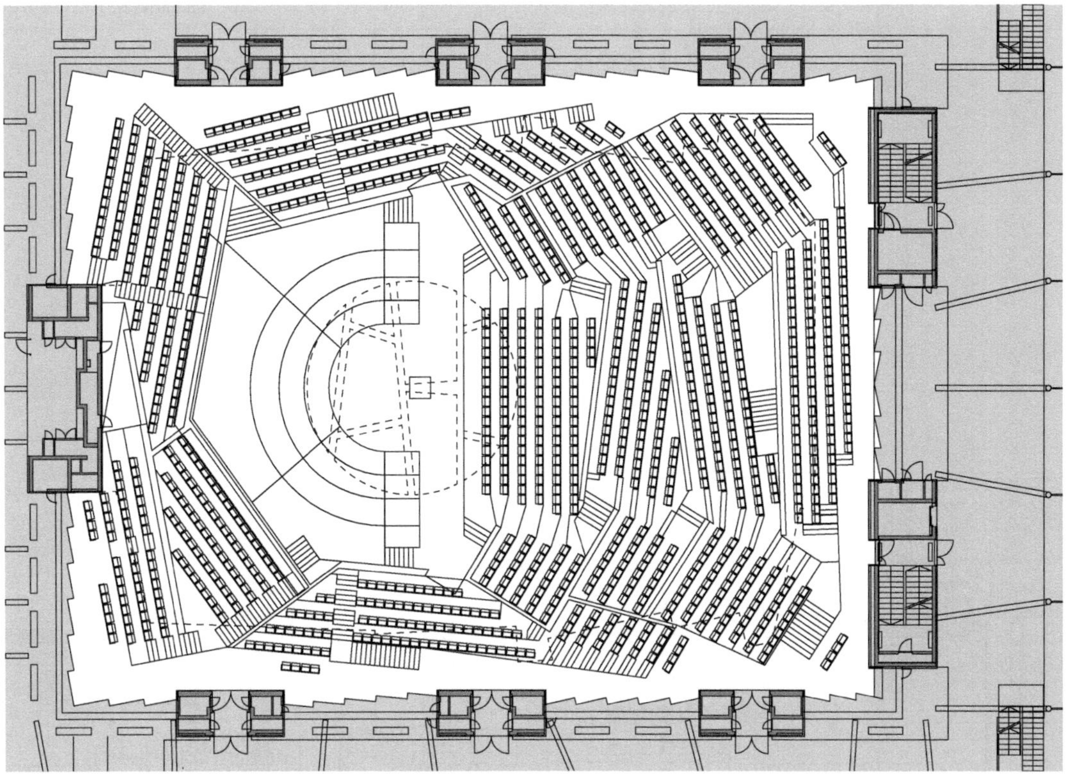

Fig. 10.1 Main level plan
Courtesy of Arkkitehtitoimisto Laiho-Pulkkinen-Raunio

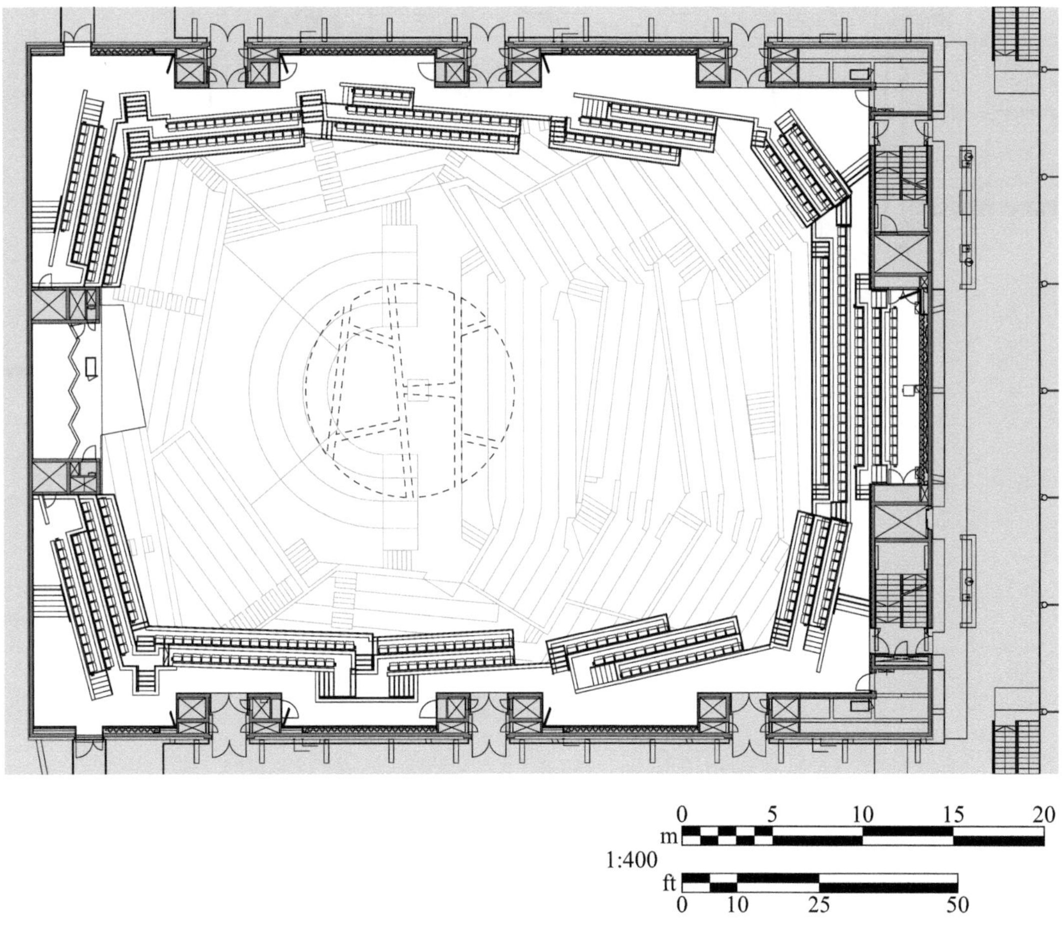

Fig. 10.2 Balcony level plan
Courtesy of Arkkitehtitoimisto Laiho-Pulkkinen-Raunio

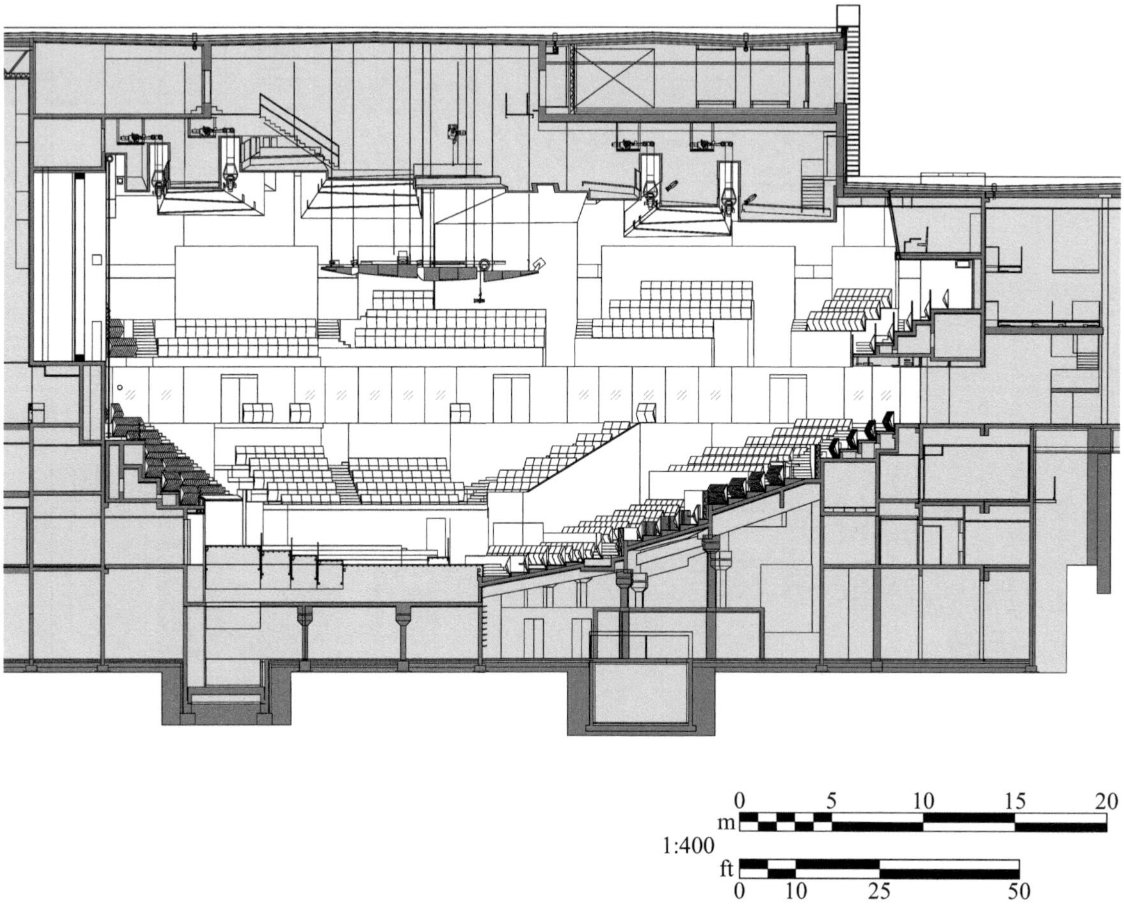

Fig. 10.3 Section
Courtesy of Arkkitehtitoimisto Laiho-Pulkkinen-Raunio

11. Kauffman Center for the Performing Arts: Muriel Kauffman Theatre

View from first balcony. The translucent balustrades are lit from within, and the artwork behind the grilles at the front three seating boxes is also lit

Y. Toyota et al., *Concert Halls by Nagata Acoustics*, https://doi.org/10.1007/978-3-030-42450-3_11

Private philanthropy and the vision of Muriel McBrien Kauffman were the driving forces behind Kansas City's Kauffman Center for the Performing Arts. In 1994, Kauffman began the exploratory process of the construction of the Center, which achieved a major milestone with the selection of Moshe Safdie as the design architect in 2000. The city, having learned from the experiences of many cities building so-called "Grand Theaters," or multi-purpose halls intended to accommodate both opera and symphonic performances, decided to build a performing arts complex with separate halls for symphony orchestra and dramatic performances such as opera and ballet. The resulting 26,500 m^2 (285,000 ft^2) facility, with a cost of \$413 million, was financed almost entirely with private donations. The 1,800-seat Muriel Kauffman Theatre for ballet, opera and similar theatrical performances, and the 1,600-seat Helzberg Hall for symphonic orchestra performances are linked by the Brandemeyer Great Hall, a glass walled and roofed atrium with sweeping views over the city to the south.

Muriel Kauffman Theatre is a traditional, proscenium-oriented venue, which was designed for opera and ballet performance. The Lyric Opera of Kansas City and the Kansas City Ballet are the main resident companies in the theater, which is also frequently used by outside presenters and for events produced by the Kauffman Center itself.

The Theatre uses the historical horseshoe typology as its base, similar to famous opera houses like La Scala (Milan, Italy), La Fenice (Venice, Italy), and the Teatro Colón (Buenos Aires, Argentina). The main floor holds 800 seats (with an additional 100 seats available on the orchestra pit lifts), and each of the three balconies holds approximately 300 seats. The balconies have rows of fixed seats at the center five segments of the horseshoe-shaped ring, and three boxes on each side close to the proscenium arch.

Using lessons learned from the design of single-purpose concert halls, the side boxes are designed in order to form the soffits which send sound reflections back to the performers, particularly to the orchestra in the pit. Particular care was taken with the boxes closest to the proscenium and the technical zone directly between the front box and the proscenium wall, since those directly adjacent to the pit are the most effective for providing the necessary supportive reflections. The work of many disciplines collides in this area, as the architecture and theater planning professionals also take keen interest in the space close to the division between audience and performers. Using this area in a novel way (for a horseshoe theater) was a delicate balancing act.

Another lesson applied from past concert hall design experiences was the ceiling height above the front of the stage and the orchestra pit. Working with Safdie's office and Theatre Projects Consultants, the proscenium opening was set at 9.1 m (30 ft) tall, resulting in a ceiling 11.9 m (39 ft) above the sunken orchestra pit. Moving towards the audience, the ceiling curves gently upwards, then makes a large step higher to give a ceiling height of 15 m (50 ft) above the front portion of the pit. The ceiling over the audience is composed of several complex curved surfaces, optimized to distribute early reflections to the entire audience.

Studies of early reflections were conducted to provide good clarity of sound in the theater, and, in the use of another lesson from concert hall design, this goal of clarity was not treated as an impediment to the longest possible reverberation time. The concave shapes inherent to a horseshoe-shaped theater were therefore not managed with sound-absorbing materials, which would reduce the reverberation time. Rather, sound-diffusing bumps line all the walls of the theater. The tall, cylindrical bumps are normally shielded from view by a grille of horizontal slats. Behind the box seats, the bumps are sometimes lit directly from behind the grille in order to expose art work by students of the Kansas City Art Institute, which was applied directly to the surface of the bumps. Balcony fronts, also a major concave reflecting surface in plan, are shaped as large bent cylindrical segments, so that detrimental focussing effects are not found in the space. The balustrade shape is made of a translucent cast resin, and is filled with a metallic foil and lighting elements to create a dynamic boundary to the inner theater volume, reminiscent of crystal chandeliers in historic opera houses.

Daniel Beckmann

View from side balcony. Box seats step down towards the stage. Multiple balconies at the rear of the hall keep all audience members close to the stage

Orchestra pit. The orchestra pit can support a full orchestra for large-scale productions

Public approach. The exterior form of the building cleverly hides the 26 m (86 ft) tall fly tower. The landscaping is reminiscent of the open fields of the local prairie

Shared lobby. The glass-roofed lobby is shared by both performance halls. From the Center's hilltop site, sweeping views to the south are active both during the day and at night. Blue carpeting denotes Helzberg Hall, and red carpeting denotes Kauffman Theatre

Kauffman Center for the Performing Arts	
Location	Kansas City Missouri, U.S.A.
Owner	Kauffman Center for the Performing Arts
Architect	Safdie Architects
Executive architect	BNIM
Acoustical consultant	Nagata Acoustics (room acoustics) *Yasuhisa Toyota, Keiji Oguchi, Motoo Komoda, Kayo Kimotsuki Kallas, Daniel Beckmann* Robert F. Mahoney & Associates (isolation and noise control)
Theater consultant	Theatre Projects Consultants
Construction cost	USD 413 Million
Design start	September 2003
Construction start	October 6, 2006
Construction end	April 2011
Building size	26,500 m^2
Muriel Kauffman Theatre	
User	Lyric Opera of Kansas City Kansas City Ballet
Opening date	September 16, 2011
Seating capacity	1,800
Room volume	9,700 m^3
Surface area	4,600 m^2
Volume/seat	5.4 m^3/seat
Volume/surface area	2.1 m
Finish material	
Ceiling	Plaster with sandblasting finish
Wall	Plaster
Audience floor	Carpet on concrete
Orchestra pit floor	Alaskan Yellow Cedar
Seat manufacturer	Theatre Solutions

Table 11.1 Muriel Kauffman Theatre—acoustical metrics at 500 Hz

RT unoccupied	1.5 s
RT occupied	1.3 s
EDT	1.4 s
C_{80}	2.8 dB
D_{50}	55%
G	2 dB

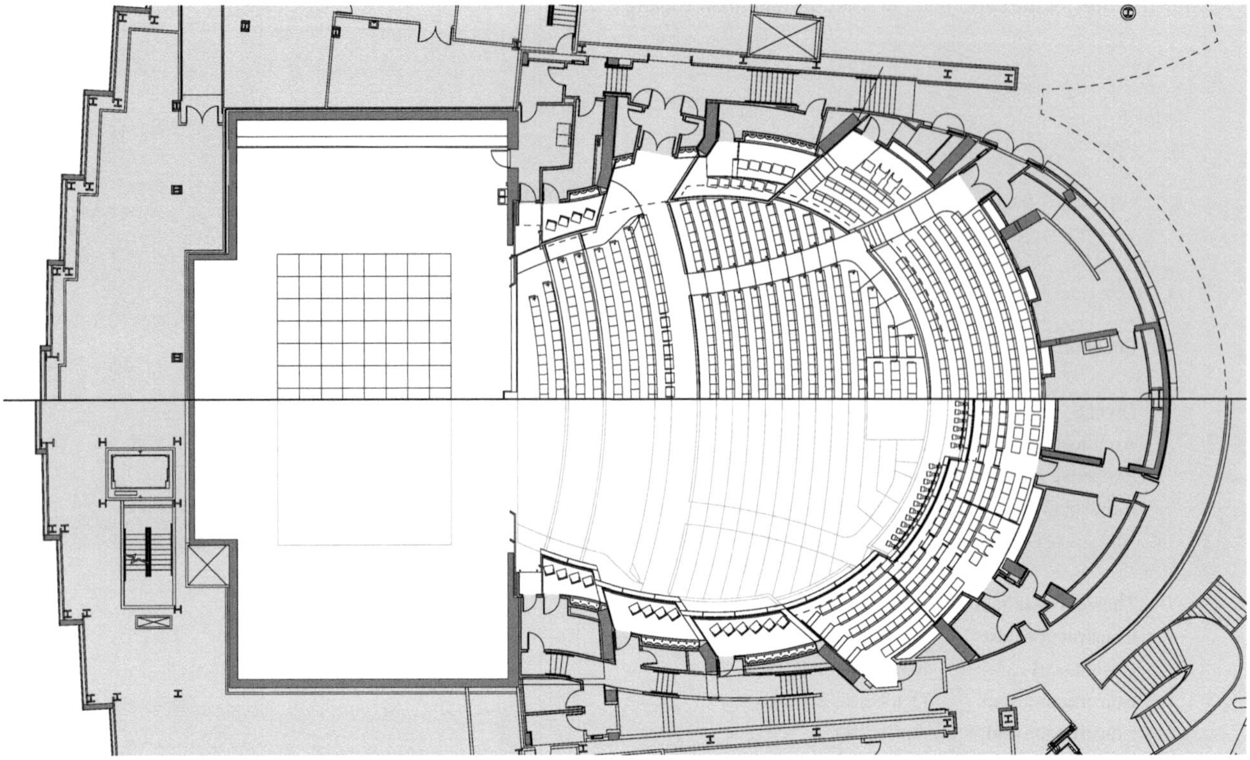

Fig. 11.1 Stage and 1st balcony plan
Courtesy of Safdie Architects

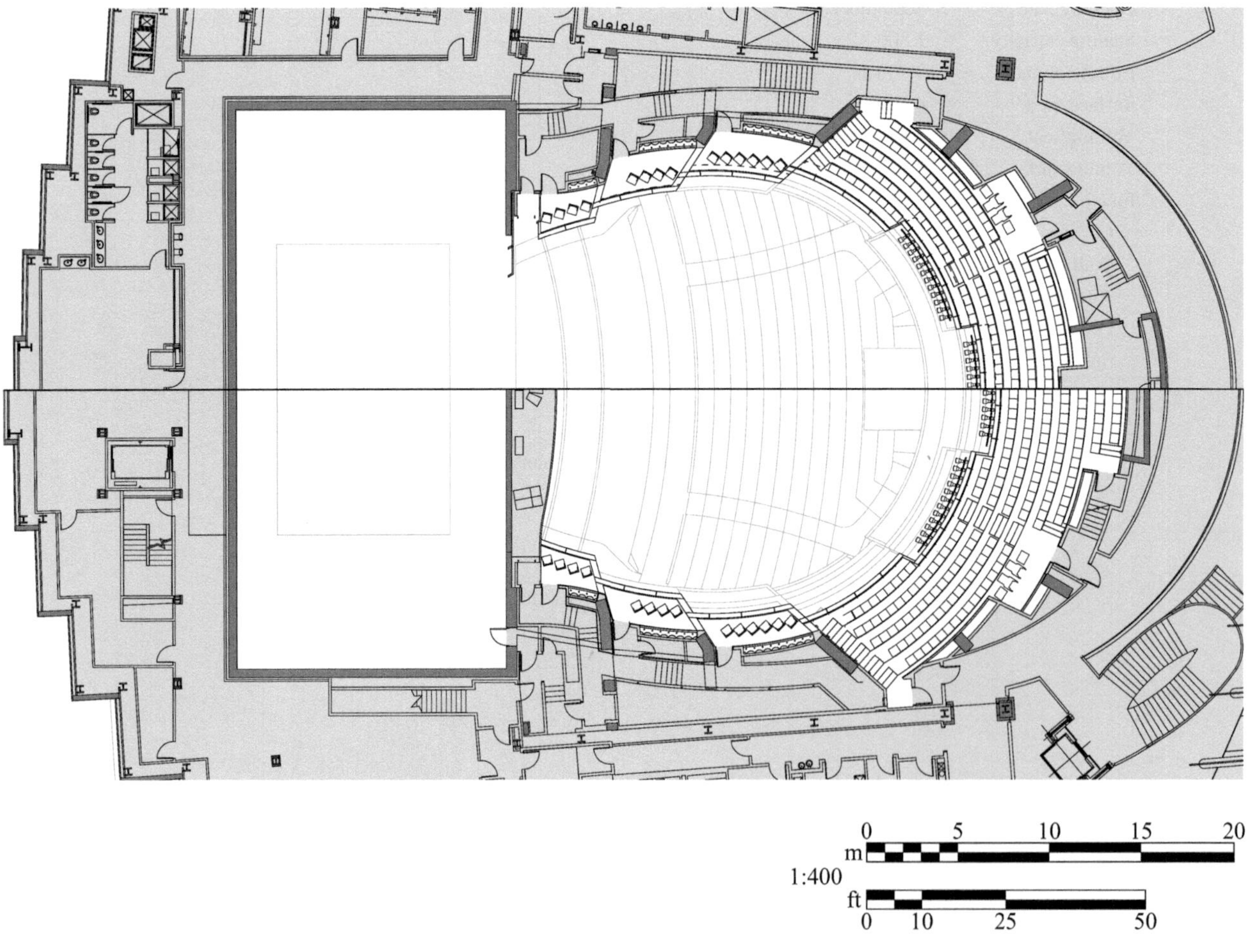

Fig. 11.2 2nd and 3rd balconies plan
Courtesy of Safdie Architects

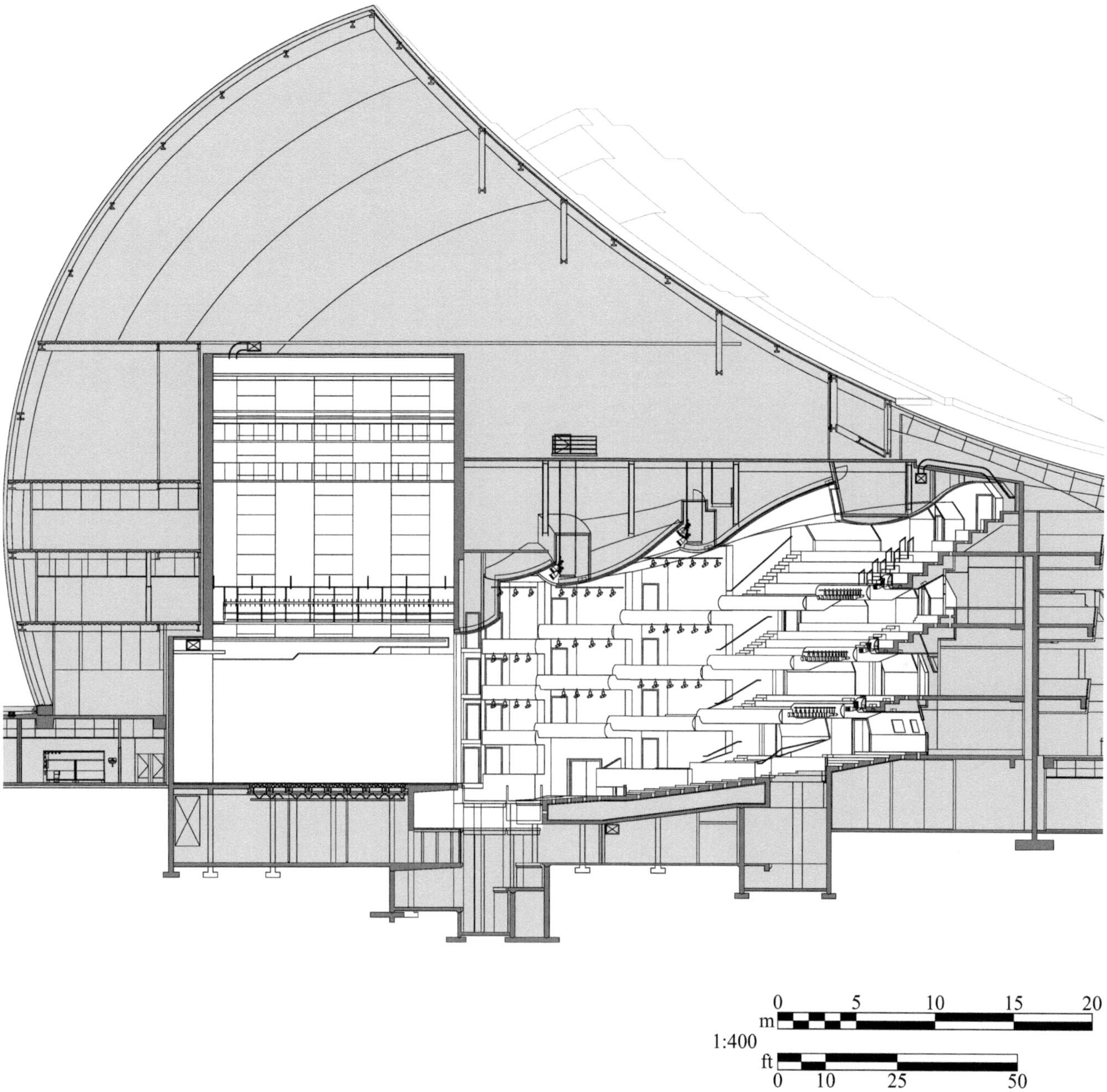

Fig. 11.3 Section
Courtesy of Safdie Architects

12. Kauffman Center for the Performing Arts: Helzberg Hall

View from main balcony. Only chorus seats and the organ are behind the stage. The acoustically transparent metal wall, mimicking the exterior shape of the building, is washed by natural light from above and hides large, sound-diffusing shapes to break up the concave shape

Y. Toyota et al., *Concert Halls by Nagata Acoustics*, https://doi.org/10.1007/978-3-030-42450-3_12

Adjacent to the Muriel Kauffman Theatre is the 1600-seat vineyard-style concert hall, named Helzberg Hall. The exterior shape of the building is clearly visible from the interior of the hall, with its dramatic 30 m (100 foot) tall wall in the shape of the "rear" wall of the building, and also the main ceiling which gently descends from the peak of the room to the farthest audience seats in the balcony. The very tall ceiling necessitated the use of an ensemble reflector above the stage, a single square GFRG (glass fiber reinforced gypsum) element measuring 11.6 m wide and 12.8 m deep (38×42 ft), bulging slightly downward. It is suspended at a height of 15 m (50 ft) above the stage. Behind the stage, the 30 m tall concave wall encloses a 79-stop Casavant Frères organ.

The stage is placed not at center of the room as is often seen in vineyard-style concert halls, rather it is closer to the tall upstage wall and organ. Only four rows of seats are behind the orchestra, and can be used for chorus or audience. Four principal audience areas face the stage in the traditional frontal configuration, all with excellent sight-lines due to the rapidly increasing rake for each section; those sitting in the last row of the balcony find the ensemble reflector above the stage exactly at their eye level. More than just the audience facing the stage directly, a variety of views are available in the hall from the three levels of balconies on the sides of the stage. Large side terraces, with five rows on each side, are closest to the stage level, and above them are two levels of balcony seats with one row each. The higher balconies also visually activate the large wall surfaces by making it an occupied space with faces to watch.

View from chorus side. The main ceiling above the stage is 30 m (100 ft) tall, and is divided roughly in half by the ensemble reflector which is at a height of 15 m (50 ft)

Helzberg Hall has several large concave elements, both subtle and obvious, even though it is challenging to accommodate these shapes in spaces for listening. The major obvious concave element in the space is the tall silver wall behind the orchestra, housing the organ and washed with natural light from hidden skylights. The visible metal wall is actually composed of a great many tiny metal bars, each measuring 3 mm in diameter (1/8 inch), that forms an acoustically transparent surface. Behind this metal wall is the acoustically reflective wall, made up of very large convex elements of varying sizes. The size of the acoustically transparent screen as well as the configuration of the large diffusing elements were both confirmed in separate 1:10 scale models, to ensure an absence of detrimental effects such as focusing (from the concave shape) or diffraction grating whistle (from the many small repetitive elements). The more subtle concave element is the overall plan itself, as the main tall side walls of the hall are bowed slightly outwards. Behind the visible structure of horizontal, acoustically transparent wood slats is a series of large, sound-diffusing, semi-cylindrical bumps running from floor to ceiling, the necessity and effectiveness of which was also confirmed in the 1:10 scale model test.

The Kansas City Symphony is resident in the hall, and the hall has become a highlight for many major American and international orchestras on tour, such as the Los Angeles Philharmonic, the Chicago Symphony, and the Mariinsky Orchestra.

Daniel Beckmann

View from chorus. The hall is symmetrical. Two levels of soffits are visible along the sides of the stage and audience, providing substantial support to the musicians on the stage

Public and musicians approach. The concert hall is at left, showing the two halls as independent structures connected by a shared glass lobby

Kauffman Center for the Performing Arts	
Location	Kansas City
	Missouri, U.S.A.
Owner	Kauffman Center for the Performing Arts
Architect	Safdie Architects
Executive architect	BNIM
Acoustical consultant	Nagata Acoustics (room acoustics)
	Yasuhisa Toyota, Keiji Oguchi, Motoo Komoda, Kayo Kimotsuki Kallas, Daniel Beckmann
	Robert F. Mahoney & Associates (isolation and noise control)
Theatre consultant	Theatre Projects Consultants
Construction cost	USD 413 Million
Design start	September 2003
Construction start	October 6, 2006
Construction end	April 2011
Building size	26,500 m^2
Helzberg Hall	
User	Kansas City Symphony
Opening date	September 17, 2011
Seating capacity	1,600
Room volume	19,000 m^3
Surface area	6,100 m^2
Volume/seat	10.9 m^3/seat
Volume/surface area	3.2 m
Finish material	
Ceiling + Canopy	Plaster with sandblasted finish
Wall	Plaster
Audience floor	Oak
Stage floor	Alaskan Yellow Cedar
Seat manufacturer	Theatre Solutions
Organ builder	Casavant Frères
Model scale	1:10

Table 12.1 Helzberg Hall—acoustical metrics at 500 Hz

RT unoccupied	2.3 s
RT occupied	2.1 s
EDT	2.3 s
C_{80}	−1.0 dB
D_{50}	33%
G	7 dB

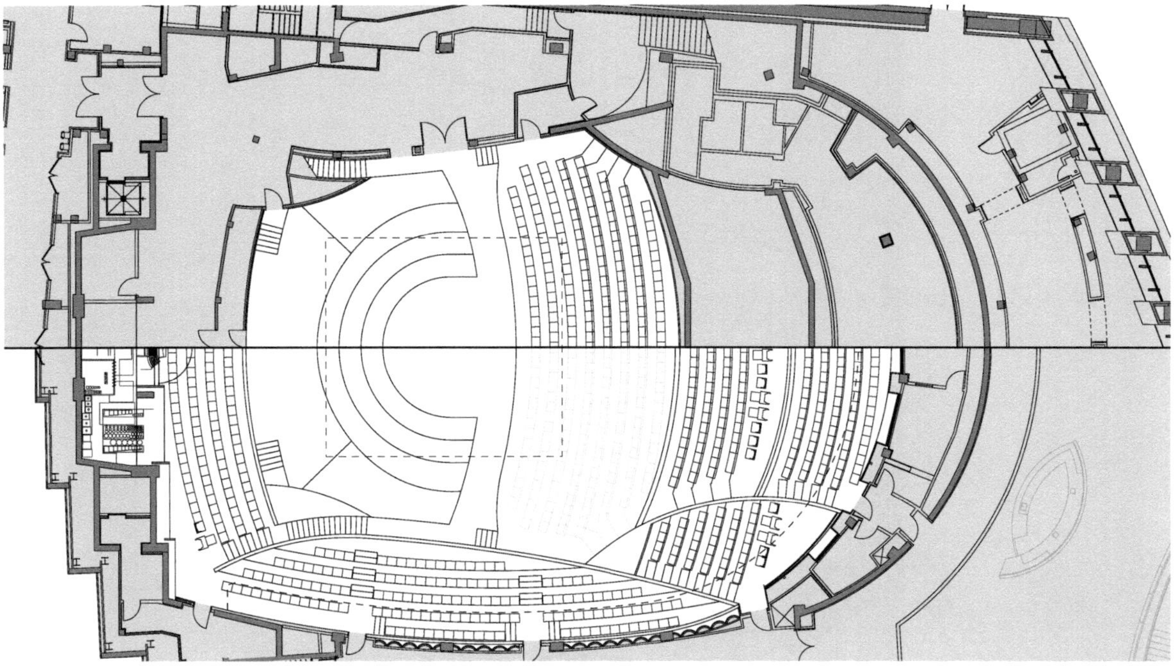

Fig. 12.1 Stage and terrace levels plan
Courtesy of Safdie Architects

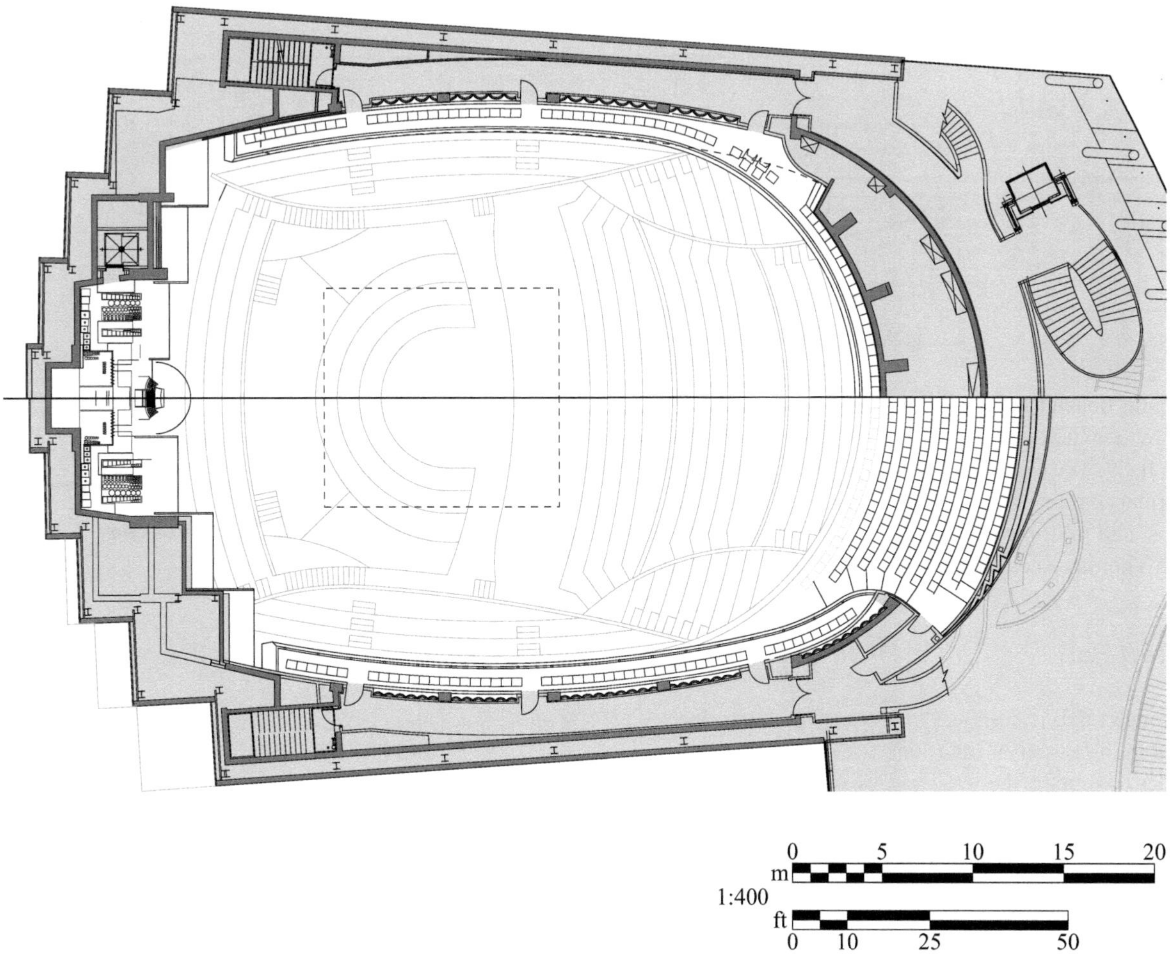

Fig. 12.2 1st and 2nd balconies plans
Courtesy of Safdie Architects

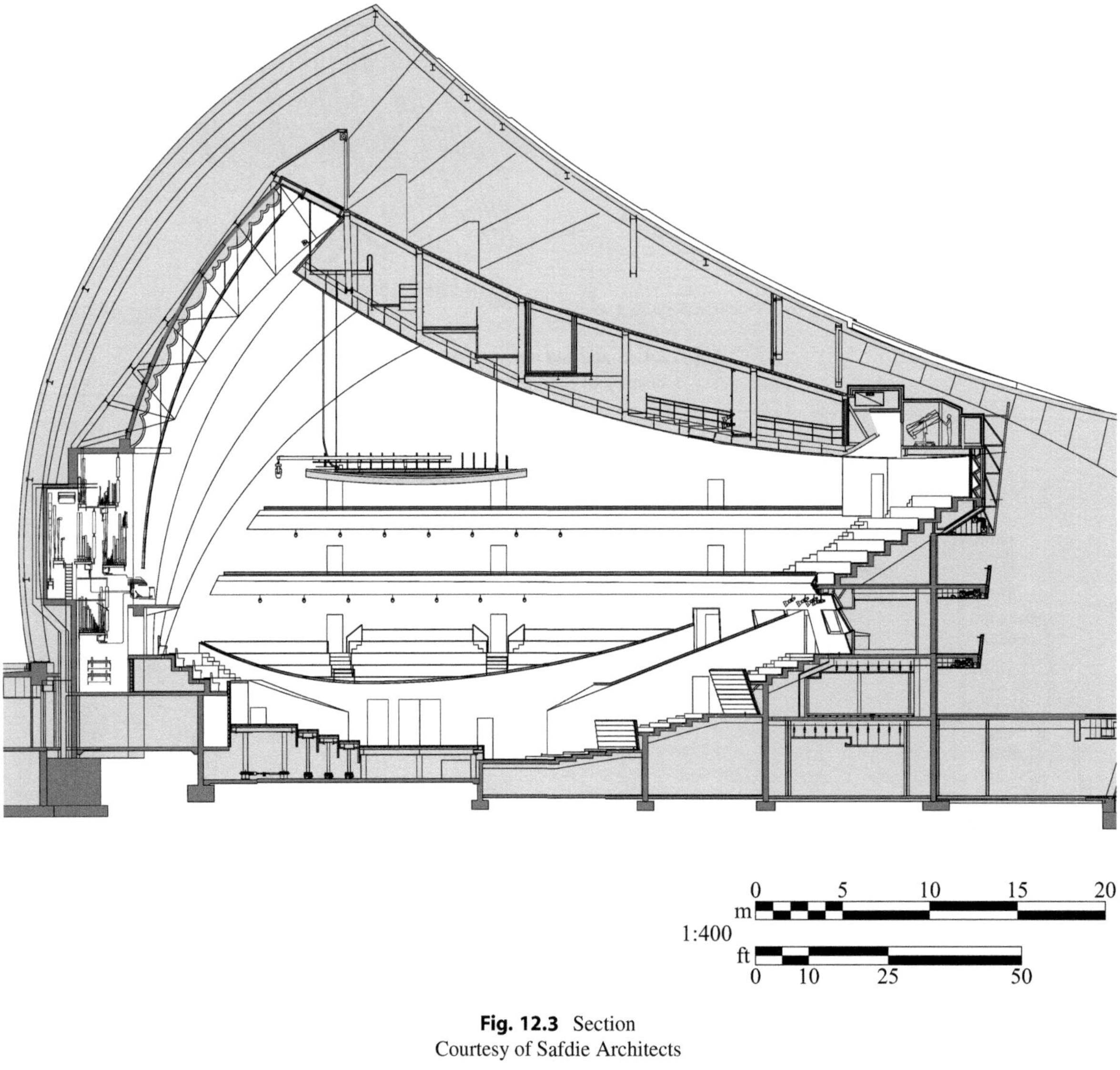

Fig. 12.3 Section
Courtesy of Safdie Architects

13. Soka Performing Arts Center

Concert stage configuration. The stage is flattened to be able to accommodate a full sized orchestra. There is no step between the stage floor and the front row of the audience so that you can experience the unique intimacy with the orchestra which is not seen in other halls

Y. Toyota et al., *Concert Halls by Nagata Acoustics*, https://doi.org/10.1007/978-3-030-42450-3_13

A new performing arts center was completed in 2011 on the Soka University America campus located in Orange County, California, USA. In the newly constructed 4-story building, a multi-purpose hall accommodating a maximum of 1,032 people, a black box theater accommodating about 150 people, practice rooms, and offices are included.

The architectural design was done by ZGF Architects (Zimmer Gunsul Frasca Architects), and the theater consultant was Auerbach Pollock Friedlander. The total construction cost was USD 73 million.

Exterior. Approach to the concert hall entrance

A Unique Style Multi-Purpose Hall

Thinking of multi-purpose halls in universities, ceremonies and lectures are generally the main usages, so an auditorium style which has a wide proscenium opening might be the first choice. However, the multi-purpose hall in this facility is quite different and has a very unique style.

Among the client team who wanted to focus on educational program related to the performing arts, there were enthusiastic classical music fans. They strongly desired to plan regular concert series inviting local professional orchestras, and several independent classical music performances. Aside from the programmatic flexibility, a unique architectural design was requested for the hall. In order to respond to such demands, a multi-purpose hall which prioritizes acoustical quality was the target of the design team.

Stage and Audience Seating

The fundamental design direction was to place the audience seats surrounding the stage, as opposed to a traditional lecture hall where the performers and audience are always facing each other. In such an arena style, the distance between the stage and audience is minimized, making it easy to achieve a sense of intimacy.

To adapt to various types of performances, concert, thrust, and semi-proscenium configurations can be realized. In the concert mode, the stage is flattened to be able to fit a full sized orchestra. There is no step between the stage floor and the front row of the audience so that you can experience the unique intimacy with the orchestra which is hard to get in a usual hall. In the thrust stage mode, the side parts of stage are sunken, and in their stead, seat wagons are installed telescoping on rails from both side of stage. In this case, an even closer relationship to the performers on stage can be appreciated. Finally, stage curtains can be suspended from batons above the stage creating a mock stage tower enclosure. Though there is not the range of stage equipment found in theater facilities with a fly tower or large stage wings, this semi-proscenium style allows for staged operas, musicals, and ballet performances.

By using the various stage configurations selectively, it is possible to accommodate a wide variety of entertainment events.

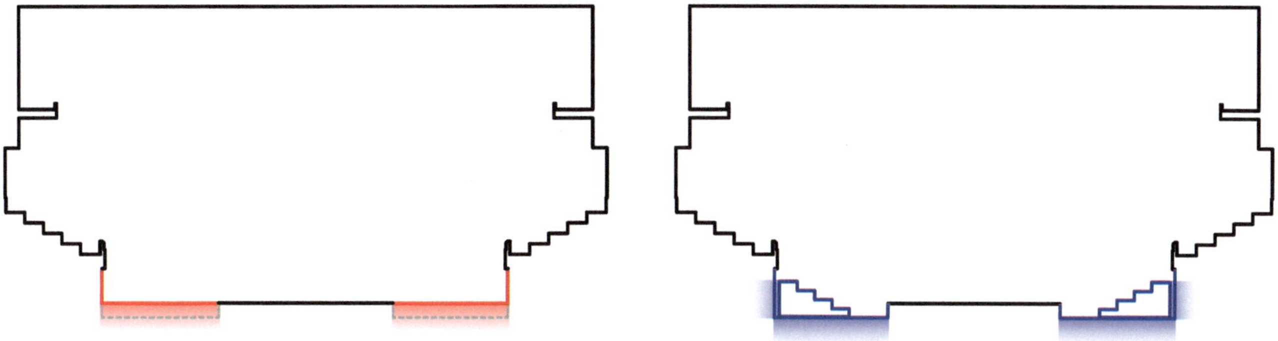

Fig. 13.1 Scope of spatial transformations

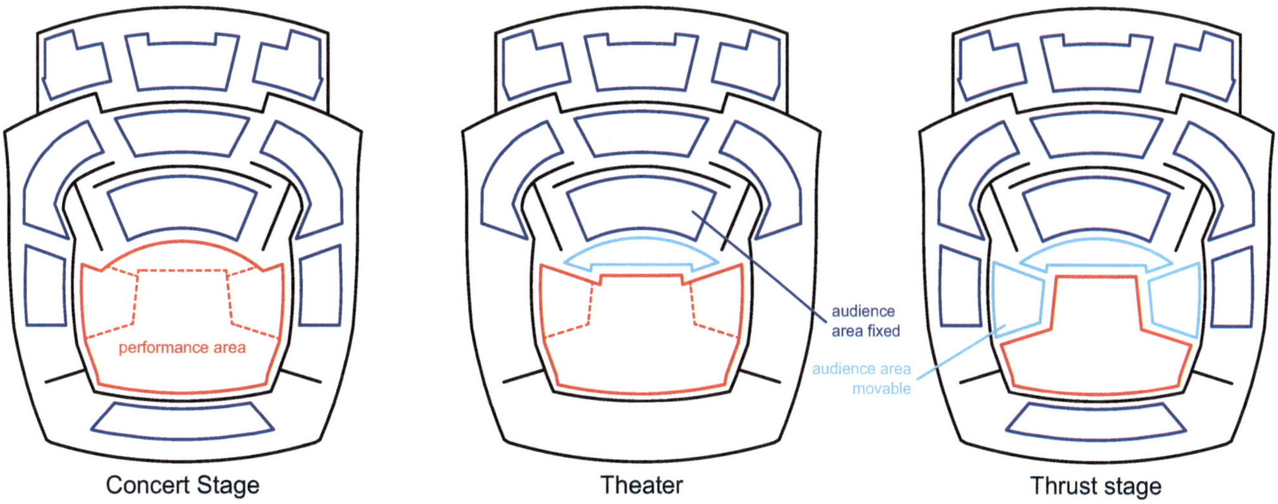

Fig. 13.2 Stage and audience configurations

Thrust stage configuration. The both sides of stage are partially sunken, and in their stead, seat wagons which are telescoped on rails are installed slided from both sides. In this case, an even closer relationship to the performers on stage can be appreciated

Theater configuration. Stage curtains suspended from batons above the stage create a semi-proscenium style and allows for operas, musicals, dance, and ballet performances, etc.

Acoustical Interior and Visual Interior

The ceiling of the hall functionally consists of two-layer structure of acoustical ceiling and visual ceiling. The acoustical ceiling is a poured concrete layer on metal deck, making it very heavy and rigid. On the other hand, the gentle curved ceiling surface that can be seen from the audience is the visual ceiling, consisting of many wooden slats which are suspended over the entire ceiling. In order to make this louver surface as acoustically transparent as possible and also to prevent excessive energy attenuation, careful considerations were made about the size and spacing of the wooden slats. Even the technical tension grid above the stage is acoustically transparent, securing the maximum ceiling height above stage, resulting in a large room air volume sufficient to obtain rich sound.

Acoustically transparent ceiling louver. The gentle curved ceiling which is suspended over the entire audience area is the visual ceiling consisting of many wooden slats, and is acoustically transparent

This double structure was also applied to the walls. Behind the louvers at terrace walls and the both side of stage balcony seats, the carefully angled heavy wall elements made of fiber reinforced gypsum were introduced. In this way, both acoustical and visual elements are nicely incorporated into the hall interior design.

Sound Absorption Curtains

Sound absorption curtains are installed in various positions in the hall to accommodate the expected variety of entertainment events. Horizontally deployed curtains are hidden in the upper part of the visual ceiling and behind the upstage louver wall, almost invisible from the stage or audience areas. These curtains can be drawn and stored mechanically with the push of a button in two to three minutes.

The measured variable range is wide and the subjective impressions are quite different as well. In the concert mode, when all the curtains are stored, the measured reverberation time in unoccupied condition is 2.4 s at 500 Hz, and when the curtains are all drawn, it drops to 1.8 s making the space more neutral for amplified performances.

Opening Series and Beyond

The opening ceremony was held on the 17th of September 2011 featuring a concert by the local Orange County orchestra, Pacific Symphony. The program included Adams's *Short Ride in a Fast Machine*, Rachmaninoff's Piano Concerto No. 2, Prokofiev's *Romeo and Juliet*, and Ravel's *Daphnis and Chloé*, all of which were wonderful. Soon after, on October 23, Polynesia-Micronesia Dance was performed, and a jazz festival was held starting on the 28th. The versatile stage and audience configurations were nicely utilized for the variety of events.

Motoo Komoda

Soka Performing Arts Center	
Location	Aliso Viejo
	California, U.S.A.
Owner	Soka University of America
Architect	Zimmer Gunsul Frasca Architects
Acoustical consultant	Nagata Acoustics (room acoustics)
	Yasuhisa Toyota, Keiji Oguchi, Motoo Komoda, Kayo Kimotsuki Kallas
	Robert F. Mahoney & Associates (isolation and noise control)
	Fred Vogler and Sonitus Consulting (sound system)
Theater consultant	Auerbach Pollock Friedlander
Construction cost	USD 73 Million
Design start	2006
Construction start	January 13, 2009
Construction end	March 2011
Opening date	September 17, 2011
Building size	9,000 m^2
Seating capacity	1,032
Room volume	13,300 m^3
Surface area	4,800 m^2
Volume/seat	12.9 m^3/seat
Volume/surface area	2.8 m
Finish material	
Ceiling	Corrugated metal/concrete, wooden louvers
Walls	Glass fiber reinforced gypsum/gypsum board, wooden louvers
Audience floor	Concrete
Stage floor	Alaskan Yellow Cedar
Seat manufacturer	Irwin Seating Company

Table 13.1 Soka University Concert Hall—acoustical metrics at 500 Hz

	Concert configuration		Thrust configuration	
	Reflective condition	Curtains deployed	Reflective condition	Curtains deployed
RT unoccupied	2.4 s	1.8 s	2.2 s	1.7 s
RT occupied	2.2 s	1.7 s	2.0 s	1.6 s
EDT			2.5 s	
C_{80}			−1.6 dB	
D_{50}			41%	
G			8 dB	

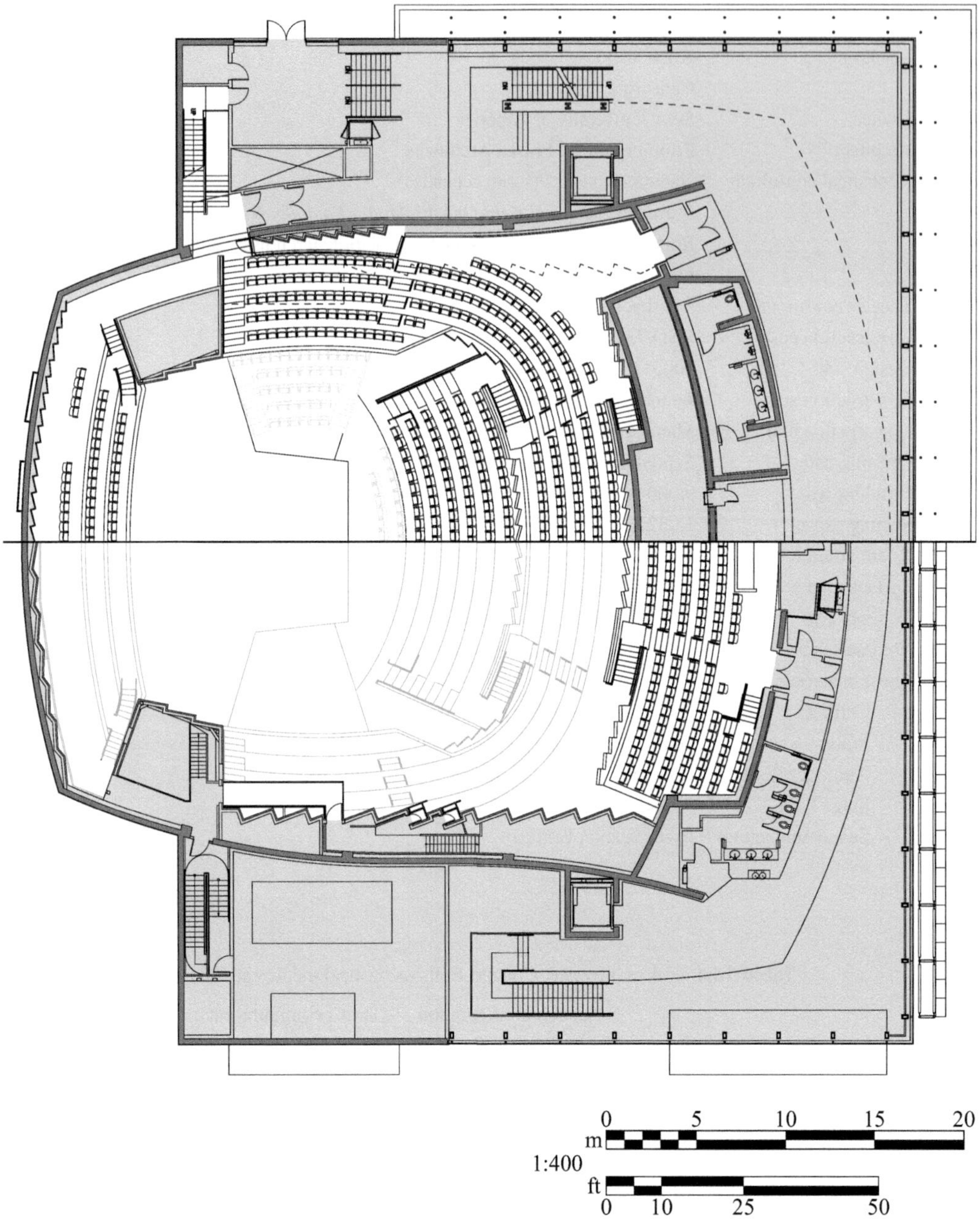

Fig. 13.3 Main and terrace levels plan
Courtesy of ZGF Architects

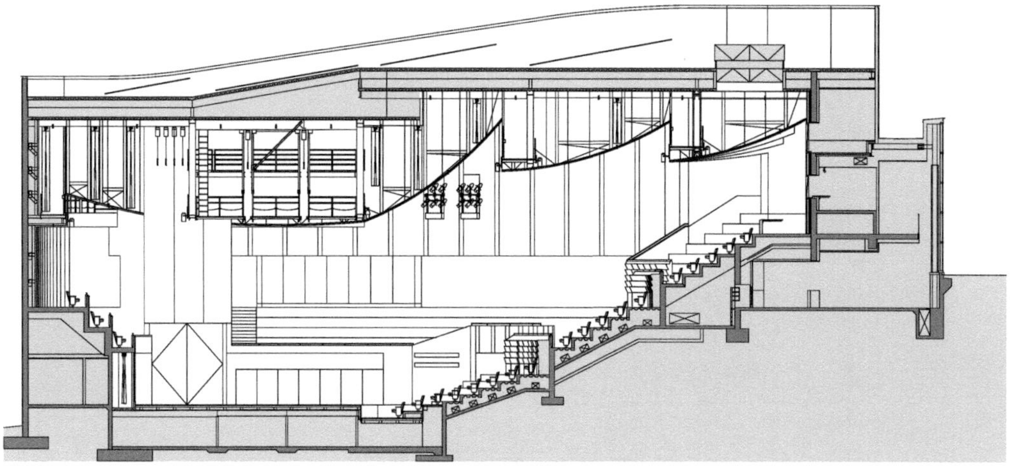

Fig. 13.4 Longitudinal section
Courtesy of ZGF Architects

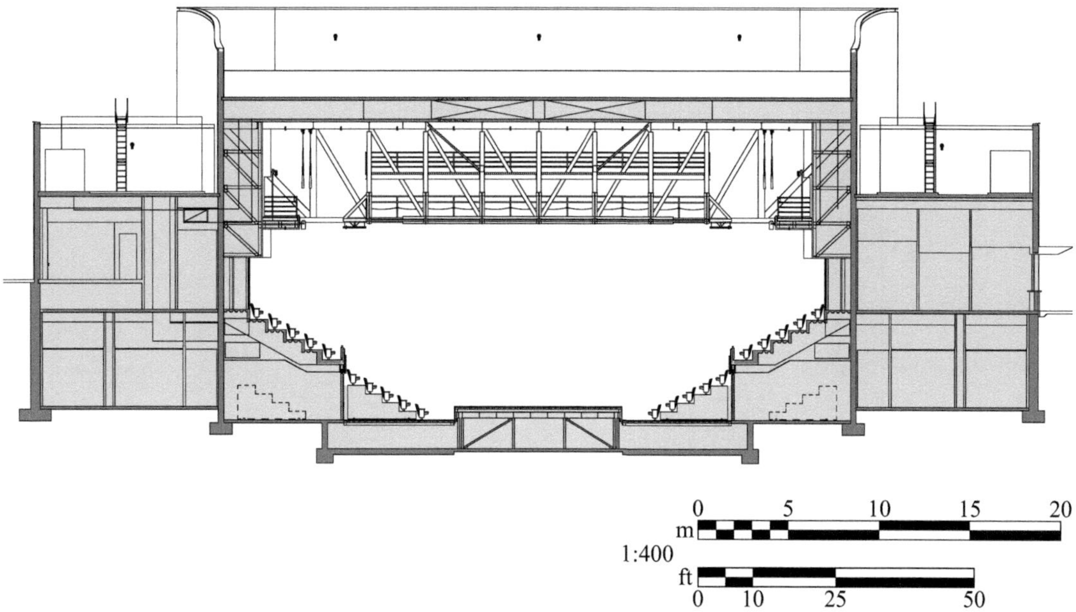

Fig. 13.5 Cross section
Courtesy of ZGF Architects

14. Isabella Stewart Gardner Museum

View from first row of seats. The hall emphasizes verticality, and the absence of any clear orientation or "front." The skylight floods the space with natural light

Y. Toyota et al., *Concert Halls by Nagata Acoustics*, https://doi.org/10.1007/978-3-030-42450-3_14

Isabella Gardner opened a museum in the form of a Venetian palazzo to the Boston public on New Year's Day, 1903. Situated in the then-wilderness of the Back Bay Fens, a part of Fredrick Law Olmsted's "Emerald Necklace" encompassing the city, the museum opened first with a concert before the galleries containing thousands of artworks laid out by Mrs. Gardner herself could be visited. One hundred years later, the museum handles two hundred thousand annual visitors, substantially more than the four thousand visitors received in each of the first years. This increase has placed a strain on the historic building, which the Board of Trustees decided to mitigate by embarking on an ambitious plan to add 6,500 m^2 (70,000 ft^2) of space to house the functions required of a museum operating in the twenty-first century. In comparison, the historic Palace encompasses 5,300 m^2 (57,000 ft^2). The board of the museum selected Renzo Piano as the architect in late 2004, based on the strength of his work at the Nasher Sculpture Center, the Menil Collection, and the Beyeler Foundation.

Piano approached the task of adding to a treasured historic building by framing it in familial terms: the new addition "must be the respectful nephew to the great aunt." The new wing of the museum would sit fifteen meters (fifty feet) away from the original structure, building a conversation between the Renaissance-inspired, inward-focused Venetian palazzo and the twenty-first century technological-modern, transparent accompaniment. Contained within the new wing are offices, a new entrance lobby and "living room" (the orientation center for the museum), a café, gift shop, educational spaces, greenhouses, artists apartments, conservation labs, a gallery space, and the music hall.

Named Calderwood Hall in recognition of major donors to the expansion project, the hall has 296 seats divided across four levels. Even though it is not difficult to achieve a feeling of intimacy with such a small audience capacity, the image and intention presented by the music director Scott Nickrenz to the design team was that of the Caio Melisso, an eighteenth-century opera house in Spoleto, Italy. This historic horseshoe-shaped space also seats only 350 viewer-listeners, and offers an atmosphere of unparalleled intimacy by allowing most of the audience to make direct visual contact with each other, since the seats are divided over four levels. Calderwood Hall takes this approach more radically, by placing the performers

Opening concert. During concerts, the view of the audience is as dynamic as the view of the performers

in the center of the square room. Viewer-listeners are seated on the same floor level with the performers, as well as on three levels of balcony surrounding the performers on all four sides of the square room. Two rows of seats share the stage level on each of the four sides of the room, totaling 116 seats on the main floor. Each balcony has only one row of 15 seats on each of four sides, or 60 per balcony, totaling 180, or almost two-thirds of the total capacity. Four equal sides give the room a groundbreaking, omni-directional character, a form that brings into the new building a reminder of the tall, glorious courtyard at the heart of the historic Palace.

Skylights. The skylight at the top of the hall is divided into nine pyramids, for a total of thirty-six 16 mm (5/8 in) thick laminated glass segments to eliminate the parallel condition with the floor. Pre-cast curved concrete panels next to the skylight diffuse sound in all directions

The first impression that greets the viewer-listener and performer upon entering the hall is one of extreme verticality. The hall is 13.5 m (44 ft) tall, crowned with a large skylight 5.5 m (18 ft) square. The skylight is divided into a three-by-three grid, with each grid space further divided into pyramids, where the glass sides of each pyramid are tilted 20° from the horizontal in order to avoid harsh reflections from the glass and to prevent flutter echoes in combination with the floor below. Flanking the skylight on each side are curved, pre-cast concrete panels. The balconies themselves are radically thin constructions, little more than pre-cast concrete panels with a wood floor on top, laid in a steel frame suspended from the ceiling. The exposed concrete ceilings have small inverse pyramids cast into the panels to diffuse high-frequency sounds. Glass sheets form the edge of the balcony overlooking the stage, making plainly visible the thinness of the balcony structure. The glass portion of the balustrade allows for good sightlines to the stage, even from the top balcony nine meters (twenty-nine feet) high. The balustrades function acoustically to hold more sound in the taller, central volume of the atrium-like space, enabling the crucial early reflections to develop in the compact room air volume of 2,500 m^3 (88,300 ft^3). All the walls in the room are clad in thin, laser-cut wood paneling, providing a visual screen that is acoustically transparent. The screen allows the wall behind to act as the heavy, acoustically effective reflecting surface, and also to allow the ductwork mounted in the cavity to act as diffusive elements to break the parallel condition of the walls. The panels themselves, at 3 mm (1/8 inch) thick with slot-shaped holes resulting in a 40% open area, are also highly effective at scattering sound so that a flutter echo is not heard.

Since the inauguration of the museum in 1903, music has played a central role in the programming of the museum. In recent decades, concerts were held in the Tapestry Room, a long, wide gallery which seated 260–300 people facing a raised stage at the end of the room. The programming of the music series invites many different chamber ensembles and solo performers, as well as the eighteen-member A Far Cry, the Chamber Orchestra In Residence at the museum. A wide variety of musical styles and eras are presented, from early Renaissance polyphony to contemporary jazz. Moving from a mono-directional performance space into the radical, omni-directional Calderwood Hall required a new approach for many musicians. The lack of a clear "main" audience area towards which to orient themselves presents a new question to performers, since the vast majority of performance spaces allow easy orientation. The solution in this space emphasizes the omni-directionality of the room by encouraging the performers also to face the center of the room, and each other directly. In this strategy, no one side of the room is given any preference over another side of the room. Ensembles can also be assisted by enhancing and easing visual communication between the individual members, since by orienting themselves in a circle, they should all have good direct eye contact with each other. The audience is drawn into the performance by the extreme intimacy between the performers, and the closeness to them, since the stage measures only 9 m (30 ft) square. The sensation of sitting together with the performers is almost like enjoying a concert in the living room at home with friends.

Calderwood Hall had its first performances at the grand opening of the museum expansion on January 12, 2012. Over four days, a roster of world-renowned musicians opened the hall: Dame Kiri Te Kanawa, Jeremy Denk, Paavali Jumppanen, Corey Cerovsek, the Jupiter and Borromeo String Quartets, and others. Cellist Yo-Yo Ma also played the Haydn Cello

Concerto with A Far Cry chamber orchestra, facing the middle of the room, together as a part of the orchestra. The hall has a relatively short reverberation time, in keeping with the low volume, and together with that an excellent clarity. Each of the balconies has a unique acoustical character, as well as the main floor. The handrails for the balcony are designed to allow the viewer-listener to easily lean forward and comfortably watch the performance while leaning on the rail. From either position, whether sitting back or leaning forward, a different and unique sound can be heard. This hall reminds the viewer-listener that she is an active participant in the performance, gathered to communicate with other people through music.

Daniel Beckmann

Audience balcony. The three balconies are only one row deep, equal on all four sides. Small articulations cast into the concrete ceiling scatters sound at the high frequencies. Perforated wood walls behind the seats also soften the parallel shape of the hall

Public approach from park. The new building stands a respectful distance away from the historic palace (brick wall visible at the right edge). The concert hall is the second cubic volume on the right, clad in green, patinated copper panels, floating above the glass lobby level

	Isabella Stewart Gardner Museum
Location	Boston
	Massachusetts, U.S.A
Owner	Isabella Stewart Gardner Museum
Architect	Renzo Piano Building Workshop
Executive architect	Stantec
Acoustical consultant	Nagata Acoustics (room acoustics)
	Yasuhisa Toyota, Keiji Oguchi, Motoo Komoda, Daniel Beckmann
	Robert F. Mahoney & Associates (isolation and noise control)
Construction cost	USD 114 Million
Design start	May, 2006
Construction start	June, 2009
Construction end	November, 2011
Opening date	January 18, 2012
Building size	6,500 m^2
	Calderwood Hall
Seating capacity	296
Room volume	2,500 m^3
Surface area	1,400 m^2
Volume/seat	8.6 m^3/seat
Volume/surface area	1.8 m
Finish material	
Ceiling	Glass, Pre-cast concrete
Wall	Perforated wood paneling, Gypsum panel
Audience floor	Wood
Stage floor	Alaskan Yellow Cedar
Seat manufacturer	Poltrona Frau
Model scale	1:24

Table 14.1 Calderwood Hall—acoustical metrics at 500 Hz

RT unoccupied	1.0 s
RT occupied	0.9 s
EDT	1.1 s
C_{80}	1.0 dB
D_{50}	41%

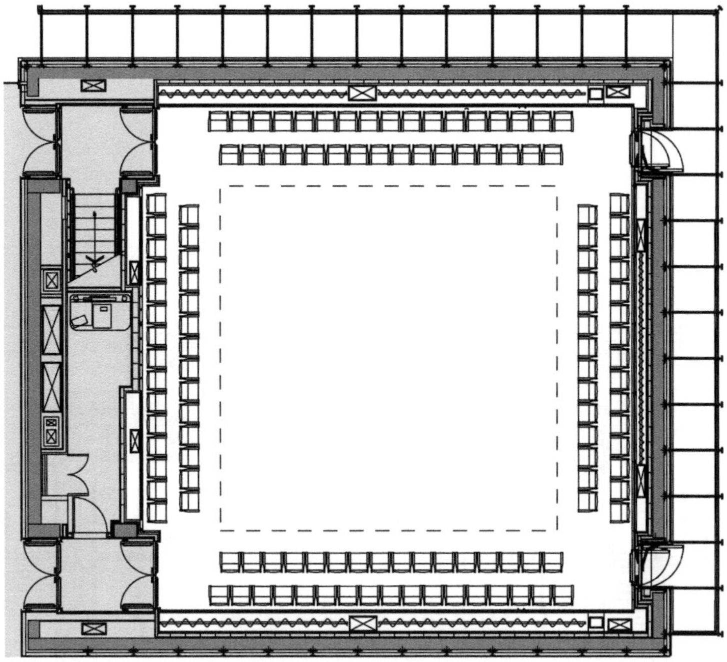

Fig. 14.1 Main level plan
Courtesy of Renzo Piano Building Workshop

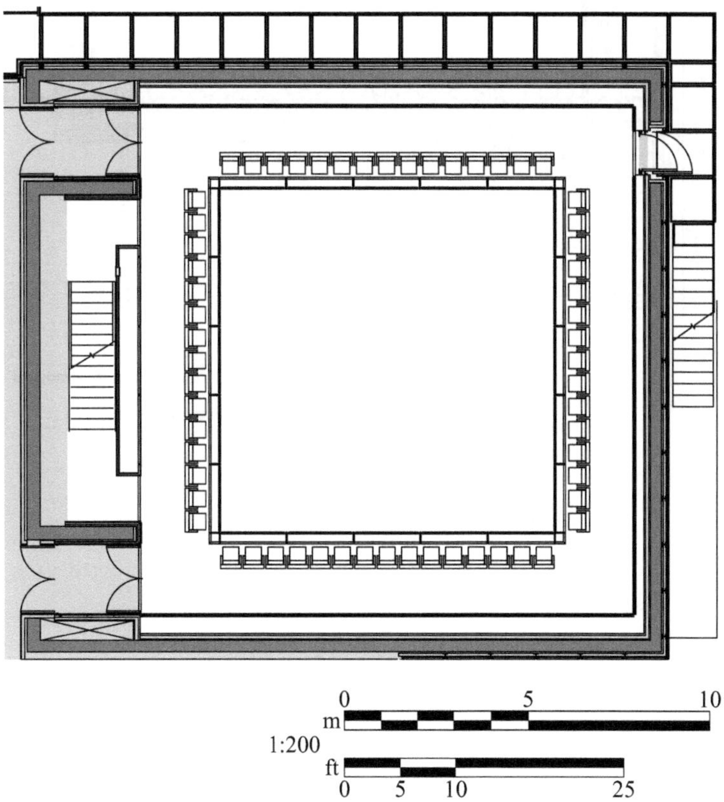

Fig. 14.2 Balcony plan
Courtesy of Renzo Piano Building Workshop

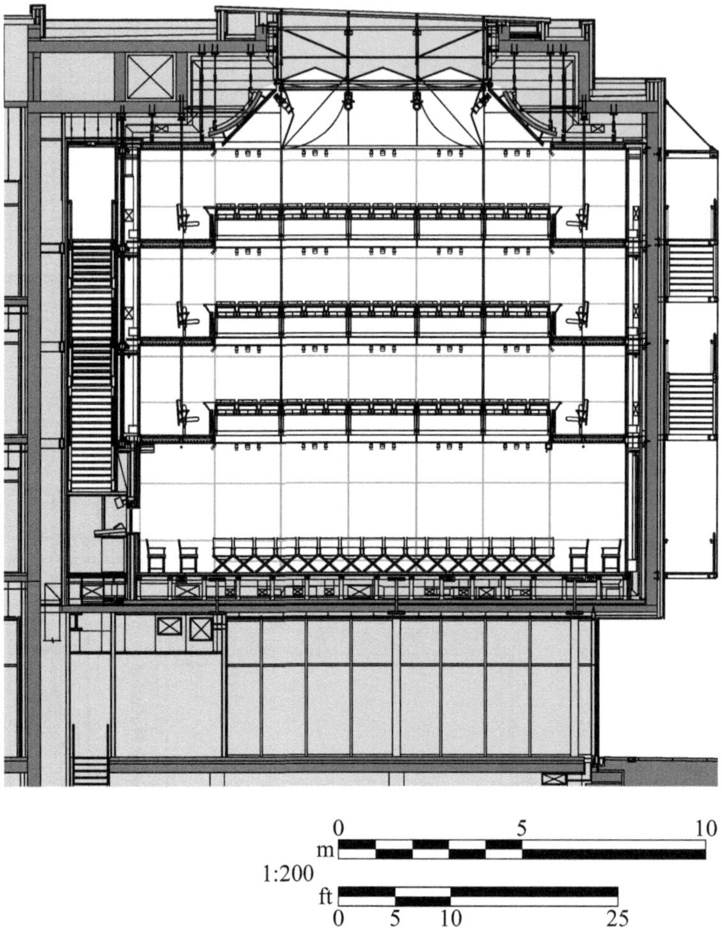

Fig. 14.3 Section
Courtesy of Renzo Piano Building Workshop

15. USC Brain and Creativity Institute

View from entrance level. The audience rake steeply descends all the way to stage level, where the front two rows wrap fully around stage left, recessed slightly into the wall. Consequently, everyone can feel the calm and intimate space

Y. Toyota et al., *Concert Halls by Nagata Acoustics*, https://doi.org/10.1007/978-3-030-42450-3_15

The University of Southern California was established in 1880 in downtown Los Angeles and is the oldest private university on the west coast. In addition to architects Frank Gehry and film director George Lucas, USC has produced numerous celebrities in the field of art, and has prestigious departments in research medicine and information engineering. Originally, the Information Sciences Institute of USC managed the IP addresses for the entire world.

The Drs. Damasio and BCI

The university's Brain and Creativity Institute (BCI), founded in 2006 by Drs. Antonio and Hanna Damasio, is literally investigating the relationship between human brain function and human creativity. The Drs. Damasio are making use of MRI, electroencephalogram, and other cutting edge medical technology for treating diseases and also for research on artistic activities.

The BCI new building project was completed in 2012 and includes a new chamber hall adjacent to the existing neuroscience research facilities which already draw experts from across the globe. In the new hall, they are continuing their ambitious attempt to try to understand the link between brain science and art. The space of just over 100 seats named Joyce J. Cammilleri Hall is designed not only as a space for classical music recitals and master classes, but also as a venue for lectures, presentations, and other experimental uses.

The brick facade, which blends in with the surrounding campus, and the bright lobby contrast with the darkened, chamber hall. We enter the hall facing away from the stage at the upper side gallery. This short, open walkway gives our eyes a chance to adjust to the black interior finishing as we spiral down to our seats. Asymmetric curvature in plan and angled walls create a more dynamic space than any simple black box chamber hall. The audience rake steeply descends all the way to stage level, where the front two rows wrap fully around stage left, recessed slightly into the wall. Consequently, everyone is unified in the calm and intimate space.

The Acoustics of a Unique, Experimental Chamber Hall

Assuming solo recital and chamber music performance, the ceiling height was set considerably high at 13.5 m at the center of the stage, and the room volume reaches 1,100 m^3. This is stunningly tall for a concert space of only 100 seats. For the stage floor, a wooden structure over an air void was engineered to obtain appropriate resonance.

The basic shape of the space was carefully designed so as not to have parallel surfaces facing each other. Above the audience seats at stage left side, a soffit is designed which can produce the reflections necessary to support the performers on stage. In addition, to avoid sound focusing caused by the gentle concave shape of some walls, irregular block shapes were added to the lower part to scatter high frequency sound. As the result, no detrimental phenomena such as flutter echoes are found.

Furthermore, for lecture or other types of events which prefer a relatively less reverberant field, the top half of the walls can be entirely covered by motorized acoustical curtains, and the upstage wall on stage can be covered by manual drawn curtains.

When construction was almost completed, we heard the first notes in the hall, performed by violinist Midori, who holds the Jascha Heifetz Chair at USC's Thornton School of Music. It was a brief performance, but it was also the moment that the marvelous rich and clear acoustics which was intended by our design work was achieved.

After the tape cutting ceremony on November 6, 2012, recitals by solo piano and cello were performed. Both were wonderful performances and all people concerned were very satisfied. Someday, some tremendous things may be discovered about the workings of our human brain and creativity in this tiny space. If the acoustics could contribute to such breakthroughs, that would be wonderful.

Motoo Komoda

View from stage. To avoid sound focusing caused by the gentle concave shape of side walls, irregular block shapes were added to the lower part to scatter high frequency sound

	USC Brain Institute
Location	University of Southern California, Los Angeles California, U.S.A
Owner	University of Southern California
Architect	Perkins+Will
Acoustical consultant	Nagata Acoustics (room acoustics)
	Yasuhisa Toyota, Motoo Komoda, David So
	Robert F. Mahoney & Associates (isolation and noise control)
Design start	Q1 2010
Opening date	November 6, 2012
Building size	1,850 m^2
	Joyce J. Cammilleri Hall
Seating capacity	101
Room volume	1,100 m^3
Surface area	790 m^2
Volume/seat	11.0 m^3/seat
Volume/surface area	1.4 m
Finish material	
Ceiling	Plaster on metal lath
Walls	Gypsum board, Glass fiber reinforced concrete
Audience floor	Concrete floor
Stage floor	Alaskan Yellow Cedar
Seat manufacturer	Krueger International

Table 15.1 USC Brain Institute—acoustical metrics at 500 Hz

	Reflective condition	Curtains deployed
RT unoccupied	1.2 s	0.8 s
RT occupied	1.1 s	0.8 s

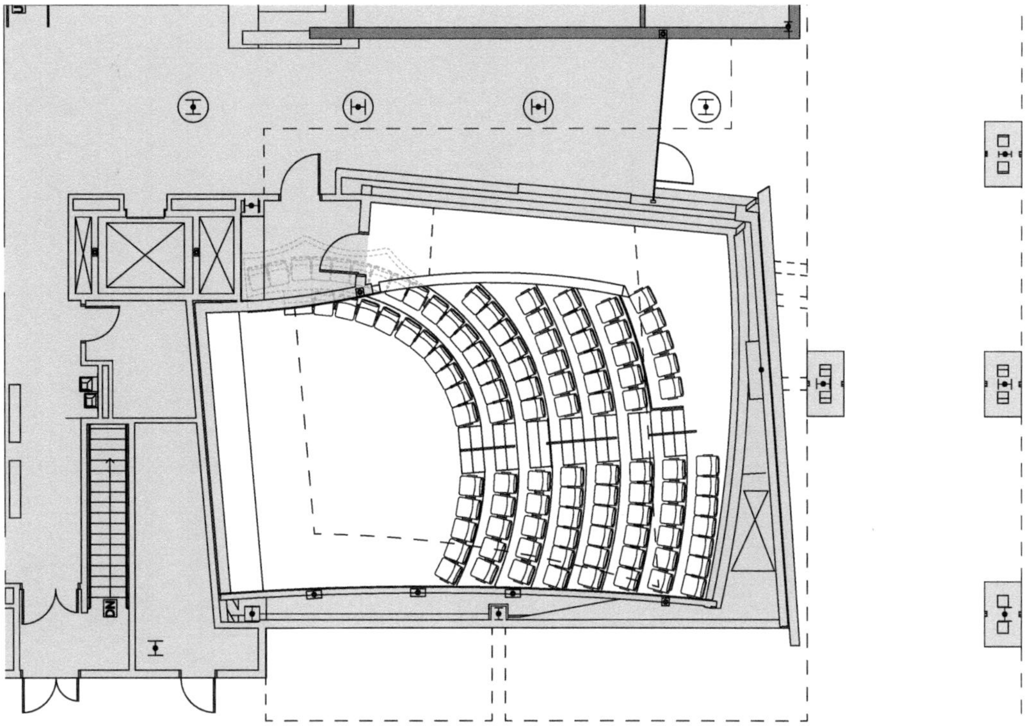

Fig. 15.1 Plan
Courtesy of Perkins + Will

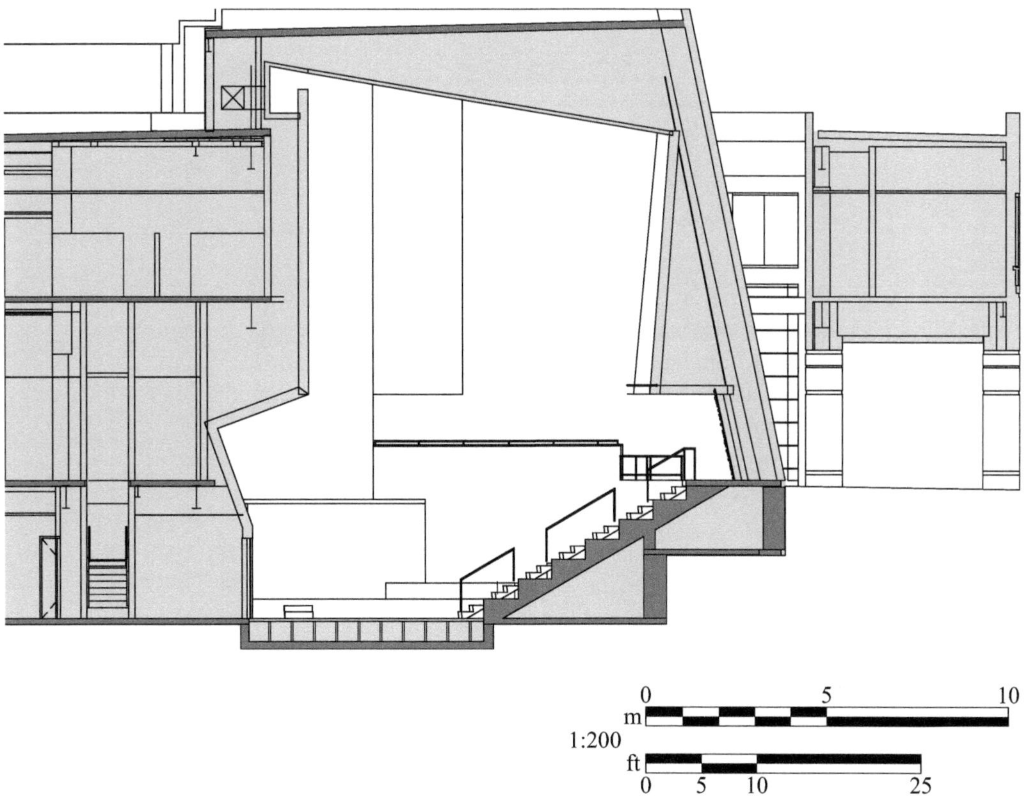

Fig. 15.2 Section
Courtesy of Perkins + Will

16. Bing Concert Hall

View from the top rear audience seats. The arena style where the stage is surrounded by blocks of audience seats was chosen. The large-sized curved wall panels created a unique entertainment space and foster a strong feeling of intimacy between the musicians on stage and the audience

Y. Toyota et al., *Concert Halls by Nagata Acoustics*, https://doi.org/10.1007/978-3-030-42450-3_16

In various fields of humanities and sciences, Stanford University's role at the center of Silicon Valley in California cannot be overstated. Among the many fields, Stanford's music department provides a broad education ranging from classical music to jazz and contemporary music, and continues into crossover research in state of the art technology, music, and acoustics.

As the production organization for performing arts, Stanford Live had been working for more than 40 years to produce concert performances and to help the activities of students and other artists in residence. Many groups of students, including Stanford Symphony Orchestra, Stanford Philharmonia Orchestra, and "New Ensemble," dedicated to performing new music, had been performing as many as 150 concert annually at Stanford's recital halls and Memorial Church. Finally, in 2013, Bing Concert Hall opened their doors as a long-awaited new space for performing arts.

In addition to the concert hall, the new facility includes a rehearsal room doubling as a small event space, dressing rooms, and offices within a 10,000 m^2 footprint. Ennead Architects (formerly Polshek Partnership) completed the architectural design of the project, and the total construction cost was USD 112 Million.

The 842 seat concert hall at the core of the building would typically be considered as a medium size performing space, and one would assume the main usages might be chamber music concerts, smaller group performances, or solo recitals. However, orchestra performances were programmed as well from the initial phase of this project. To this end, a large-size stage which can accommodate a full-size orchestra was required as one of the most important design conditions.

As the rough room shape, the arena style where the stage is surrounded by blocks of audience seats was chosen. The large-sized curved wall panels created a unique performance space and foster a strong feeling of intimacy between the musicians on stage and the audience. This feeling of dialogue between the musicians and the audience is totally different from the traditional layouts like the shoe-box style.

Notably, the height of stage floor here was designed to be at the same level as the first row of audience seating, while most other existing halls have height difference at the stage edge. In other words, the stage floor is the lowest position in the hall space. This arrangement provides direct sound energy effectively and made for better visual sight lines for the front rows of seating. While this feature was a relatively new challenge for the design team, the result is that the intimate feeling in the hall space is emphasized.

View from stage side audience seats. To further pursue the intimate atmosphere, the hall has an oval shape in plan. To prevent undesirable sound focusing caused by the oval shape, the walls consist of large, three-dimensionally convex shapes later called sails by the design team were introduced

To further pursue the intimate atmosphere, the hall has an oval shape in plan. As easily imagined, however, the risk of acoustical focusing phenomena is much higher than in other conventional hall shapes, such as a rectangular plan. To prevent such undesirable focusing effects, three-dimensionally convex shapes—later dubbed "sails" by the design team—were introduced. The sails are made of fiber reinforced polymer filled with heavy concrete. The soffit surfaces which are created from the overhang of these large-scale panels create the beneficial early sound reflections which are directed towards the stage and the front audience area. The ceiling was designed as heavy cement with gently curved convex shape as well. During the design phase, a 1:24 scale acoustical model of the hall was built and used to test the general shape of hall and, after some minor adjustments, no possible detrimental acoustic effects such as focusing or echoes were found.

Various surface textures on the walls. Wavy shapes were introduced below the sails. On the terrace walls, similar wave-like wooden panels were installed. These relatively small textured elements promised sound scattering effects at higher frequency bands

Furthermore, to prevent strong and harsh sound reflections from large-sized smooth surface, various types of interior surface textures were introduced. On the terrace walls which separate the audience in blocks and provide very early sound reflections to each closer audience areas, wave-like wooden panels were installed on a glass-fiber reinforced concrete (GFRC) base. Similar wavy shapes were introduced on the walls below the sails. The surface of these walls panels has been sprayed with a thin layer of powdered marble. On the main ceiling, in addition to the wave geometry, the cement surface was sand-blasted. These relatively small textured elements promised sound scattering effects at higher frequency bands.

The full-size stage surrounded by the audience seats and the tall ceiling of the hall resulted in a luxurious room volume of 17,000 m^3. The resulting room air volume per seat reaches over 20 m^3. The room volume and heavy interior materials help to achieve the long and rich reverberation in the hall.

Variable curtains in between sails. In the recessed portions between the sails, acoustical curtains area can be changed with one-push buttons to adjust the reverberation, not for the natural acoustics but only for the programs with amplified sound

In order to accommodate a variety of music performances, including not only classical but also modern amplified music, several systems of variable acoustics were introduced in the hall. In the recessed areas between the sails, acoustical curtains can be deployed or stored with one-push buttons to adjust the reverberation. Next, under the large-scale convex shape panels, the acoustical curtains can be set on the rails or be stored in the boxes hiding within the walls. Additionally, a motorized vertical system of sound absorptive panels was introduced behind the grilled wall surfaces on stage to adjust the reflections from the hard reflective walls as needed. Finally, portable, self-standing baffles and carpets on stage were made specifically for amplified music performances.

On January 10, 2013, the Bing Concert Hall building was officially handed over to Stanford University. The next day, the resident St. Lawrence String Quartet and the San Francisco Symphony conducted by Michael Tilson Thomas performed a superb opening night concert. In the following weeks, pianist Emanuel Ax, cellist Yo-Yo Ma, and violinist Midori each took the stage to celebrate the birth of the brand new space. Through the listening experiences of orchestral, chamber music, and solo concerts which were performed in the hall, the excellent sound clarity and the rich sound in the intimate space were confirmed. Stanford University and the main donors Helen and Peter Bing expressed their joy at the successful completion of concert hall project.

Motoo Komoda

Bing Concert Hall	
Location	Stanford University, Stanford
	California, U.S.A.
Owner/User	Stanford University
Architect	Ennead Architects
Acoustical consultant	Nagata Acoustics (room acoustics)
	Yasuhisa Toyota, Motoo Komoda, Kayo Kimotsuki Kallas, Daniel Beckmann
	Robert F. Mahoney & Associates (isolation and noise control)
	Fred Vogler and Sonitus Consulting (sound system)
Theater consultant	Fisher Dachs Associates
Construction cost	USD 112 Million
Design start	January 2008
Construction start	May 2010
Construction end	December 2012
Opening date	January 11, 2013
Building size	10,400 m^2
Seating capacity	842
Room volume	17,000 m^3
Surface area	6,300 m^2
Volume/seat	20.2 m^3/seat
Volume/surface area	2.7 m
Finish material	
Ceiling	Concrete, fiber reinforced polymer, sand-blast
Walls	Concrete, fiber reinforced polymer, sprayed with a thin layer of powdered marble
	Glass-fiber reinforced concrete, wood rib fascia
Audience floor	Beech on concrete
Stage floor	Alaskan Yellow Cedar
Seat manufacturer	Ducharme Seating
Model scale	1:24

Table 16.1 Stanford Bing Concert Hall—acoustical metrics at 500 Hz

RT unoccupied	2.2 s
RT occupied	2.1 s

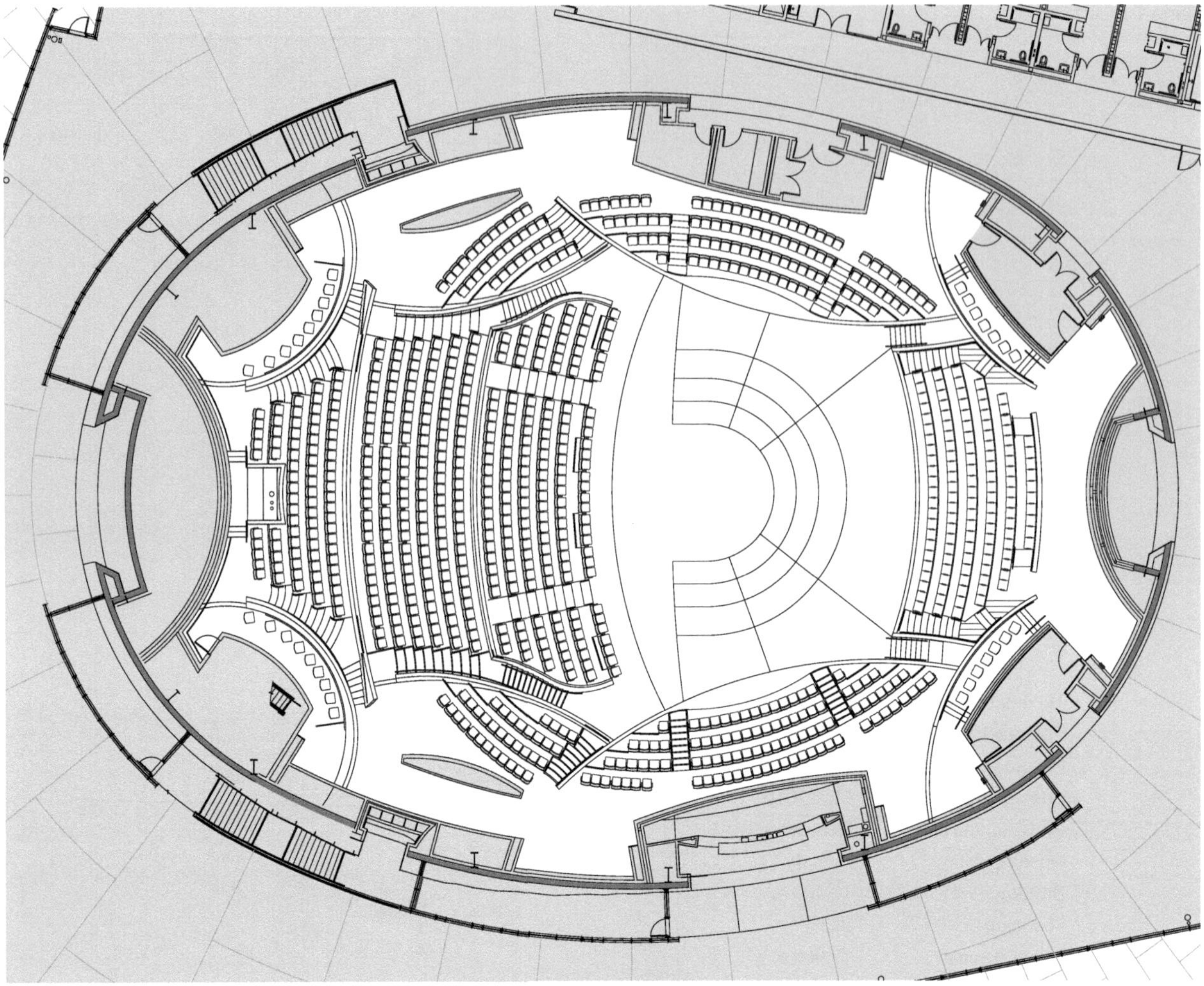

Fig. 16.1 Plan
Courtesy of Ennead Architects

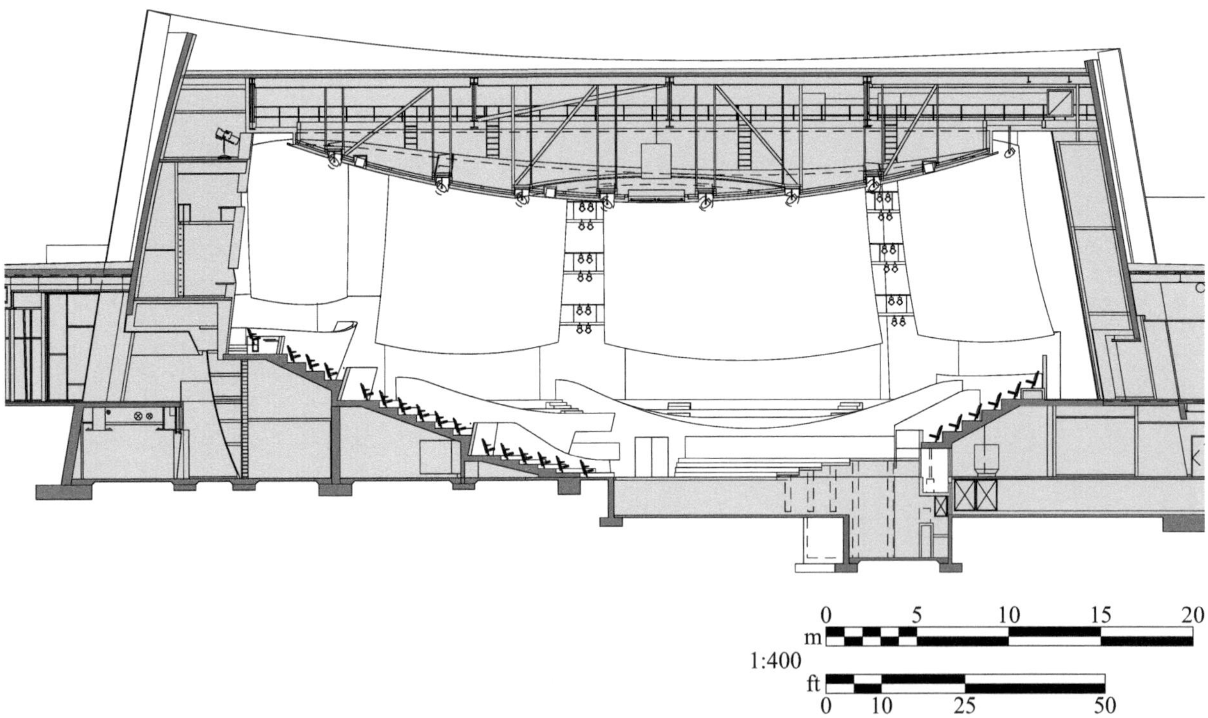

Fig. 16.2 Section
Courtesy of Ennead Architects

17. Charles Bronfman Auditorium

View from sloped side seating in Lowy Concert Hall. Visible changes introduced by the renovation include a new stage layout and a set of mechanical risers, new titled wall reflectors and terraced seating in the central tier. The acoustical ceiling height was significantly increased but the original ceiling was recreated with acoustically transparent expanded metal and wood grill

Y. Toyota et al., *Concert Halls by Nagata Acoustics*, https://doi.org/10.1007/978-3-030-42450-3_17

Located in central Tel-Aviv (Israel), Frederick R. Mann Auditorium was inaugurated in 1957. Planning began as early as 1951, just a few years after the founding of the state of Israel in 1948. Also known to Israelis as Heichal Hatarbut (Culture Palace), it serves as a multipurpose hall for a wide range of performances and events, and has been the main residence of Israel Philharmonic Orchestra (IPO) since its opening. Upon its completed renovation in 2013, the building was renamed Charles Bronfman Auditorium, in recognition of its principal contributor. In the years following the inauguration, the renovated main hall was named Lowy Concert hall.

Original Design of the Hall and Plans for Renovation

The building was designed by architects Dov Karmi and Zeev Rechter, both prominent figures of Israeli architecture, in association with Rechter's son, Yaakov. Its design is representative of the style unifying Tel-Aviv's White City, with strong Bauhaus and Modernist influences. Acoustical design for the main hall was provided by Bolt Beranek and Newman, and the project is featured in Leo Beranek's seminal publication "Concert Halls and Opera Houses - Music, Acoustics, and Architecture."

Lowy Concert Hall (previously Frederik R. Mann Auditorium) before renovation. In spite of the need for acoustical improvements to the design of the hall, preservation of the visual appearance was paramount for the project. An acoustically transparent ceiling recreated the original ceiling, while wall paneling was either refurbished or rebuilt identical and chair design was kept

With an original seating capacity of 2,715, the audience layout followed a fundamental fan shape with the stage located on the narrow end, in front of three rows of niched chorus seating. The central seating section, approximately 20 m deep, is framed on both sides by sloped terraces 7–9 m wide. The terraces climb almost 7 m to reach an extensive balcony covering the full width at the rear of the hall and extending over 15 rows deep.

With an average height of less than 14 m from the stage level, and as low as 8 m above the stage, the original main concrete ceiling was visually concealed by largely open wood mesh around the periphery and pyramidal reflectors above the stage and central seating section. These elements contributed in lowering the visual and acoustical ceiling height.

Plans to refurbish the building and renovate the hall were initiated in 2001. More than 40 years after its opening, the intense programming had outgrown the technical capabilities of the hall and both the interior and exterior were in need of refurbishment. Additionally, the acoustics of the auditorium were judged overly dry and lacking in intimacy, and Music Director Zubin Mehta and the orchestra had long voiced their difficulties on stage.

Initial design strategies intended to modify the room shape and audience layout, and to raise the roof of the building. However, when UNESCO declared Tel-Aviv White City District—to which the building belongs—a World Heritage Site, any change to the building exterior and overall visual appearance suddenly became impossible. Additionally, local historical preservation groups assembled to object any significant change even to the interior, and in particular to the visual appearance of the concert hall.

Architects Kolker Kolker and Epstein (KKE), from Jerusalem, were appointed in 2004 to oversee the renovation project but the project was once halted before being revived in late 2007. The ambitious plan they developed involved excavations under the large public square in front of the building to add up to 5,000 m^2 to the building without altering its appearance. This also allowed moving some services away from the concert hall, therefore freeing up space around the concert hall within the building envelope.

How to Improve Without Changing?

Nagata Acoustics was invited in October 2007 to consult on the acoustical renovation of the hall. We started our work by visiting the hall to experience concerts and rehearsals, discuss with Maestro Mehta and IPO musicians, as well as orchestra and concert hall management. We could confirm the most critical shortcoming to be the generally low ceiling, especially above the stage, resulting in a constrained volume and difficulty of ensemble hearing and projection to the hall. In addition, the seats on the central seating section, although closer to the stage, lacked in acoustical intimacy and clarity. Finally, the thin paneling cladding the walls resulted in an excess in low frequency absorption.

A computer model of the hall was subsequently built from plans and photos to conduct geometrical studies and acoustical simulations. Initial studies confirmed the preliminary assessments, and the potential for improvement by raising the ceiling of the hall. But most importantly, they also showed that improvements could be obtained without altering the building roof and exterior. We then developed a series of conceptual proposals, which were discussed and developed into a formal design in close collaboration with Opher Kolker and his team at KKE. Above all other considerations, the fundamental room shape, audience layout, and relation between stage and viewer-listeners had to be maintained. Therefore, the end-stage fan shape was conserved in full, in its proportions and dimensions.

Thanks to the displacement of air-handling units previously located above the stage, a new interior concrete ceiling was designed at a raised height of up to 14 m above the stage and 16 m on average across the hall. This resulted in an increase of the room air volume by close to 3,000 m^3. In order to maintain the visual aspect of the original suspended reflectors, the pyramids were built out of expanded metal mesh combining acoustical transparency with sufficient visual opacity.

The central seating section was divided into three blocks by low tilted walls providing the previously lacking early reflections to this area, while the increased rakes facilitated sightlines to the stage. Row depths were also increased to improve viewer-listener comfort. Along the side walls, new tilted panels were added to serve much needed early reflections to the side terraces. Both the new walls and panels, and the renewed side walls were built with massive construction approximately 50 kg/m^2 to enhance low frequency response.

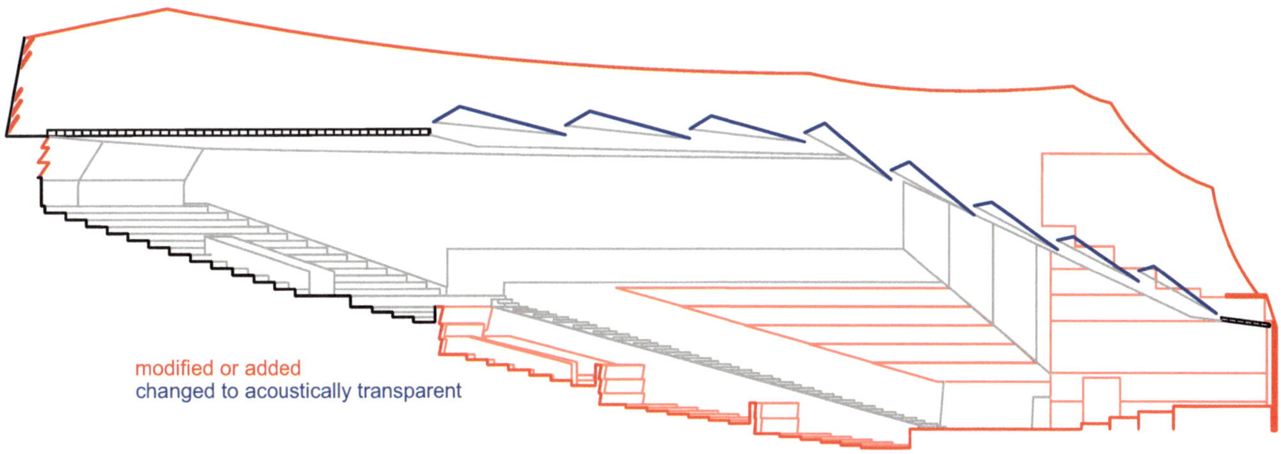

Fig. 17.1 Scope of renovation in Lowy Concert Hall

The stage area was remodeled to reach a total depth of 16 m suitable for large ensembles as well as smaller ensembles with choir. It was equipped with semi-circular mechanical risers, complemented by portable risers downstage, to offer both more compacity and more flexibility to the ensemble layouts. The risers also participate in the enhanced acoustics on stage, facilitating both communication and sound projection. Additionally, horizontal soffit reflectors were created around the stage to provide effective reflections to the musicians, but they remain invisible behind acoustically transparent cladding.

Construction began on site in August 2011 at the end of the artistic season, with plans to re-open in October 2012, but construction works eventually completed in early 2013.

First Notes in the New Acoustics

The first rehearsal in the renovated hall was scheduled in March 2013. At that time, construction was not fully completed and parts of the visually suspended ceiling were still missing. Both the building as a whole and the concert hall itself still felt as a construction site, with exposed scaffolds, harsh projector lights, and half-painted surfaces. Nevertheless, the first notes quickly confirmed how much the hall had gained in liveliness and richness of sound, coupled with a new sense of proximity and intimacy with the stage. Over the following weeks leading to the opening, the musicians gathered more time on stage. Gradually becoming accustomed to the new hall and stage environment, they confirmed, along with Maestro Mehta, the improved acoustical conditions on stage and a new ease of ensemble hearing.

The inaugural concert took place on May 25, 2013, with the Israel Philharmonic Orchestra conducted by Music Director for Life Zubin Mehta and violinist Itzhak Perlman, featuring Beethoven's Violin Concerto and Mahler's Fifth Symphony.

A Chamber Hall in the Basement

The underground building extension devised by KKE was completed in October 2017. It was mainly destined to create facilities dedicated to the orchestra, including a large rehearsal hall which also offered the opportunity for a smaller concert space tailored for recitals or chamber ensembles, as well as amplified performances or lectures. Named Zucker Hall, it can seat an audience of up to 480 people for concerts or accommodate a full size orchestra for rehearsals.

Although detailed design and actual construction of the new hall were delayed, excavation works were carried out during construction on the renovated building. It was therefore essential to secure early on the fundamentals of the new room dimensions. Importantly, the depth of excavation was scaled to allow for an appropriate ceiling height of approximately 14 m in the new hall.

The fundamental footprint is rectangular, with a long side of 31 m and short side of 17 m. On the lower level, a flat stage area 20 m wide and 14.5 m deep was reserved to accommodate large ensemble in rehearsals. Audience can be seated on two banks of four rows on each side of the rehearsal area, and an additional ten rows of retractable seating can be deployed on one side. Finally, chairs can be placed on the remaining stage area around smaller ensembles. A technical gallery surrounds the hall, at a height of approximately 8 m from the lower stage level, providing sound reflections to the musicians on stage. To prevent detrimental echoes or favor an even distribution of sound reflections throughout the hall, the ceiling surface is scalloped and the walls are cladded with a grid of curved panels.

In order to tune the overall reverberance of the hall to the needs of the programs or occupancy, as well as adjust ensemble balance in the rehearsal area, retractable acoustical draperies were planned in front and behind the orchestra, as well as around the upper levels.

The orchestra started rehearsing in the hall in the summer of 2017, and the hall was inaugurated on October 13, 2017 as part of the celebrations of the 60th anniversary of the overall facility and opening of the new concert season that year.

Marc Quiquerez

Audience view from the last row of Lowy Concert Hall. The renovated hall kept its fundamental fan shape footprint, accommodating over 2,400 seats. This leads to a maximum width of almost 50 m at the rear, and a maximum distance from stage to last row of close to 40 m

Orchestra rehearsing in Zucker Hall. The new hall was initially designed primarily as a rehearsal hall for Israel Philharmonic Orchestra. Its tall ceiling, large volume, and ample stage area make it suitable for accommodating large ensembles

Zucker Hall in recital configuration. By deploying retractable seating and fitting additional rows of chairs on the flat stage, the hall can accommodate intimate chamber music concerts or recitals

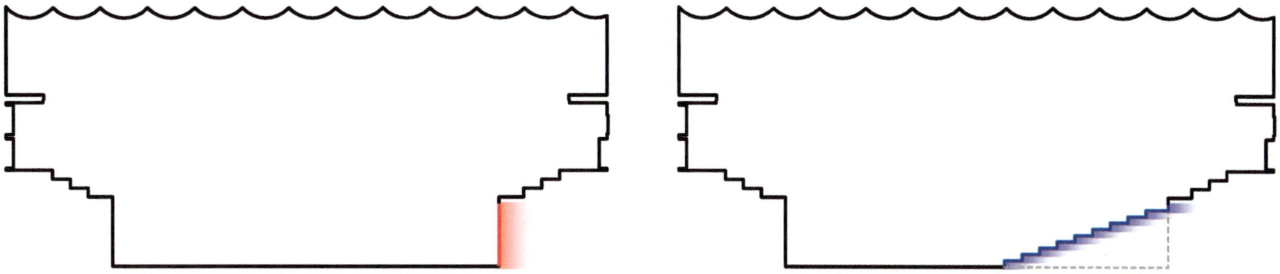

Fig. 17.2 Scope of spatial transformations in Zucker Hall

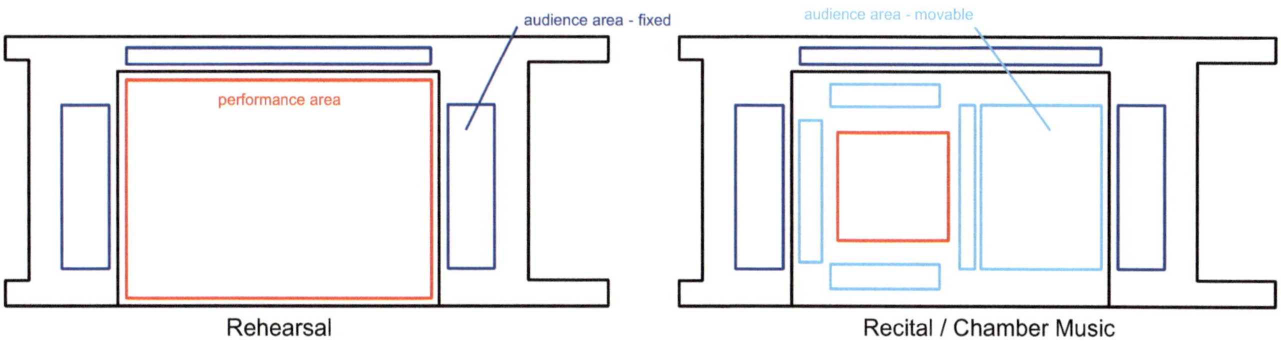

Fig. 17.3 Stage and audience configuration in Zucker Hall

Charles Bronfman Auditorium	
Location	Tel Aviv, Israel
Owner	Heichal Hatarbut
Original name	Frederick R. Mann Auditorium
Architects	Dov Karmi, Zeev Rechter and Yaakov Rechter (original, 1957)
	Kolker Kolker Epstein Architects (renovation, 2013)
Construction cost	ILS 165 Million (renovation)
Building size	17,500 m^2 (after renovation)
Lowy Concert Hall	
User	Israel Philharmonic Orchestra
Acoustical consultant	Bolt Beranek Newman (original, 1957)
	Nagata Acoustics (renovation, 2013) (room acoustics)
	Yasuhisa Toyota, Marc Quiquerez
Design start	October 2007
Construction start	August 2011
Construction end	May 2013
Opening date	May 25, 2013
Seating capacity	2,430 (after renovation)
Room volume	24,400 m^3 (after renovation)
Surface area	7,250 m^2
Volume/seat	10 m^3/seat
Volume/surface area	3.4 m
Finish material	
Ceiling	Shotcrete, Expanded metal mesh, Wooden grille
Walls	Plywood, Plaster
Audience floor	Wood over concrete
Stage floor	Alaskan Yellow Cedar
Seat manufacturer	Irwin Seating Company
Zucker Hall	
User	Israel Philharmonic Orchestra
Acoustical consultant	Nagata Acoustics (room acoustics)
Design start	Q1 2013
Construction start	Q3 2015
Construction end	September 2017
Opening date	October 13, 2017
Seating capacity	480
Room volume	6,100 m^3
Surface area	2,600 m^2
Volume/seat	13.3 m^3/seat
Volume/surface area	2.3 m
Finish material	
Ceiling	Painted shotcrete
Walls	Plastered concrete, medium-density fiberboard and steel plates
Stage floor	Alaskan Yellow Cedar
Seat manufacturer	Ezcaray Internacional

Table 17.1 Charles Bronfman Auditorium, Lowy Concert Hall (after renovation)—acoustical metrics at 500 Hz

RT unoccupied	2.2 s
RT occupied	1.9 s

Table 17.2 Charles Bronfman Auditorium, Zucker Hall—acoustical metrics at 500 Hz

	Rehearsal mode	Concert mode
RT unoccupied	1.8 s	1.7 s
RT occupied		1.6 s

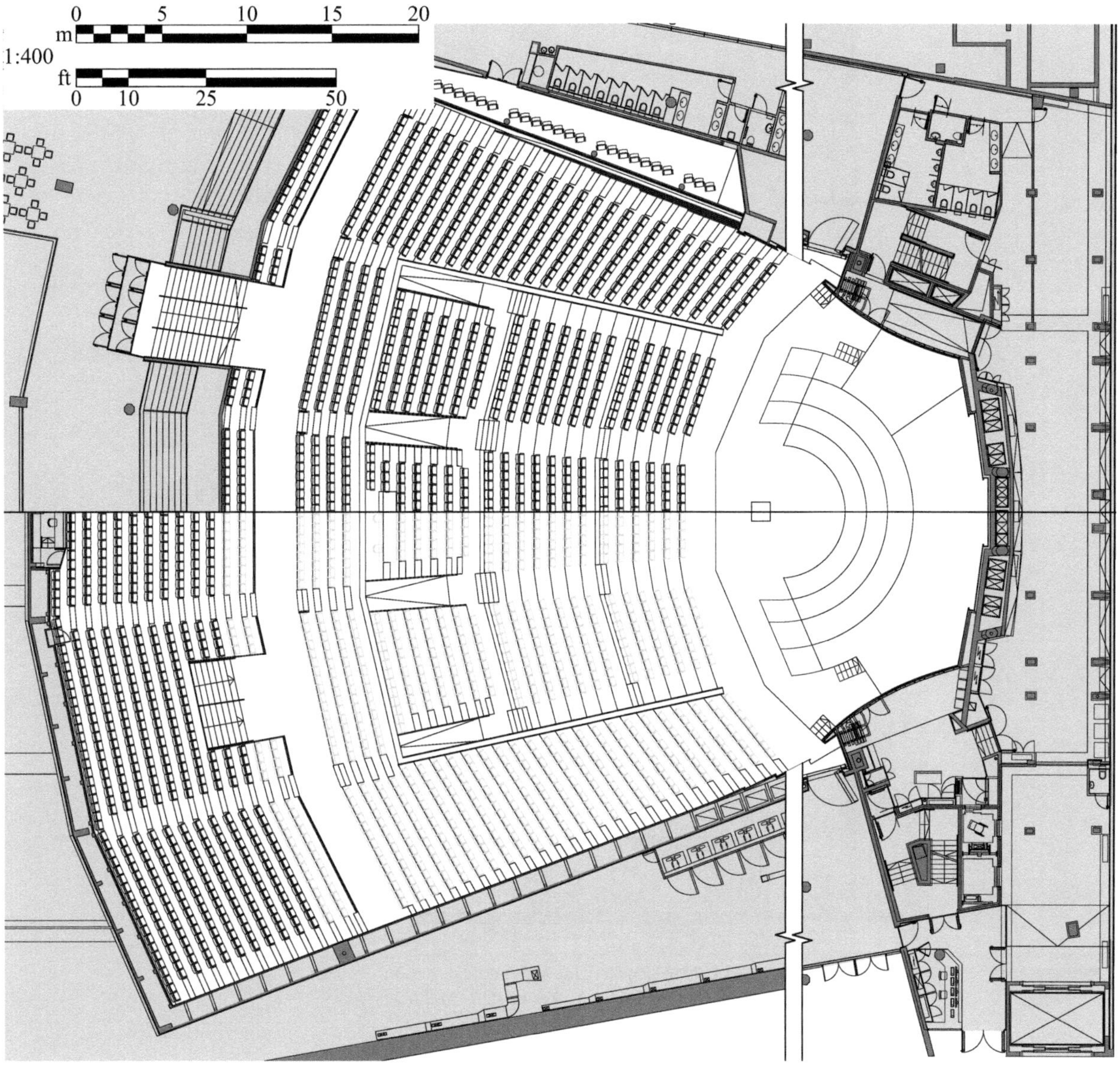

Fig. 17.4 Lowy Concert Hall stage and main levels plan
Courtesy of Kolker Kolker Epstein Architects

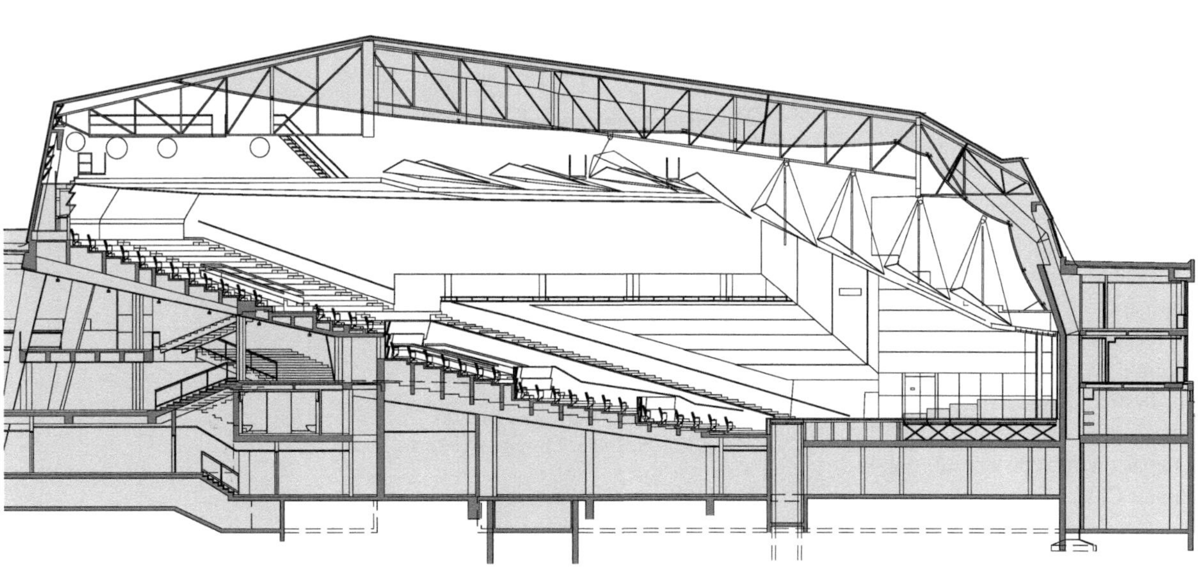

Fig. 17.5 Lowy Hall section
Courtesy of Kolker Kolker Epstein Architects

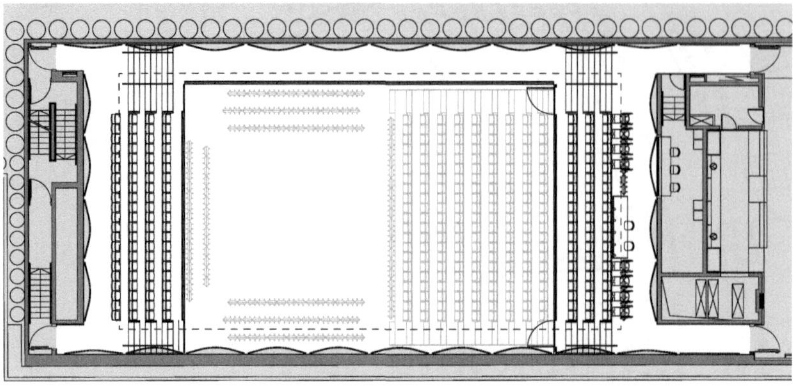

Fig. 17.6 Zucker Hall plan
Courtesy of Kolker Kolker Epstein Architects

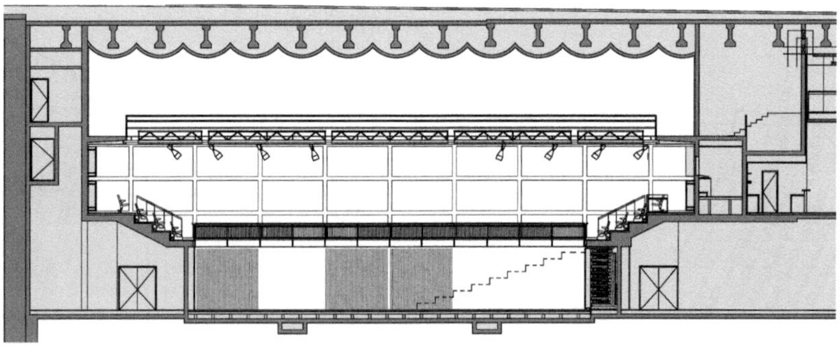

Fig. 17.7 Zucker Hall longitudinal section
Courtesy of Kolker Kolker Epstein Architects

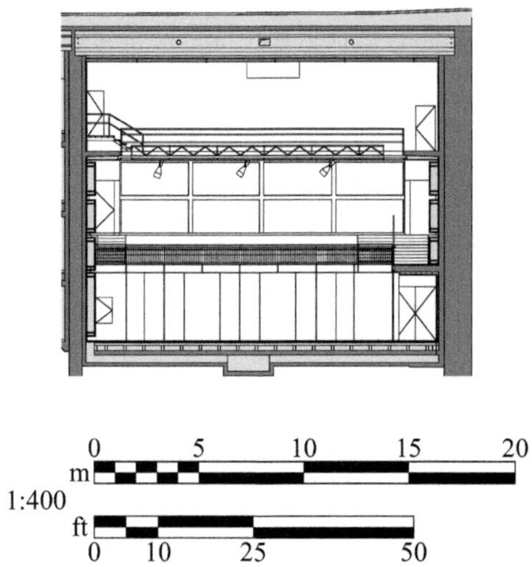

Fig. 17.8 Zucker Hall cross section
Courtesy of Kolker Kolker Epstein Architects

18. Auditorium Giovanni Arvedi

View from upstage seating. The height of the hall was increased by digging below the existing floor, and placing the stage at the lowest point. The single row of seats overlooking the stage from the side is at the pre-renovation floor level

Y. Toyota et al., *Concert Halls by Nagata Acoustics*, https://doi.org/10.1007/978-3-030-42450-3_18

The Museo del Violino (Museum of the Violin) is the adaptive re-use of the former Palazzo dell'Arte in Cremona, built in the 1930s by the Italian Rationalist architect Carlo Coccia. The building, located on Piazza Marconi within the historic city center, had seen multiple uses and was unused when Giovanni Arvedi, a local entrepreneur, decided to renovate it and convert it into the current Museum. The art and craft of violin-making was perfected in Cremona beginning in the 1500s, reaching its zenith in the early 1700s with the iconic work of Antonio Stradivari and Guiseppe Guarneri (del Gesù). The museum will exhibit several major collections of historical instruments, as well as contemporary prize-winners of the triannual violin-making competition, and a collection of Stradivari's tools.

Common sense would dictate that a museum dedicated to a musical instrument should contain a performance space in which to enjoy the music produced by these excellent instruments. A large, rectangular room, once used as a gymnasium by a local girls' school, seemed to offer an ideal space for a chamber music hall, with its measurements of 14 m wide, 36 m long, and 10 m tall. Given the limited footprint available for the hall, it was clear that it would serve primarily as a chamber music hall, with a smaller stage for solo performances and ensembles such as quartets or octets. The ceiling height of the room was considered too low, though, to produce a truly excellent space for natural acoustics. As a historic building, there were many aspects which could not be changed, including the exterior envelope of the building (a unique, highly textured, and delicate brick design) and the many historical doors and windows with their sculptural, modernist frames lining three sides of the space. To provide sufficient ceiling height in order to produce excellent natural acoustics for the "Auditorium Giovanni Arvedi," on-site investigations revealed that the floor of the hall could be lowered to the base of the footings for the main walls, or 2.8 m lower than the existing floor level. The floor at the center of the room, for the stage, is lowered an additional meter to give a final ceiling height of 13.7 m.

The stage is placed near the center of the room, in order to give a variety of different views to the performers on the stage. This enables more seats to be closer to the stage, further enhancing the intimacy in the 476-seat hall. The furthest seat from the stage is only 15.9 m away, and the closest seats form a complete ring on the stage floor itself, with forty seats on all four sides of the 65 m^2 oval stage. With viewer-listeners seated directly on the stage floor, the communication between performers and audience is free of any physical barriers.

Good sightlines to the stage are ensured by a very steep rake for all of the audience seats, both in front of and behind the stage. At the original floor level, one row of seats remains to look directly down on the stage from the sides, and gives another visual connection between the main and upstage audience areas. The largest segment of the audience area, containing 243 seats, sits in a sculptural structure, reaching from the stage almost to the ceiling. It is separated from the walls in order to allow the space under and around the seats to join the acoustic volume. Gaps in the floor at the edge of the seating element also allow the underground level, newly formed as a result of digging below the existing floor level, to contribute to the acoustic volume and increase the reverberance of the space.

The architects, locally based Giorgio Palù and Michele Bianchi, chose to show the contrast between the existing historic structure and the new constructions for the chamber hall by leaving the historic structure primarily white plaster, which emphasizes the dark wood of the historic doors and windows, and by cladding all of the new work in honey-colored wood. The new insertions into the space subtly evoke the curved forms of a violin.

Circulation. The lower-level circulation space is still connected to the overall auditorium volume, by a gap between the structure of the highest audience seats and the main level perimeter circulation

Once the construction was completed, a period of six months elapsed before the grand opening ceremony. During that time, the hall invited many ensembles to rehearse in the hall to learn the acoustics at a leisurely pace. The hall is unique for a "chamber hall," since it places the stage almost in the middle of the room. Performers have been confused about which orientation to take when first encountering the stage. Small ensembles are encouraged to arrange themselves in a circle, thereby not showing particular favor to one segment of the audience over another. The ensemble can also be improved when all members have eye contact with each other easily and directly as a result of sitting in a circle, rather than in a line as would normally occur in a more mono-directional hall. The extended rehearsal period before opening allowed the local ensembles ample opportunity to acclimate to this unique feature and the hall in general, ensuring a successful opening concert which highlighted the distinctive qualities of this small auditorium.

Daniel Beckmann

View from stage level seats. At the stage, seats are placed very close to the performers, and there is no clear border between audience and ensemble. Perforated walls behind the three rows of seats at the sides of the stage add diffusion in order to break the parallel condition of the walls

Historic facade. The original building is composed of highly sculptural brick designs, as shown here on exterior side of the stage-right wall of the hall, which forms the back wall of the overall museum building

Museo del Violino	
Location	Cremona, Italy
Owner	Museo del Violino
User	Commune de Cremona
	Fondazione Stradivari
Architect	ARKPABI|Giorgio Palù e Michele Bianchi Architetti
Building size	6,500 m^2
	Auditorium Giovanni Arvedi
Acoustical consultant	Nagata Acoustics
	Yasuhisa Toyota, Daniel Beckmann
Design start	March, 2010
Construction end	June, 2012
Opening date	September 14, 2013
Seating capacity	475
Room volume	5,330 m^3
Surface area	3,000 m^2
Volume/seat	11.2 m^3/seat
Volume/surface area	1.8 m
Finish materials	
Ceiling	Painted plaster
Walls	Plaster, wood, historic doors
Stage floor	Alaskan Yellow Cedar
Seat manufacturer	CALOI

Table 18.1 Museo del Violino—acoustical metrics at 500 Hz

RT unoccupied	1.6 s
RT occupied	1.4 s
EDT	1.5 s
C_{80}	2.0 dB
D_{50}	44%
G	9 dB

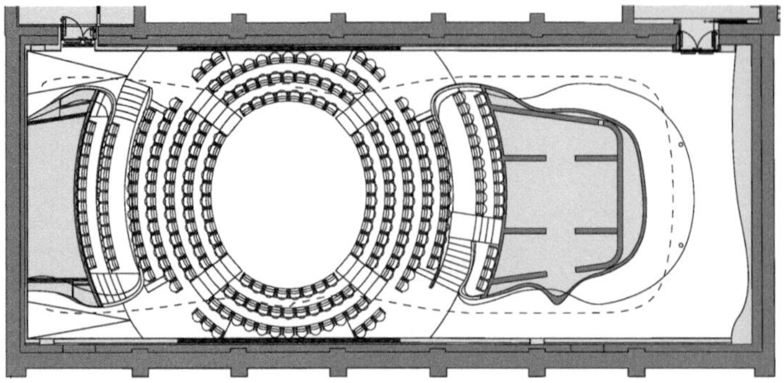

Fig. 18.1 Stage level plan
Courtesy of ARKPABI

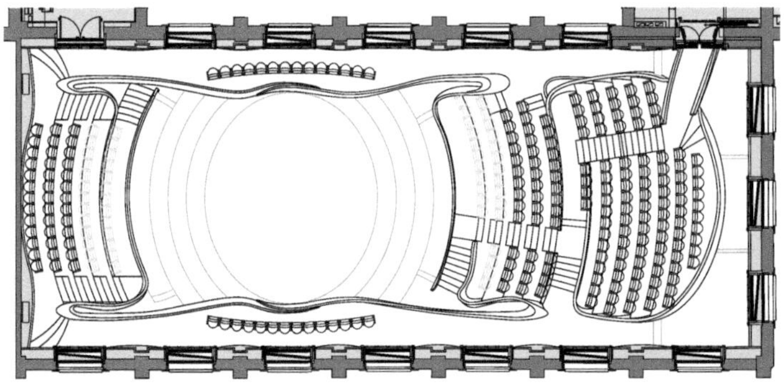

Fig. 18.2 Balcony plan
Courtesy of ARKPABI

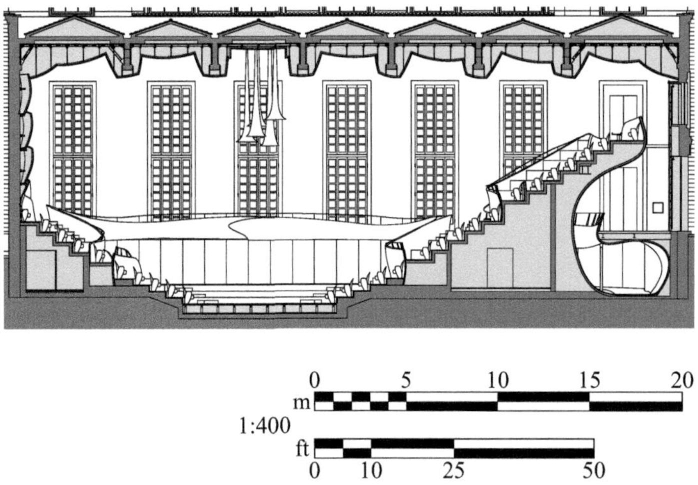

Fig. 18.3 Section
Courtesy of ARKPABI

19. Shanghai Symphony Hall

Main Hall, View from rear seating. The audience seating is arranged in terraces without overhangs. Basket-woven panels hang suspended within the smooth, outer acoustical volume

Y. Toyota et al., *Concert Halls by Nagata Acoustics*, https://doi.org/10.1007/978-3-030-42450-3_19

Shanghai Symphony Orchestra (SSO) is the oldest, extant, Chinese symphony, founded in 1879. In 2008, the city of Shanghai intended to create a new rehearsal for the orchestra, coincident with the arrival of Long Yu as Music Director. At the time, the SSO performed in the Shanghai Grand Theater located in People's Square and considered it impossible to build another dedicated concert hall. The Grand Theater seats 1,800, but is definitively a multipurpose theater resulting in difficult acoustics for symphonic performances. Gradually and strategically, the rehearsal hall was expanded into a campus containing a Main Hall, Chamber Hall, and management offices in the heart of the former French Concession, less than three kilometers from the Grand Theater.

Nagata Acoustics was invited by the city as acoustical consultant in 2008 and sat on the jury during the architectural competition. While there was some political risk of choosing both a Japanese architect *and* Japanese acoustician, the design by ISOZAKI+HuQian Partners (the Shanghai Office of Arata Isozaki and Associates) was the stand out winner, followed by Tongji Architectural Design who became the local architect. Originally, the goal was to complete the project by the "Expo 2010 Shanghai," however, it was clear that such a short timeline was unrealistic. The building opened on September 9, 2014, still a respectably quick schedule.

The Main Hall is accessed from two foyer levels but every seat can be projected onto a single, rectangular plan. This combination of a steep rake and lack of balcony overhangs ensures that every seat has excellent sightlines and reception of direct sound as well as the possibility to receive acoustic reflections from all directions. Small walls textured with horizontal ridges divide the audience into 18 symmetrical seating terraces. Smooth, white plaster forms the acoustic envelope. Within this space, seven large panels define the architectural shape and acoustic reflections: two on either side, one upstage, one at the rear hiding the control room, and one as the ceiling. The basket-weave motif in ochre wood provides acoustical scattering and hides lighting and loudspeakers. Diffuse lighting on the outer envelope gives the impression that the panels float in space.

More than simply an ensemble reflector, the ceiling panel extends a full two-thirds of the total length of the hall. The space above the panel was designated as a space for a one-time adjustment to the reverberation during acoustical tuning at the end of construction. However, after several rehearsals and opening night, it became clear that no adjustment was needed. While the reverberation is longer—2.7 s at 500 Hz in the unoccupied condition—than what is generally thought to be acceptable, the excellent clarity allows for the added richness without any compromise.

Unlike many concert halls, the Main Hall has no variable acoustics illustrating the intent of the space as primarily for classical performances.

Adjacent to the Main Hall sits a medium-sized Chamber Hall. The hall is rectangular in footprint with a control room and several rows of raked, fixed seating at the far end. The ceiling is doubly curved and covered with a wooden basket-weave pattern connecting the architectural language with the Main Hall. To avoid echoes resulting from the parallel walls, the profile on each level is slightly tilted upwards.

Moveable audience chairs on a 3×4-grid of platforms each 4.35 m square form the main floor. Each of the twelve risers can be controlled independently, creating a huge variety of possible stage and seating layouts which can seat a maximum of approximately 600. Lowering the second row of risers and gradually incrementing the next rows up to the fixed seating creates a conventional end-stage layout. For small chamber ensembles or soloists, the audience is often arranged in the round with a single, central riser raised for the musicians. When not in use, the moveable seats can be stored underneath the fixed seating.

The incredibly flexible stage layout results from the diverse programming requirements, focusing on chamber concerts and rehearsals, but ranging to lectures and even light opera. While there is no fly tower, three balconies can be outfitted with technical equipment. The first balcony can accommodate audience or can be fully or partially cleared of seats to make room for lighting, theatrical equipment, or scenery. The second and third balconies are strictly technical. The third functions almost as a catwalk, with its metal grate floor and open, ribbed facing making it acoustically transparent so acoustically absorbent curtains can be deployed at this level.

Chamber hall, Second balcony. Panels on the balcony retract to allow for theatrical equipment and variable acoustical curtains. Even when the panels are in their forward position, the vertical louvers are acoustically transparent

The backstage areas are shared between the two spaces making it easy to move instruments and equipment between the halls. As we walk from one hall to the other, we cross over two sets of resilient joints which are the only visible indication of the Herculean effort to isolate the halls from their busy, urban environment.

Vibration isolating springs. In order to isolate the halls from subway vibration, the entire structure of each hall is floated on a system of giant springs

It should come as no surprise that the location of the SSO campus in the center of Shanghai meant that vibration isolation would be of particular concern. One of the busy metro system lines runs only 14 m from the foundation of the main hall. In order to prevent unwanted vibrations from disturbing concerts every few minutes, drastic measures were necessary. Since airborne noise was not of outstanding concern, a box-in-box system was not needed, and instead, a system of springs floats both halls separately in their entirety. The huge springs manufactured by GERB Vibration Control carry approximately 900 kN each and have a resonant frequency of less than 5 Hz. The system is successful and no structural vibration can be perceived in either hall.

Erik Bergal

Main Hall, View from stage. From stage, the central panel seems to take up the entire ceiling, but the volume above it is still valuable as a reverberant space. An extruded block at the rear of the hall houses the various control and spot light rooms

Main Hall, Terrace wall texture. Horizontal ridges in the terrace walls provide acoustical diffusion

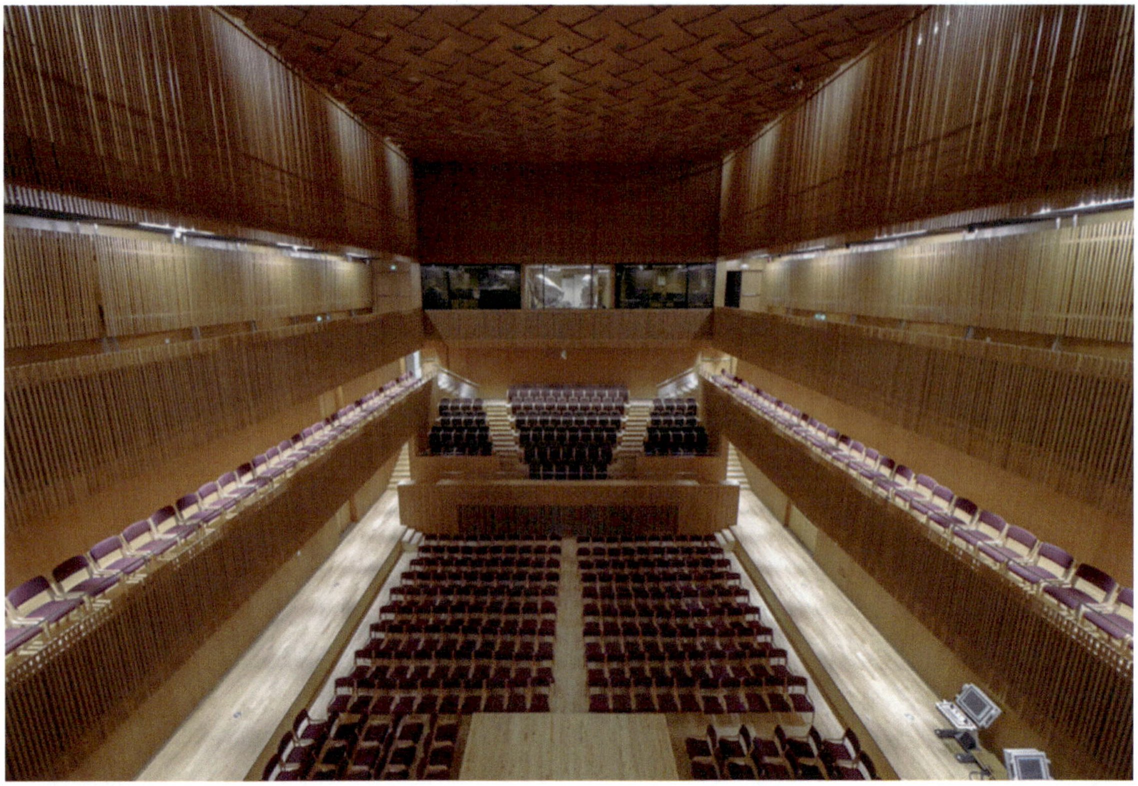

Chamber Hall, View from second balcony. The chamber hall features a mix of fixed and movable seating. The fixed seating is pushed to one end, along with the control rooms. The basket-weave ceiling texture connects the architectural language with the Main Hall

Chamber Hall, View from fixed audience. The movable seats are places on twelve mechanically operated platforms. The first balcony features audience seats, but the second and third are used as technical and lighting positions

Exterior. A vine covered pergola runs the length of the Shanghai Symphony site, with the audience entrance and artists entrance on either end. The bowed roof has earned the building its nickname, "The Little Dumpling"

Shanghai Symphony Hall	
Location	Shanghai, China
Owner/user	Shanghai Symphony Orchestra
Architect	ISOZAKI + HuQian Partners
Acoustical consultant	Nagata Acoustics
	Yasuhisa Toyota, Keiji Oguchi, Motoo Komoda
Theater consultant	Theatre Projects Consultants
	Theatre Workshop
Construction cost	RMB 630 million
Design start	June 2008
Construction start	October 2009
Construction end	December 2013
Opening date	September 9, 2014
Main Hall	
Seating capacity	1,123
Room volume	20,000 m^3
Surface area	7,600 m^2
Volume/seat	16.6 m^3/seat
Volume/surface area	2.6 m
Finish material	
Ceiling + wall panels	Wood on concrete
Walls	Plaster on glass fiber reinforced concrete
Audience floor	Wood flooring on concrete
Stage floor	Hinoki
Seat manufacturer	Dafeng
Scale model	1:10
Chamber Hall	
Seating capacity	400–600
Room volume	6,500 m^3
Surface area	2,900 m^2
Volume/seat	10.8 m^3/seat
Volume/surface area	2.2 m
Finish material	
Ceiling + Walls	Wood on concrete
Stage/audience floor	Hinoki

Table 19.1 Shanghai Symphony Hall, Main Hall—acoustical metrics at 500 Hz

RT unoccupied	2.7 s
RT occupied	2.3 s
EDT	2.3 s
C_{80}	0.1 dB
D_{50}	39%
G	3.3 dB

Table 19.2 Shanghai Symphony Hall, Chamber Hall—acoustical metrics at 500 Hz

RT unoccupied	1.8 s
RT occupied	2.3 s
EDT	1.7 s

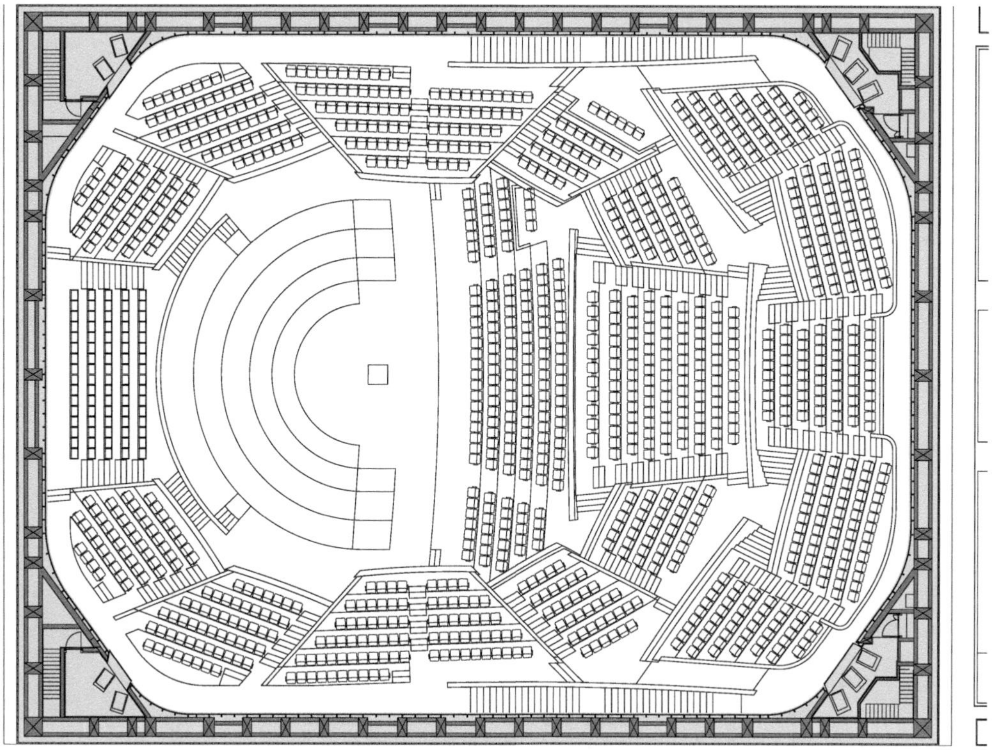

Fig. 19.1 Main Hall, plan
Courtesy of Arata Isozaki & Associates

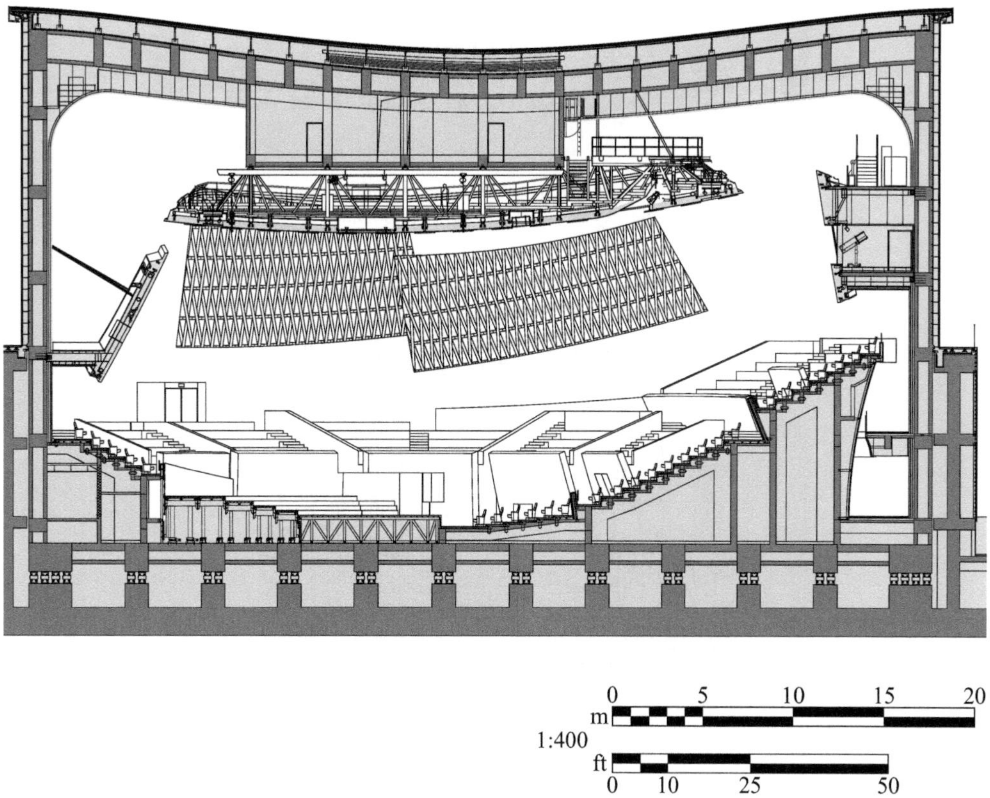

Fig. 19.2 Main Hall, section
Courtesy of Arata Isozaki & Associates

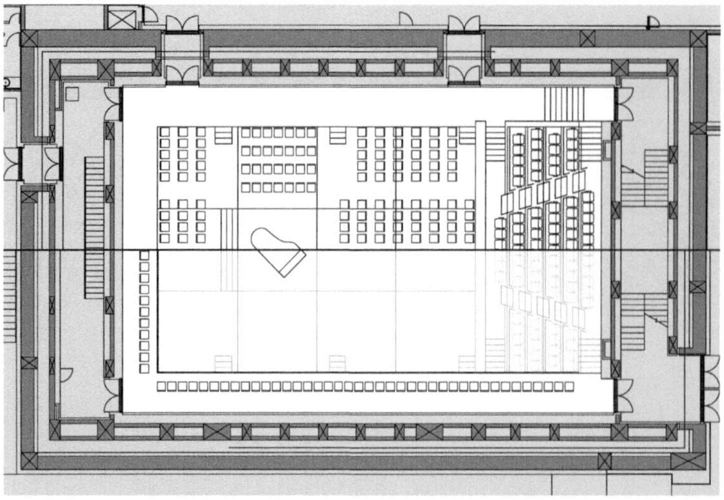

Fig. 19.3 Chamber Hall, stage and first balcony plan
Courtesy of Arata Isozaki & Associates

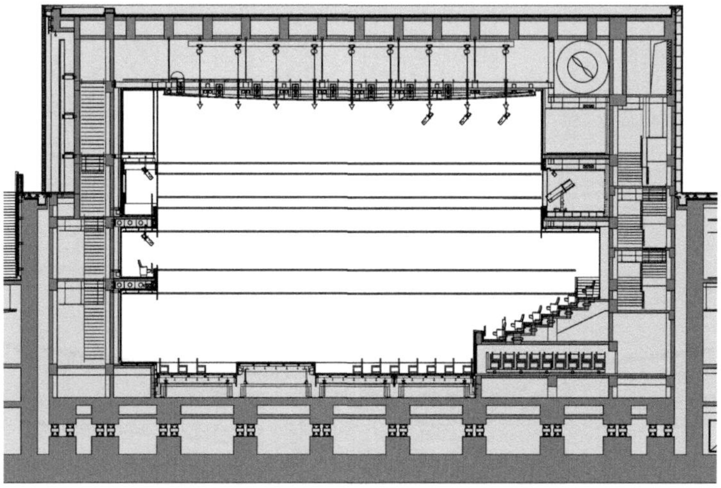

Fig. 19.4 Chamber Hall, longitudinal section
Courtesy of Arata Isozaki & Associates

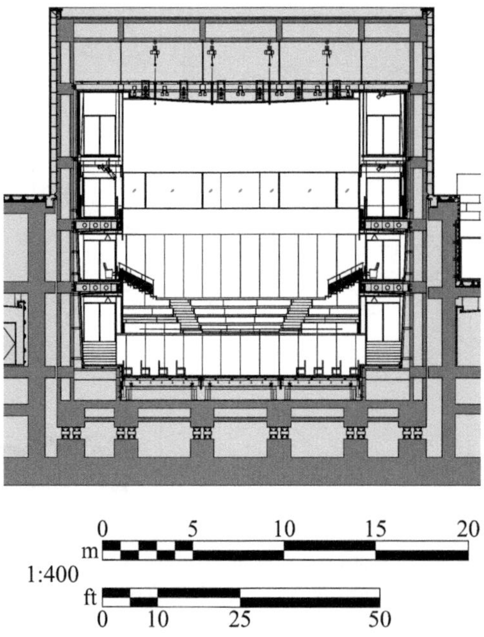

Fig. 19.5 Chamber Hall, cross section
Courtesy of Arata Isozaki & Associates

20. NOSPR Katowice

View from the first balcony. The stage can accommodate a large symphonic orchestra, or smaller ensembles with choir on stage. Above the stage area, a circular convex reflector is suspended from the ceiling to provide the necessary early reflections to musicians and surrounding audience. ©Daniel Rumiancew

Y. Toyota et al., *Concert Halls by Nagata Acoustics*, https://doi.org/10.1007/978-3-030-42450-3_20

Narodowa Orkiestra Symfoniczna Polskiego Radia (Polish National Radio Symphony Orchestra), often referred to by its initials NOSPR, moved into its new headquarter building in Katowice on October 1, 2014. The new building provided them with a 1,780-seat concert hall as well as a 300-seat chamber hall and a range of practice and rehearsal rooms. Founded in 1935 in Warsaw, the ensemble was revived in 1945 after World War II in the city of Katowice, where it has maintained its managing and artistic residence ever since. Today, it is one of the country's two national symphony orchestras, alongside Warsaw National Philharmonic Orchestra.

The new facility is located in an area that once belonged to a large mining corporation. While the city of Katowice is still an important industrial and coal-mining hub, a heritage of the boom of its development in the mid-nineteenth century, it has transformed itself in recent years into an active business and trade fair center and continues to develop its cultural activities. As evidence of this dynamic and ambitions, the nearest neighbors of the new building are the new campus of Silesian Museum to the east, the new International Congress Center and the imposing Spodek, or "Saucer," multipurpose arena to the west.

The building is surrounded by a park lush with trees and outdoor activities. The audience can approach both from the west and east across large open plazas. The facade alternates massive vertical stripes of red bricks, a hint to local building techniques, and thin glazed cuts opening to the public spaces as well as private offices and functions. The mass-dyed concrete shell of the hall protrudes visibly on top of the rectangular building and continues inside the main foyer in a gentle curve, making the overall volume of the concert hall immediately apparent.

Beyond the Shoebox

The competition held in 2008 awarded the design of the project to Katowice-based architect Tomasz Konior and his firm Konior Studio. From the onset, renowned pianist Krystian Zimerman, who was born in the area and studied in Katowice, supported the project and acted as an advisor. Beyond his extraordinary talent and lengthy artistic career, Zimerman is also known for his acute knowledge of his instrument and his keen interest in concert hall acoustics. In 2009, on his recommendation, Nagata Acoustics was invited to take over the role of acoustical consultant on the room acoustics of the concert hall.

The original concept developed for the competition followed a traditional shoebox typology, with a narrow width and mainly frontal seating layout. Upon our appointment, we worked closely with Tomasz Konior team and the client on adapting the original design to enhance the visual and acoustical intimacy by reducing the physical and perceived distance between audience and performers. While maintaining a fundamentally rectangular footprint, this meant gradually increasing the overall width of the hall and bringing audience seating to the sides and back of the stage. This direction was strongly supported by the design architect, who was enthusiastic about the potential to create a more intimate concert experience. In spite of some initial reluctance on the part of the project client to seating listeners to the sides of the stage, first-hand experiences in rehearsals and concerts at the recently opened Danish Radio Concert Hall in Copenhagen finally convinced them of the merits of such an approach, which they fully embraced.

The Design of the Concert Hall

From the stage, the central seating section gradually widens in a gentle curve as the rake steepens. To create sufficiently early sound reflections, two tilted terrace walls were introduced that divide the section into three blocks. To the rear, this central section reaches the same level as a ring of terrace seating which wraps up to five rows around the stage.

The two levels of balconies are the most literal evidence of the original shoebox layout of the design. The first balcony is inscribed in the same footprint as the lower level and does not exceed three rows so as to limit the overhang and maintain direct visibility to the ceiling for the seats below. The second balcony is slightly pushed outward, but the seating area extends towards the rear with nine rows facing the stage.

Steep rakes all around facilitate unobstructed sightlines to the stage. In spite of the mainly frontal seating arrangement, the terraced and surround seating offers over 20% of seats on first rows, emphasizing the sense of proximity with the stage and intimacy throughout the hall.

The large stage area, with near constant width of 21 m, offers more than 300 m^2 for large orchestras and choir. The platform is divided by a total of 18 mechanical risers creating a semi-circular layout around the conductor.

Above the stage, a large circular ensemble reflector appears as if pushing the central part of the ceiling up. Suspended at a height of 14.5 m above the performers, its large diameter of 15.5 m and gentle curve serve early sound reflections to the stage as well as surrounding seating blocks.

Behind the stage, the balconies are interrupted to make way for a pipe organ. Although no instrument was installed upon completion of the hall, mock pipes were put in place to visually and acoustically mimic the presence of the instrument. In May 2018, a consortium lead by Anton Škrabl was commissioned to finally design and build the instrument, which is still expected to retain the same facade design.

The hall's visual atmosphere is dominated by the deep blacks of the painted concrete walls, ceiling paneling, and chair upholstery, the dark honey tones of lacquered beech plywood of the balcony fascias and terrace walls, and lighter colors of the wooden floors.

Side balcony and concrete walls. The interior concrete walls of the concert hall were cast-in-place with a silicone form creating irregularities of varying depth, width, and orientation. This uneven texture helps scattering high-frequency sound reflections. The balcony overhangs were limited to maintain direct visibility to the ceiling providing first-order early reflections

The most distinct feature of the hall may be the main walls surrounding the space, which consist of cast-in-place concrete shaped with silicone form panels to create a random pattern of irregularities with varying widths and depths. This not only creates massive reflective surfaces with high-frequency sound diffusion, but also the visual illusion of smooth hills and valleys. The pattern and dimensions of these irregularities were confirmed through testing of a physical 1:10 scale model. Fixed absorptive treatments were also introduced on the rear walls under the balconies to prevent long path echoes.

The ceiling shape follows a profile similar to the smooth landscape-like illusion of the wall texture. It is shaped by plywood panels resting on transverse steel beams. The lower part of the beams is exposed and cladded in wood so as to visually emphasize the shape and create some surface irregularities. Concrete was poured over the decking to create a continuous massive surface.

First Rehearsal and Inaugural Concert

The orchestra played their first notes in the hall in late August 2014 with their newly appointed music director at the time, Alexander Liebreich. Although the tension was palpable, the musicians quickly sounded at ease on their new stage, and their praises matched our own appreciation not only of the qualities that the hall immediately exhibited, but also of the quality of the ensemble and overall balance they achieved in this first rehearsal.

The inaugural concert took place on October 1, 2014, with the NOSPR orchestra led by Alexander Liebreich, the Bavarian Radio Choir, and pianist Krystian Zimerman. The program featured pieces from Polish composers Witold Lutosławski, Krzysztof Penderecki and Wojciech Kilar, as a tribute to their influence and relationship with the orchestra. The evening also included Johannes Brahms's Piano Concerto No. 1 and concluded with Ludwig van Beethoven's Symphony No. 9.

Marc Quiquerez

View from chorus seating. The room shape follows the elongated box footprint inherited from the competition design, but introduced a terraced main floor, and increased the width to sit more audience to the sides and back, with the objective to improve sightlines and favor a more compact seating layout all around the stage. ©Daniel Rumiancew

Concert hall entrance in the main foyer. The volume of the concert hall sits in the middle of a large, full-height foyer receiving daylight from the facade windows and skylight. The wooden finishes evoke the interior of the hall and draw audience to the doors carved into the dark dyed concrete monolith. ©Daniel Rumiancew

Building facade and approach. ©Daniel Rumiancew

NOSPR Katowice Concert Hall	
Location	Katowice, Poland
Owner/user	Narodowa Orkiestra Symfoniczna Polskiego Radia (NOSPR) (National Polish Radio Symphony Orchestra)
Architect	KONIOR STUDIO Tomasz Konior
Construction cost	PLN 270 million
Design start	September 2009
Construction start	2012
Construction end	August 2014
Opening date	October 1, 2014
Building size	35,100 m^2
Concert Hall	
Acoustical consultant	Nagata Acoustics (room acoustics) *Yasuhisa Toyota, Motoo Komoda, Marc Quiquerez* Pracownia Akustyczna (building acoustics)
Seating capacity	1,800
Room volume	22,000 m^3
Surface area	8,100 m^2
Volume/seat	12.2 m^3/seat
Volume/surface Area	2.7 m
Finish material	
Ceiling	Concrete over plywood
Canopy	Birch plywood
Walls	Painted concrete
Balcony fronts	Birch plywood
Audience floor	Wood flooring on concrete
Stage floor	Alaskan Yellow Cedar
Seat manufacturer	Nowy Styl
Organ builder	Anton Škrabl & Thomas Jann Orgelbau
Model scale	1:10

Table 20.1 NOSPR Katowice Concert Hall—acoustical metrics at 500 Hz

	Reflective condition	Curtains deployed
RT unoccupied	2.3 s	1.8 s
RT occupied	2.1 s	1.7 s

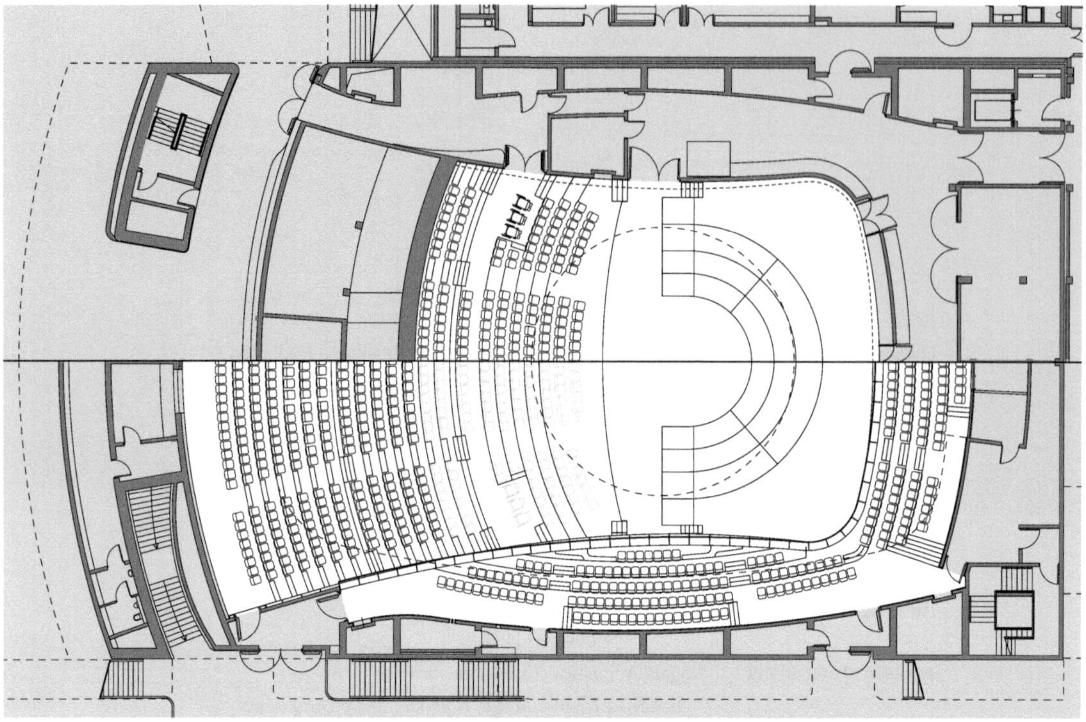

Fig. 20.1 Stage and terrace levels plan
Courtesy of KONIOR STUDIO Tomasz Konior

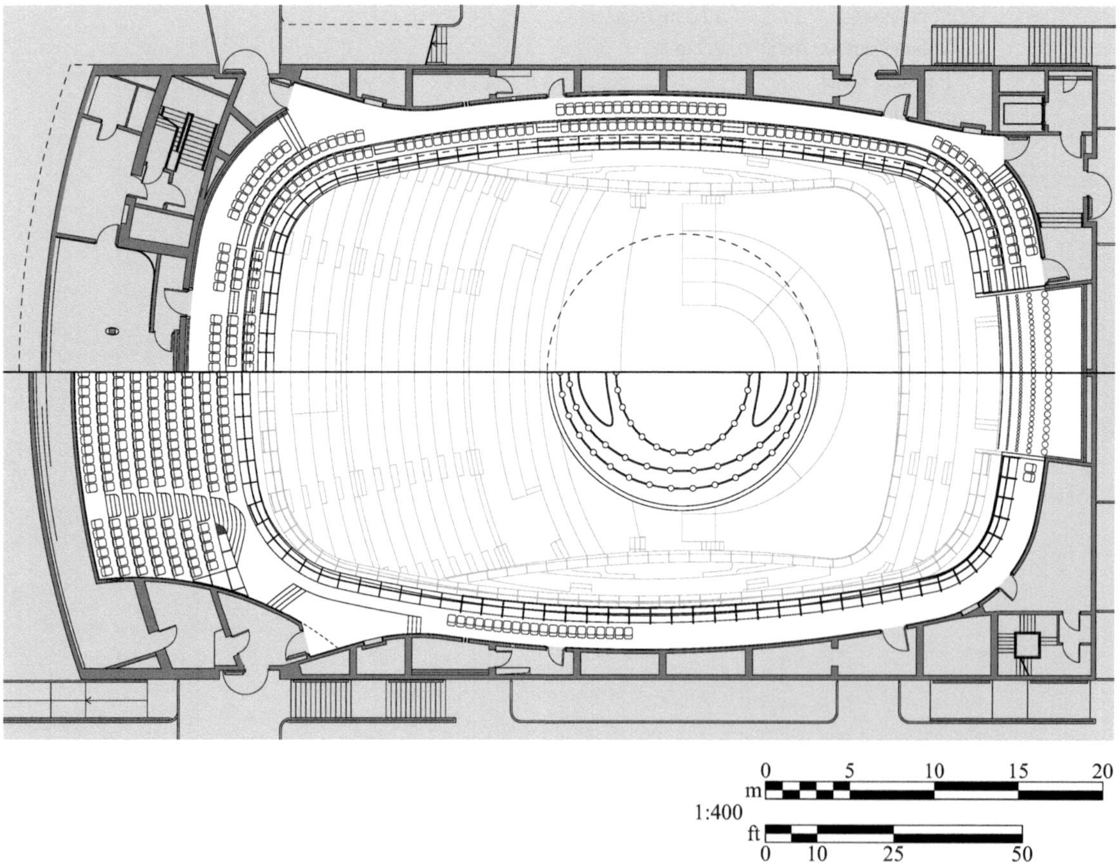

Fig. 20.2 First and second balconies plan
Courtesy of KONIOR STUDIO Tomasz Konior

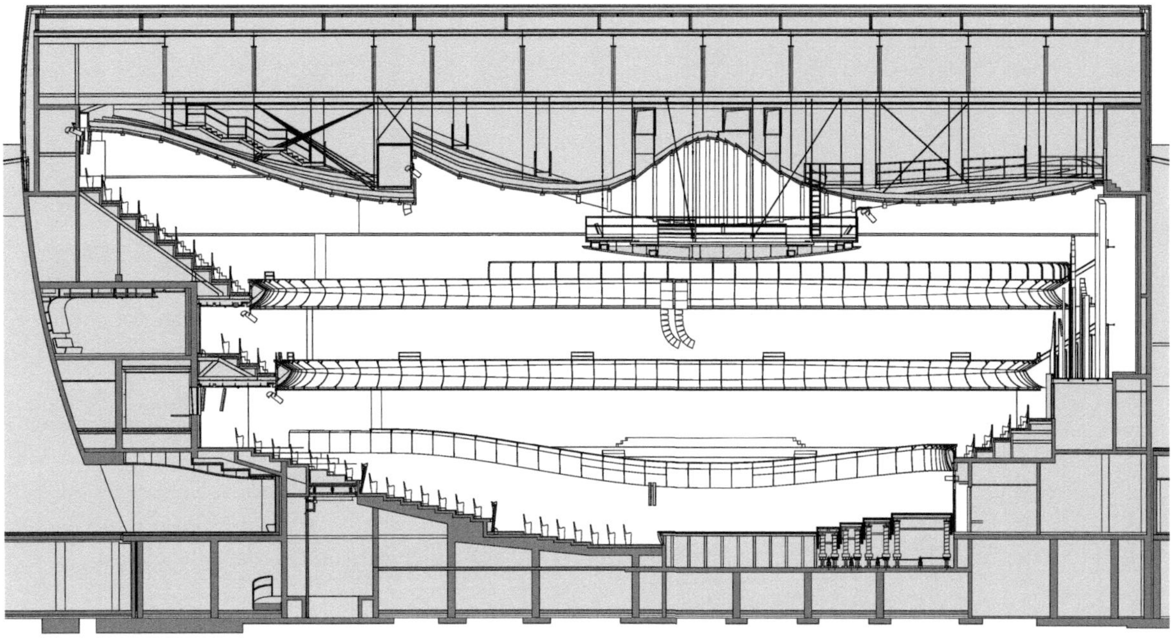

Fig. 20.3 Longitudinal section
Courtesy of KONIOR STUDIO Tomasz Konior

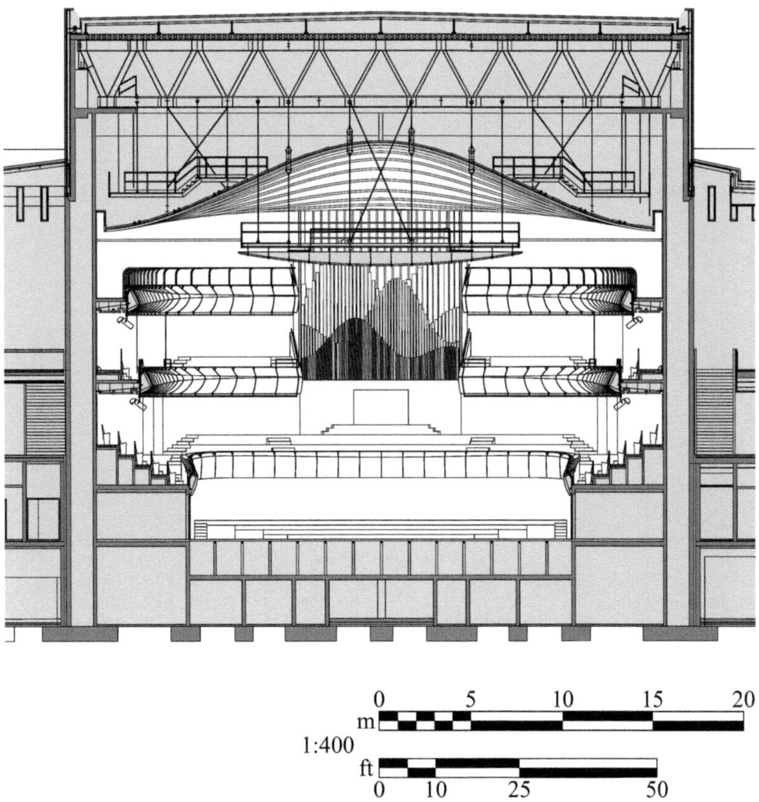

Fig. 20.4 Cross section
Courtesy of KONIOR STUDIO Tomasz Konior

21. Auditorium, Fondation Louis Vuitton

View from the balcony. Curved glazed panels were introduced at the bottom of the glazed walls to diffuse sound reflections from the mostly hard and smooth surfaces. Massive reflectors can be lowered behind and above the stage to help the sound projection to the audience and provide support to performers

Y. Toyota et al., *Concert Halls by Nagata Acoustics*, https://doi.org/10.1007/978-3-030-42450-3_21

Designed by Frank Gehry and opened in October 2014, Fondation Louis Vuitton is a sculptural building of glass, steel, wood, and white concrete tiling. Nestled in the trees of Bois de Boulogne, a large public park on the western edge of Paris, it borders the nineteenth century Jardin d'Acclimatation amusement park.

Financed by LVMH Moët Hennessy Louis Vuitton group, the new museum is dedicated to contemporary art, and features eleven galleries across over 3,800 m^2 of spaces presenting collections, permanent exhibitions, and temporary artistic interventions. Additionally, the building houses a performance space, the Auditorium, with modular configurations for standing audience or up to 350 seats. On the invitation of Frank Gehry, Nagata Acoustics served as acoustical consultant for the room acoustical design of the Auditorium.

The Design of the Auditorium

Upon the start of design in 2006, initial programming directions described the Auditorium, or "Forum" as it was once referred to, as a multifunctional event space for lectures, projections, or amplified music performances. Secondary uses were to include fashion shows as well as occasional musical performances in natural acoustics.

To accommodate the flexibility of uses, the main floor of the auditorium is divided into individual mechanized rows with adjustable heights and reversible seating. The hall can therefore be configured with a fully flat floor or a variety of rakes, seating up to 350 people. At its steepest, the seating meets the two-row balcony at the rear of the hall to create a seamlessly continuous slope. Similarly, the stage floor consists of a grid of motorized lift platforms which can be configured in a variety of options, thus complement the spatial flexibility of stage and audience layouts.

Although rather compact in footprint, the hall features a comparatively tall ceiling height, at 15 m across the entire space, in order to create suitable dimensions and volume for classical music performances without the use of electronic sound enhancements or amplification. The walls are shaped into large blocks and towers sculpting the volume with mainly sharp edges and straight lines. Audience typically enters the hall from the balcony level, descending to the main floor via a single staircase on the right.

One of the principal acoustical design challenges for the space was the ample use of glazed surfaces, continuing the overall building concept of transparency and openness. The facade walls give the hall its unique atmosphere, combining the intimacy and proximity of a small venue with spectacular views opening to the broader surroundings and the slowly moving water descending from the stepped fountain behind the stage, but glass is an inherently difficult material for an acoustical designer. Being particularly hard and smooth, glazed surfaces create sound reflections distinctly colored by abundant higher frequencies and an extremely sharp character. The solution was to introduce curved glazed panels where the glass facade was the most exposed to stage and seating, and served direct sound reflections to performers or listeners. Consequently, sound reflections off of glazed surfaces are either diffused by the curved geometry, or reflected on at least one other surface, thus avoided strong direct reflections.

By contrasts, the walls and ceiling surfaces were designed to distribute and alternate solid reflective area with soft absorptive patches. This acoustical checkerboard is visually unified by an openwork finish comprised of a milled artificial stone grid backed by nylon mesh fabric, either glued or stretched over the acoustical finish. The grid also creates a textured finish to the surfaces, thus favoring further high-frequency sound scattering.

Two massive sound reflectors can be lowered above and behind the stage to project the sound to the audience and improve communication for musicians. Shaped as crumpled sheets of papers, they are in fact made of cast aluminum, reaching a weight of approximately 30 kg/m^2.

View from stage right. In spite of a compact footprint, the hall features a tall ceiling to accommodate performances in natural acoustics. The seating layout can vary from a fully flat floor to a steep rake meeting the balcony level in a single slope

In order to suit the various possible functions and occupancies of the space, elements of acoustical variability were introduced in the design. They consist in a series of retractable banners embedded in the technical gallery and ceiling, and curtains covering the upper rear surfaces of the hall, around the control room. These acoustical draperies largely leave the glass facade exposed and visible, but additional curtains can be deployed to completely black out the space when required by the program.

Featured Art

Several contemporary artists were commissioned by the Fondation to create pieces and installations to compliment and inhabit the architecture of the building, inside and outside. For the Auditorium, American painter and sculptor Ellsworth Kelly proposed a set of five large rectangular canvases, each with a single bright solid color, distributed on the wall surfaces and punctuating the boxes and towers of the architectural design. For the design and creation of these pieces, careful considerations were given to the canvas fabric and backing, so as not to match the intended acoustical properties of the covered surfaces. The installation, titled "Colored Panels (Red, Yellow, Blue, Green, Violet)," is complimented by an additional piece for the stage curtain, featuring twelve colored vertical lines and titled "Spectrum VIII" in continuation of a long running series of work and variations started in 1953.

Installation by artist Ellsworth Kelly. American painter and sculptor Ellsworth Kelly was commissioned to create an original installation for the Auditorium. He designed a set of five large rectangular canvases of single bright solid colors. Each was carefully studied depending on the intended acoustical properties of the covered surfaces

The installations are left on permanent display in the Auditorium. When no performance is scheduled in the space, it is open to the visitors of the museum.

The Concert Hall in the Museum

During the construction of the building, the management of the Fondation and future operator of the Auditorium decided to pivot the primary programming of the space towards classical music and performances in natural acoustics.

Less than 2 years later, and a few months before the building's inauguration, a group of classical musicians were invited to test the hall and its acoustics. Pianists, violinists, cellists, quartet, clarinetists, and vocalists, musicians unequivocally praised the acoustics of the auditorium, its clarity, its richness, and the quality of its reverberance, in spite of a relatively short reverberation time by traditional standards and expectations for a dedicated classical music venue.

Ever since its inaugural concert by star-pianist Lang Lang in October 2014, the Auditorium has hosted a wide variety of artists, performances, and events, ranging from hip-hop and electro musicians, to semi-staged contemporary operas, solo performers, and larger chamber ensembles. The concert program alternates between thematic productions in relation with concomitant exhibitions, and distinct concerts or series. But the primary focus has been in presenting classical music in a supremely intimate setting, and featuring both established artists and rising talents, including training workshops and masterclasses.

Alongside the wonderful success of Fondation Louis Vuitton as a cultural destination in Paris, and once imagined as a multifunctional event space, the Auditorium has since established itself as a genuine concert hall.

Marc Quiquerez

Acoustical banners deployed. Acoustical banners can be lowered from the technical catwalk and ceiling to introduce additional sound absorption and reduce the reverberance of the space, while still maintaining the visual transparency of the glazed facade

Example of flexible configuration, set up for contemporary opera performance. Dedicated black-out curtains can be deployed to block daylight or ensure privacy. The independent rows of seating can be lowered to extend to stage area as desired, while still maintaining a raked audience layout

Building approach

Fondation Louis Vuitton	
Location	Paris, France
Owner	Fondation Louis Vuitton
Architect	Gehry Partners, LLP
Executive architect	STUDIOS
Design start	Q3 2006
Construction start	Q3 2008
Construction end	Q2 2014
Opening date	October 27, 2014
Building size	11,700 m^2
Auditorium	
Acoustical consultants	Nagata Acoustics (room acoustics)
	Yasuhisa Toyota, Kayo Kimotsuki-Kallas, Marc Quiquerez
	Lamoureux Acoustics (building acoustics)
Theater consultant	dUCKS scéno
Inaugural concert	October 28, 2014
Seating capacity	350
Room volume	5,900 m^3
Surface area	3,000 m^2
Volume/seat	16.9 m^3/seat
Volume/surface Area	2.0 m
Finish material	
Ceiling	Medium-density fiberboard and steel plate, glass wool
Wall	Medium-density fiberboard and steel plate, melamine foam
Surface finish	HI-MACS open grid
Stage reflector	Cast aluminum
Glass facade	Curved panes at bottom
Seat manufacturer	Poltrona Frau
Model scale	1:20

Table 21.1 Auditorium—acoustical metrics at 500 Hz

	Reflective condition	Curtains deployed	Acoustical and black-out curtains deployed
RT unoccupied	1.3 s	0.9 s	0.7 s
RT occupied	1.2 s		

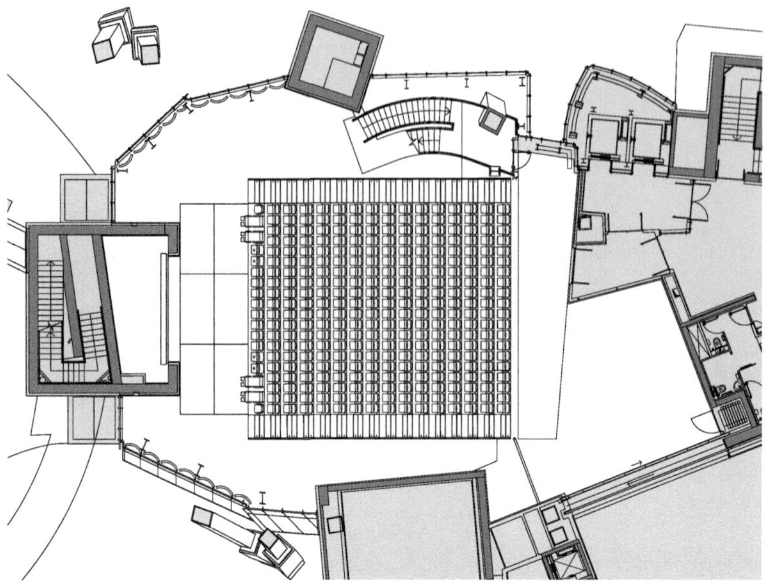

Fig. 21.1 Main level plan
Courtesy of Gehry Partners, LLP

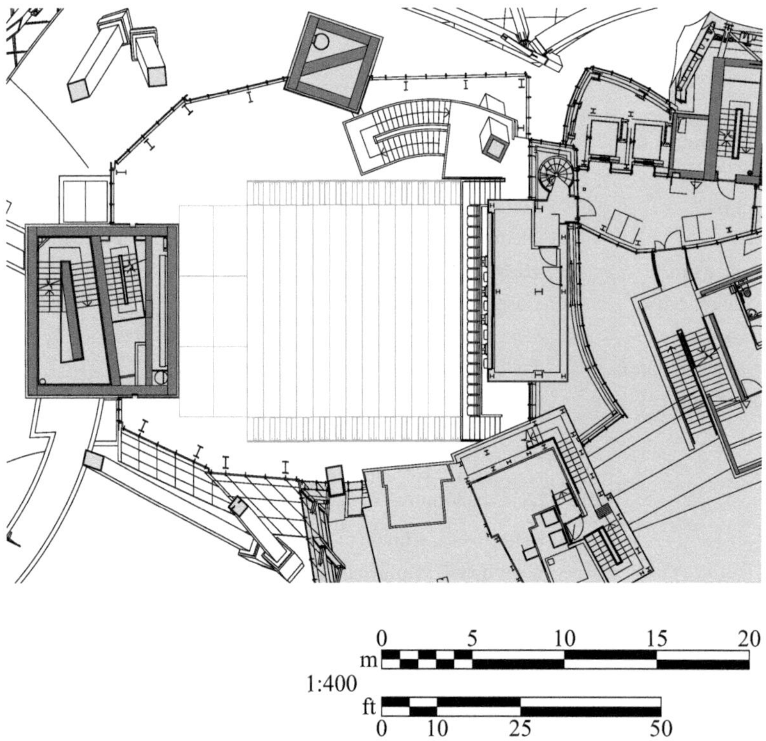

Fig. 21.2 Balcony plan
Courtesy of Gehry Partners, LLP

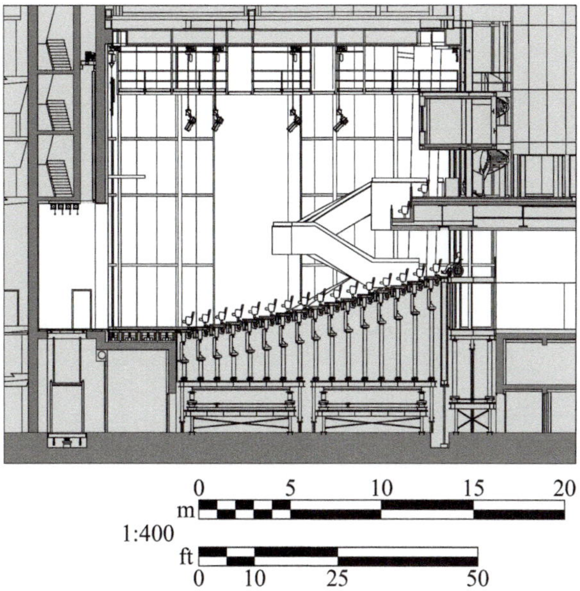

Fig. 21.3 Section
Courtesy of Gehry Partners, LLP

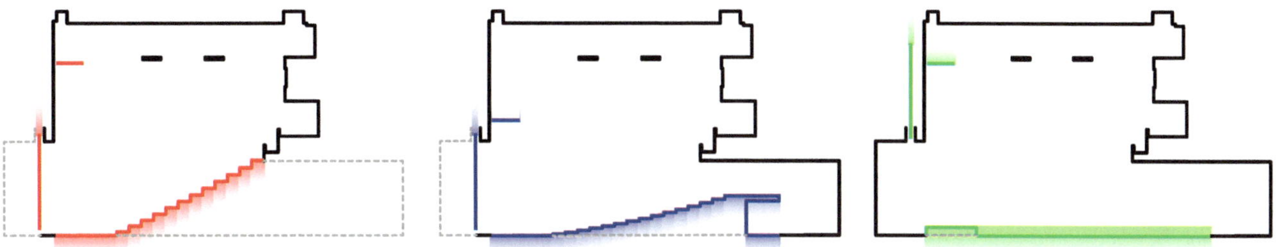

Fig. 21.4 Scope of spatial transformations

22. Radio France Concert Hall

View from upper stalls. With seating surrounding the stage on three levels, the Auditorium offers a supremely intimate and shared concert hall experience, in a very immediate yet warm acoustics

Y. Toyota et al., *Concert Halls by Nagata Acoustics*, https://doi.org/10.1007/978-3-030-42450-3_22

Maison de la Radio in Paris was inaugurated in 1963. Designed by French architect Henry Bernard, it served as the headquarter of the French national public radio and television company ORTF until 1975, when only radio remained in the building under the "Radio France" structure. Often referred to as "Maison Ronde," or "round house," the building is organized in two concentric rings—the outer most of which measures 500 meters in circumference—and a central tower 68 m high. Once described as a "radio factory," the building houses more than 3,000 people working on site across more than 140 different trades. Among them are no less than four permanent music ensembles. Its two orchestras Orchestre National de France and Orchestre Philharmonique de Radio France, professional choir Chœur de Radio France, and 150-strong children and youth choir Maîtrise de Radio France make Radio France a prominent music institution in France and in the world.

In 2003, parts of the building were deemed unsafe by local authorities and evacuated, which triggered discussion on the future and conservation of the building. An ambitious rehabilitation of the entire edifice was soon decided and became an opportunity to rethink its organization and operations. In late 2004, an architectural competition was announced, which set the clear goal to create more openness to the surroundings and to the public. As part of this plan, a brand new 1,500-seat concert hall devoted to the resident music ensembles was to be created. Paris-based AS.Architecture-Studio was appointed in April 2005 as the design architect for the overall project. Construction work on the overall renovation broke ground in December 2009. Works on the Auditorium effectively started in 2012, after the demolition of previously standing studios, and completed in Fall 2014.

The Design of the Auditorium

Nested in the outer ring of the building, the Auditorium remains invisible to the visitor on the outside or in the surrounding foyers and circulations, with only its domed roof hinting at its presence when viewed from above. Within the footprint where two smaller studios once stood, the hall is constrained on all sides in an outline resembling a rounded hexagon approximately 40 m in diameter.

The original client brief described a hall which could seat 1,500 patrons in a layout surrounding the stage. But accommodating a stage suitable for a full-size orchestra meant that the stage would already occupy half of the width and one third of the length of the hall. The design naturally developed into a quasi-cylindrical hall with the conductor's podium at its center. Since each level exhibited the same outer boundary, dictated by the surrounding building, the hall could not spread in plan. Instead, the audience layout grew vertically with two levels of overhanging balconies. This presented two major difficulties in acoustical design, with a circular shape prone to detrimental sound focusing and the risk of stacked balconies overshadowing the seats underneath.

Balconies and surface texture. The terraced seating familiar to vineyard-style halls was here transferred into stacked balconies divided into blocks. The large balcony fronts serve as reflectors, and their surfaces are textured with variations of depths and width. Behind the seats, convex panels help diffuse sound reflections to prevent focusing from the fundamental round shape of the hall

One key solution was to divide the walls and balconies into small blocks of audience, reminiscent of the terraces found in most vineyard-type concert halls. This allows for the fundamental circular and concave geometry to be broken down into smaller linear segments arranged in "scales" which also create a peripheral circulation and distribute audience access to each block. Each scale is lined with convex cylinders which further disperse sound reflections and prevent detrimental focusing and echoes. Additionally, the balcony levels gradually recede to limit overhangs and secure maximum visibility from each row of seats to the main ceiling above. Although the undersides of balconies are tilted to maximize the rake and allow for improved sightlines in this mostly vertical layout, the soffits on the sides of the stage were locally adjusted to maintain a right angle with the wall and therefore function as a second-order reflector for both the musicians on stage as well as audience.

This resulted in an extraordinarily compact and intimate audience layout, where almost half of the audience is seated to the sides or behind the stage, and over one out of four viewer-listeners is seated on a first row. Not one seat is located further than 22 m from the conductor's podium, comparable to the most distant pair of musicians in the orchestra on stage. To maximize the acoustical volume of the hall, large overlapping ceiling panels rise to a peak height of 20 m above the stage level. In order to serve early enough reflections to the stage and nearby audience, a large curved reflector is suspended above the musicians at a fixed height of 14 m.

View from chorus seating. A large reflector is suspended above the stage to serve necessary early reflections to the musicians and surround audience, while large ceiling panels raise the ceiling up to 20 m from stage level

The unique arena-style arrangement of the hall and the acoustically sensitive nature of its circular organization required the help of a physical model to fully assess the overall geometry of the hall and secure the complete elimination of any detrimental echoes or focusing. A physical 1:10 scale model was built and thoroughly tested for this purpose, and local adjustments were made to the design, in the form of supplementary convex shapes or thin absorptive treatments. These treatments were concealed with acoustically transparent finishes to blend with the overall visual design of the hall.

Though the initial design distributed audience all around the stage on all levels, a pipe organ was later added to the project and slightly reduced the capacity to 1,450 seats. Sitting on the first balcony behind the stage, the instrument was designed by Barcelona-based, German-born organ builder Gerhard Grenzing and features 86 stops and 5 manuals. The facade pipes are flanked on each side by swell boxes followed by acoustically transparent wall opening to the larger instrument volume behind, visually blending in with the mainly horizontal patterns of the walls finishes.

As one walks into the hall, the immediate image is one of warm and vibrant changing wood tones embracing the light-colored stage in the center. The walls are indeed lined with three different wood types (cherry wood, birch, and beech), but only in the form of a thin veneer. The main underlying construction is composed of medium-density fiberboard backed by fiber-reinforced gypsum boards to realize the massive complex required for sound-reflecting surfaces. The different wood types draw thin lines on the walls and balcony fronts, which are emphasized by grooves of varying sizes and depths to create small irregularities for high-frequency sound diffusion. To reach double the surface density of the walls, the gypsum backing is replaced on the ceiling and ensemble reflector by steel plates over 10 mm thick.

The stage floor is made of lightly oiled Port Orford Cedar decking supported by soft wood sleepers and beams over an airspace. The stage is fitted with 17 mechanically operated risers which allow for a stepped arrangement of musicians in a wide range of configurations.

First Notes and Inaugural Concert

Just a week before inauguration day, the two resident orchestras took turns on the stage to sound their first notes and prepare for the opening concert. This offered a unique opportunity to witness this process happen simultaneously for two distinct ensembles in the same space. Day after day, even in this short and most certainly stressful period, we could hear each group gradually adjust their ensemble playing to the new space and take advantage, each in their own ways and characters, of the clear yet rich acoustics of the space, where even the softest notes reach the audience almost as if heard from the stage.

The opening concert on November 14, 2014 featured both orchestras and the choir and celebrated a milestone in the history of Radio France's music ensembles in a hall they can now all truly call home.

Concert Hall or Studio?

With its exceptional proximity to the stage, surround layout and steep rakes, the Auditorium offers a visually striking experience at every seat. The hall can feel solemn yet inclusive, intimate yet grand. Acoustically, this translates into a sound that is always very immediate and close, precise yet warm. In spite of the relatively large volume, which allows it to shine even with full-size ensembles or organ music, one almost feels as if hearing the performance from within the orchestra. As such, the venue creates a one-of-a-kind concert experience, one that blurs the line between concert hall and studio, presenting all viewer-listeners at once with glimpses at the sonic inner workings of a musical ensemble.

Marc Quiquerez

Maison de la Radio	
Location	Paris, France
Owner	Radio France
Architects	Henry Bernard (original, 1963)
	AS.Architecture-Studio (renovation)
Auditorium de Radio France	
Users	Orchestre National de France
	Orchestre Philharmonique de Radio France
	Chœur de Radio France
	Maîtrise de Radio France
Acoustical consultants	Nagata Acoustics (room acoustics)
	Yasuhisa Toyota, Keiji Oguchi, Motoo Komoda, Marc Quiquerez
	Lamoureux Acoustics (building acoustics)
Theater consultant	Changement à Vue
Design start	Q2 2005
Construction start	December 2009
Construction end	November 2014
Opening date	November 14, 2014
Seating capacity	1,461
Room volume	14,500 m^3
Surface area	6,200 m^2
Volume/seat	9.9 m^3/seat
Volume/surface Area	2.3 m
Finish material	
Ceiling and canopy	Steel sheets and medium-density fiberboards
Walls	Fiber-reinforced gypsum boards and medium-density fiberboards
Audience floor	Wood over concrete
Stage floor	Port Orford Cedar
Seat manufacturer	Quinette Gallay
Organ builder	Gerhard Grenzing Orgelbau
Model scale	1:10

Table 22.1 Radio France Concert Hall—acoustical metrics at 500 Hz

RT unoccupied	2.0 s
RT occupied	1.8 s

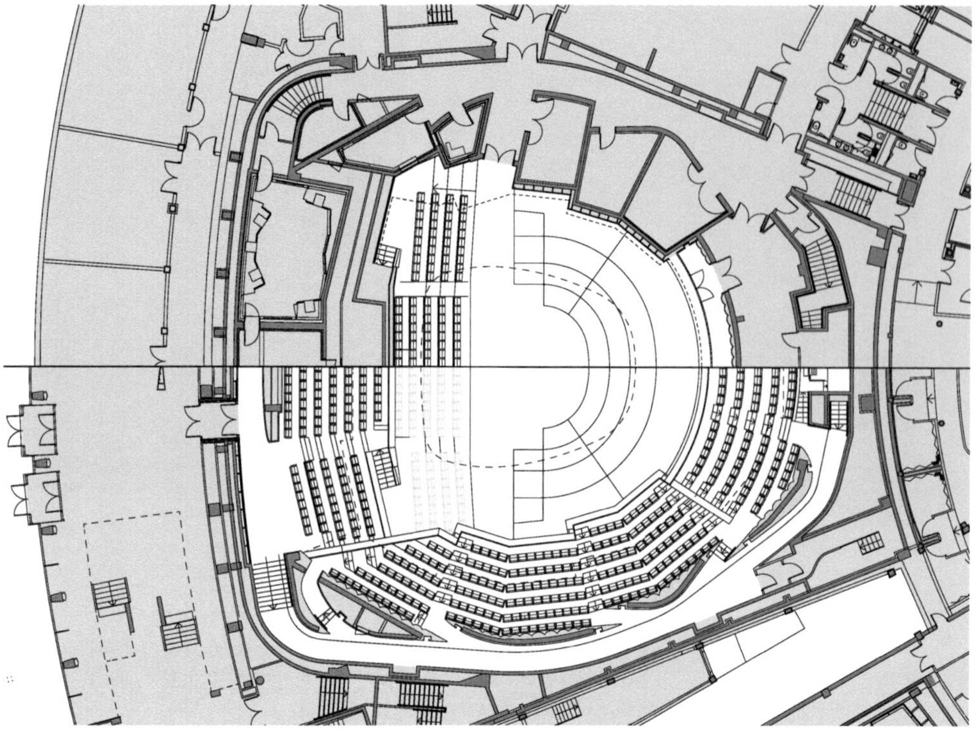

Fig. 22.1 Stage and terrace levels plan
Courtesy of AS.Architecture-Studio

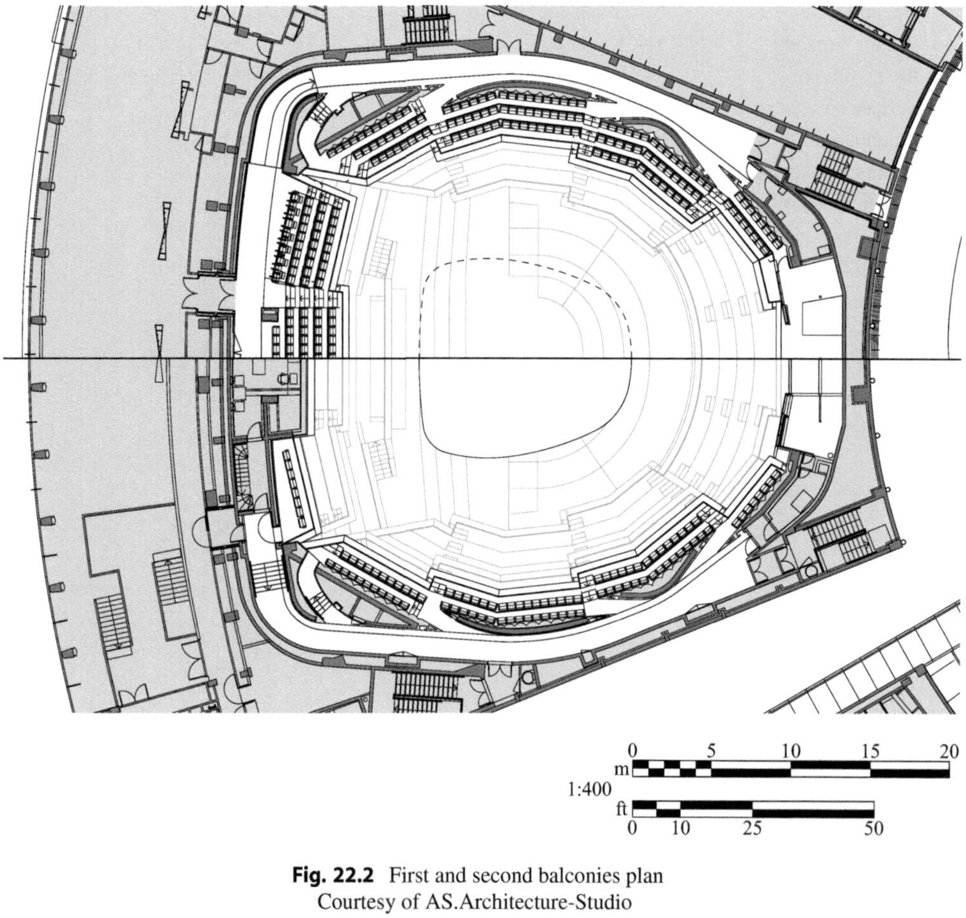

Fig. 22.2 First and second balconies plan
Courtesy of AS.Architecture-Studio

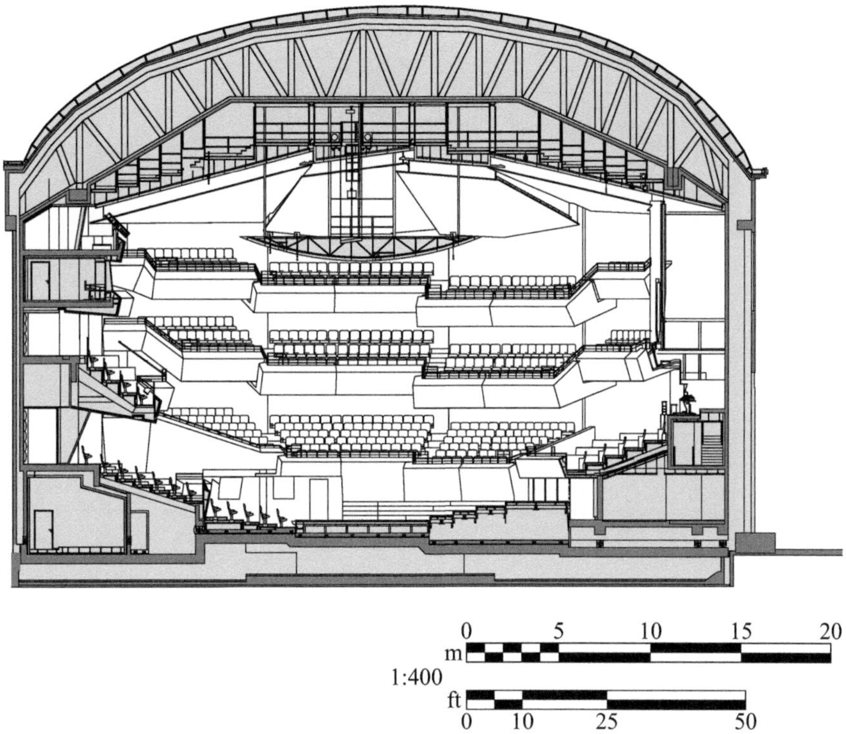

Fig. 22.3 Section
Courtesy of AS.Architecture-Studio

23. Philharmonie de Paris

View from the first balcony. In spite of the large seating capacity of the hall, the design achieved a great intimacy for all seats with its compact balconies and seating wrapped around the stage

Y. Toyota et al., *Concert Halls by Nagata Acoustics*, https://doi.org/10.1007/978-3-030-42450-3_23

The idea to create a brand new purpose-built concert hall for orchestra music in Paris had been circulating for quite a long time, and dates back as far as the 1970s. It was only in the mid-2000s that a firm plan finally crystalized, gathering the joint political and financial support of the French national government, the city of Paris, and the regional government, under the name "Philharmonie de Paris."

The building finally opened to the public in January 2015, after eight full years of development and construction not devoid of challenges.

A Building in Context

Philharmonie de Paris stands on the east edge of Parc de la Villette, along the ring freeway encircling Paris, in the north-east of the city. With its swirling shiny core of aluminum and steel set among the angular edges of large planes cladded with flocks of bird-shaped tiles in grey shading, the imposing silhouette does not give away its content or inner forms. It does, however, hint at the fundamental organization of its functions: daytime activities such as exhibitions, seminars, and educational workshops are located on the lower ground level as an extension of the park, while the hall is lifted two floors up and can be reached by a large inclined ramp and a grand staircase.

The construction is integrated within the dense cultural hub formed by a variety of institutions in and around the park, such as the National Music and Dance Conservatory, the Zenith indoor arena, the Grande Halle, and Trabendo music club among others. Even more significant was the organizational merge of the new facility with the adjacent Cité de la Musique (opened in 1995) to create a global institution overseeing two buildings, three halls (for a total of 3600 seats), eighteen rehearsal spaces of various sizes, twenty-four dedicated educational spaces, a museum, and a multimedia library. Five different musical ensembles are resident or associate to the complex: Orchestre de Paris, Ensemble Intercontemporain, Orchestre de Chambre de Paris, Les Arts Florissants, and Orchestre National d'Île-de-France.

A Complex Design Team Organization for an Innovative Concept

The international competition launched at the end of 2006 awarded the design to French architect Jean Nouvel in the spring of 2007. During the competition, Nouvel enlisted acoustical consultants Marshall Day Acoustics (MDA) from New Zealand and theater planners dUCKS scéno out of Lyon, France. Nouvel later invited Nagata Acoustics to act as personal acoustical advisors for the design of the main hall. In total, no less than six acoustical consulting firms participated in the project, from briefing and support to the client, to design and commissioning, to construction.

No stranger to the design of large performing arts facilities and cultural buildings at large, and encouraged by an ambitious brief, Nouvel embarked upon the project with the determination to explore beyond the formal frontiers of concert hall typologies. This also meant tackling the challenge of coordinating varied inputs from the brief, which called for a wide array of physical transformations and programs, and the various components of the design team with a holistic approach aiming at merging all the goals and contributions into a truly unique and innovative design. The fundamental acoustical principle established during the competition was to independently control the audience layout and early sound reflections on the one hand, and the late reverberant field on the other.

The Design of the Concert Hall

As one passes through the doors to the hall, their first encounter is an abstract volume bounded by the white envelope of the hall and the rounded back of the balconies. The vast and intricate womb-like space barely feels as if being part of a performance space, and yet already evokes the sense of envelopment that the hall will soon reveal to the eyes and ears.

Large surrounding volume. Detaching the balconies from the outer walls created a large outer volume which contributes to the rich reverberance of the hall

Through the entry ways concealed in the balcony walls, the hall suddenly appears, opening to the large inner volume with its shiny surfaces and unusual colors. One can suddenly embrace the entirety of the hall and audience surrounding the stage. And yet, it is not an overwhelming sense of gigantism and space that dominates. Rather, what immediately strikes is the proximity of all seats and one's eyes are naturally drawn to the stage which stands out of the surrounding dark walls in its natural bright wooden tones. It is as if everything was physically within reach.

Two levels of balcony gently sweep around the hall in smooth curves. In the back, ribbon-like walls continuously vary in height and tilt to serve early reflections to the rows of audience immediately in front. On each side of the stage, the walls fold over to create a shallow overhang and exact right angle to direct sound reflections to the performers on stage. Reminiscent of the terraced design of vineyard halls, the balconies remain compact, with any given seat never further than seven rows away from the surface of the ribbon walls. Detached from one another and only connected to the main outer envelope by cantilevering bridges, they hover in the grand volume of the space and moderately overhang to maintain the proximity to the stage and keep ceiling and overhead reflectors always visible. Steep rakes, at times vertiginous, secure open sightlines to the stage and enhance the sense of proximity and intimacy.

Cantilevered balconies. Balconies are cantilevered from the outer walls, as if suspended in the overall volume of the hall. This maintains the proximity and intimacy between stage and audience ©Lida Guan ©2020 Jean Nouvel / Artists Rights Society (ARS), New York / ADAGP, Paris

By comparison to the intricate balconies, the lower seating tier and chorus area may appear rather simple in form. Their outline shape was, however, driven by the programming request to make them fully transformable. Each row on the parterre, up to the terrace wall, consists of its own lift: chairs can disappear and the overall floor can become a large flat area for standing audience or various layouts of performers. The entire stage platform can also be lowered to meet this level, effectively extending the potential flat area to a total length of almost 30 m. The chorus area comprises two sections of telescopic seating rows on a large lift platform. This allows to create a flat, raised platform for end-stage configurations. A wide range of intermediate configurations is also available, with an array of possible performance types ranging from symphonic music to end-stage pop-rock to staged performances or movie projections.

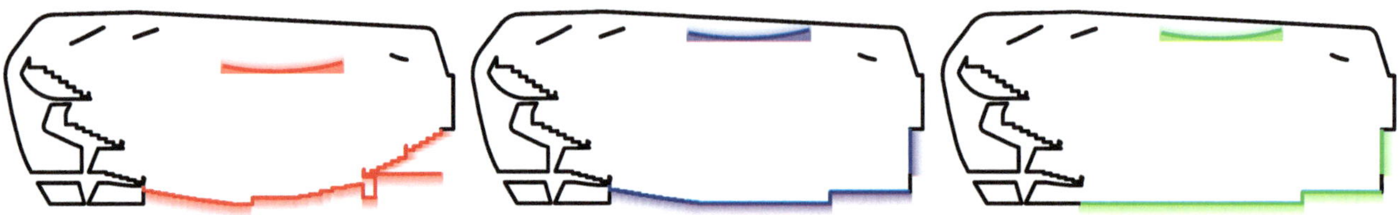

Fig. 23.1 Scope of spatial transformations

A large reflector hangs over the stage and a series of petal-like reflectors, or "clouds," cap the audience chamber. Though partially obscured by these reflectors throughout the hall, the main ceiling above still contributes effective early reflections—including first order reflections—to the audience on the balconies.

The undeniable modernity of the shapes is matched by the chosen expression of the materiality of the space. Parting with a more traditional wooden atmosphere, Nouvel created color variations on the balconies' veneers and chair upholstery, with a progression from the strict black surrounding the stage and spreading over the lower seating area from ochre nuances of amber and golden yellow to solid white matching the envelope. Rather than strictly defining zones or sections in the hall, the tones follow the flow of the balconies and surfaces and participate in the overall sense of unity and envelopment.

Although an abundance of surfaces throughout the hall are perfectly smooth, the white plaster envelope is sprinkled with square dents of varying sizes and depths, while protruding blocks are scattered across the balcony walls and fascias. Designed to scatter high-frequency sound reflections from surfaces close to the audience, the various irregularities also participate in preventing detrimental echoes, as measured and studied with a physical 1:10 scale model.

In spite of the exceptionally large seating capacity of 2,400 seats, the ample total room air volume of more than 30,000 m^3 and a maximum ceiling height of almost 23 m from the platform level, the design manages to sit only a fraction of viewer-listeners beyond a 30 m distance to the stage. Yet, the abundance of record-breaking numbers hardly paints a complete picture of the success of the design. The result is a concert experience which is consistently intimate and balanced, with a warm yet precise sound, bathed in a rich reverberance enhanced by sound flowing behind and between the balconies, which only truly exhibits an obvious double decay at full stops.

Beyond the Inauguration

Visiting the site in the early days of January 2015, one could hardly imagine that inauguration day was only a few days away. With so many chairs missing and the others wrapped in plastic, unfinished plywood boards and stage decking still under construction, there was little evidence that the hall could be ready to shine shortly.

Much has been written about the turbulent circumstances surrounding the inaugural concert of the hall on January 14, 2015. The hall and the building were still visibly and undeniably unfinished, and the principal resident ensemble, Orchestre de Paris, and its music director at the time Paavo Järvi only had a mere few hours to spend on stage before the ribbon was cut and the proverbial "curtain" raised. Yet, from the night of the opening concert onward and in the following months, the myriad of programs by varied ensembles, conductors, and artists was continuously met with the enthusiasm and praises of performers and attendees alike, cementing the dawning status of Philharmonie de Paris as a truly world-class venue.

In October 2016, the hall officially took the name "Grande Salle Pierre Boulez" in honor of the late French composer and conductor who had been an instrumental and ardent supporter of the project and overall plans to create a new venue for orchestra music in Paris.

Marc Quiquerez

View from upstage balcony. Ribbon-like walls behind balcony seating direct early sound reflection to the stage and audience, complimented by large reflectors suspended from the ceiling

Building approach from Parc de la Villette.

Philharmonie de Paris	
Location	Paris, France
Owner	Cité de la musique-Philharmonie de Paris
Architect	Jean NOUVEL—Ateliers Jean Nouvel
Design start	Q2 2007
Construction start	February 2011
Construction end	January 2015
Opening date	January 14, 2015
Building size	42,000 m^2
Grande Salle Pierre Boulez	
Users	Orchestre de Paris
	Choeur de l'Orchestre de Paris
	Ensemble Intercontemporain
	Les Arts Florissants
	Orchestre de Chambre de Paris
	Orchestre National d'Île-de-France
Architects	Jean NOUVEL—Ateliers Jean Nouvel
	Métra + Associés (Associated Architects for the conception and the realization of the concert hall)
Acoustical consultants	Nagata Acoustics (room acoustics, advisor to Jean Nouvel)
	Yasuhisa Toyota, Keiji Oguchi, Motoo Komoda, Marc Quiquerez
	Marshall Day Acoustics (room acoustics)
	Studio DAP (building acoustics)
Theater consultant	dUCKS scéno
Seating capacity	2,400
Room volume	37,700 m^3
Surface area	13,700 m^2
Volume/seat	15.7 m^3/seat
Volume/surface Area	2.7 m
Finish material	
Ceiling	GF boards
Canopy and clouds	Wood on fiber-reinforced gypsum boards
Envelope	Plaster
Walls	Wood on fiber-reinforced gypsum boards
Audience floor	Wood on fiber-reinforced gypsum boards
Stage floor	Scots pine
Seat manufacturer	Figueras Seating Solutions
Organ builder	Rieger Orgelbau
Model Scale	1:10

Table 23.1 Philharmonie de Paris—acoustical metrics at 500 Hz

RT unoccupied	3.1 s
RT occupied	2.6 s

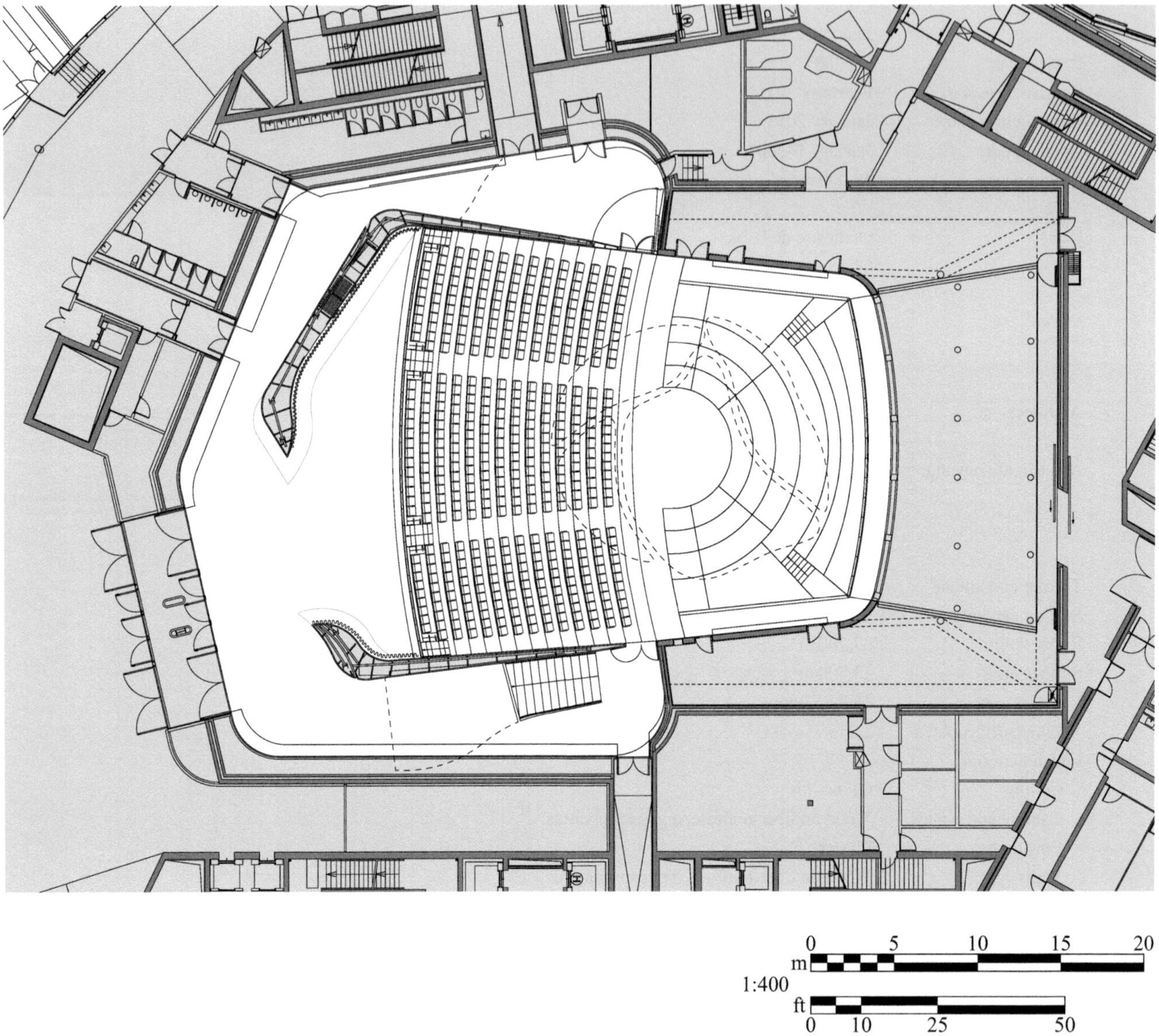

Fig. 23.2 Stage level plan

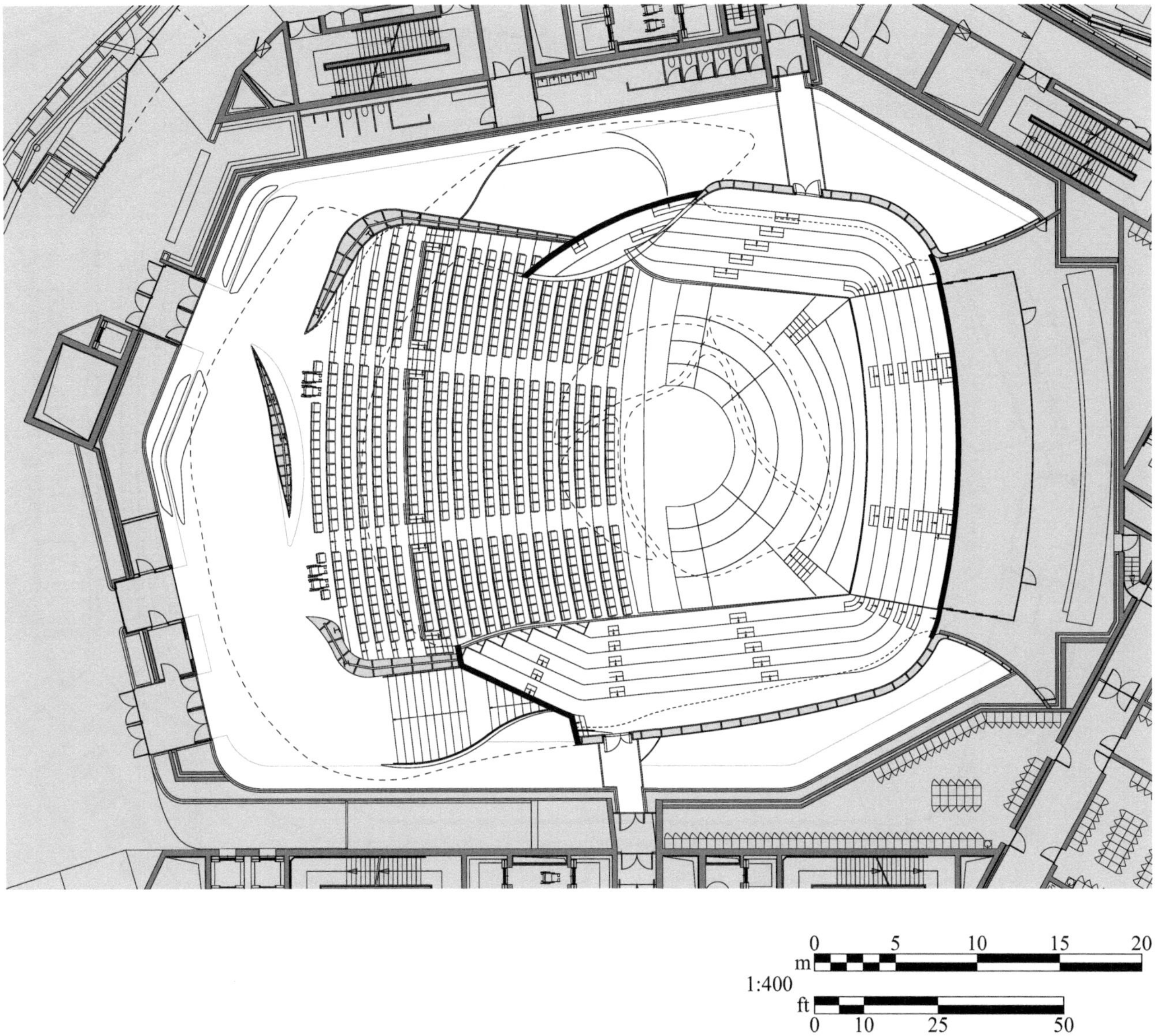

Fig. 23.3 Terrace level plan

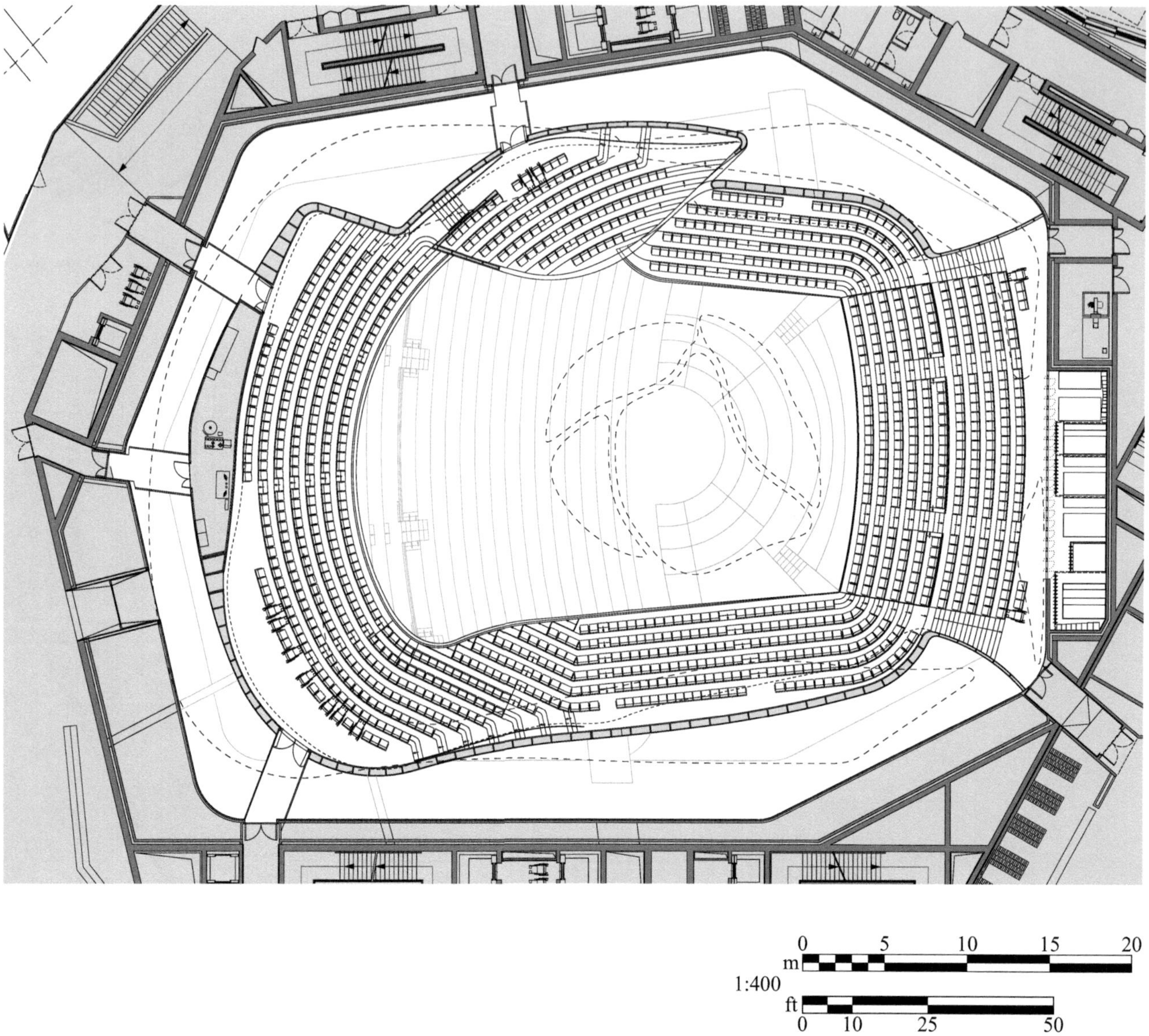

Fig. 23.4 First Balcony plan

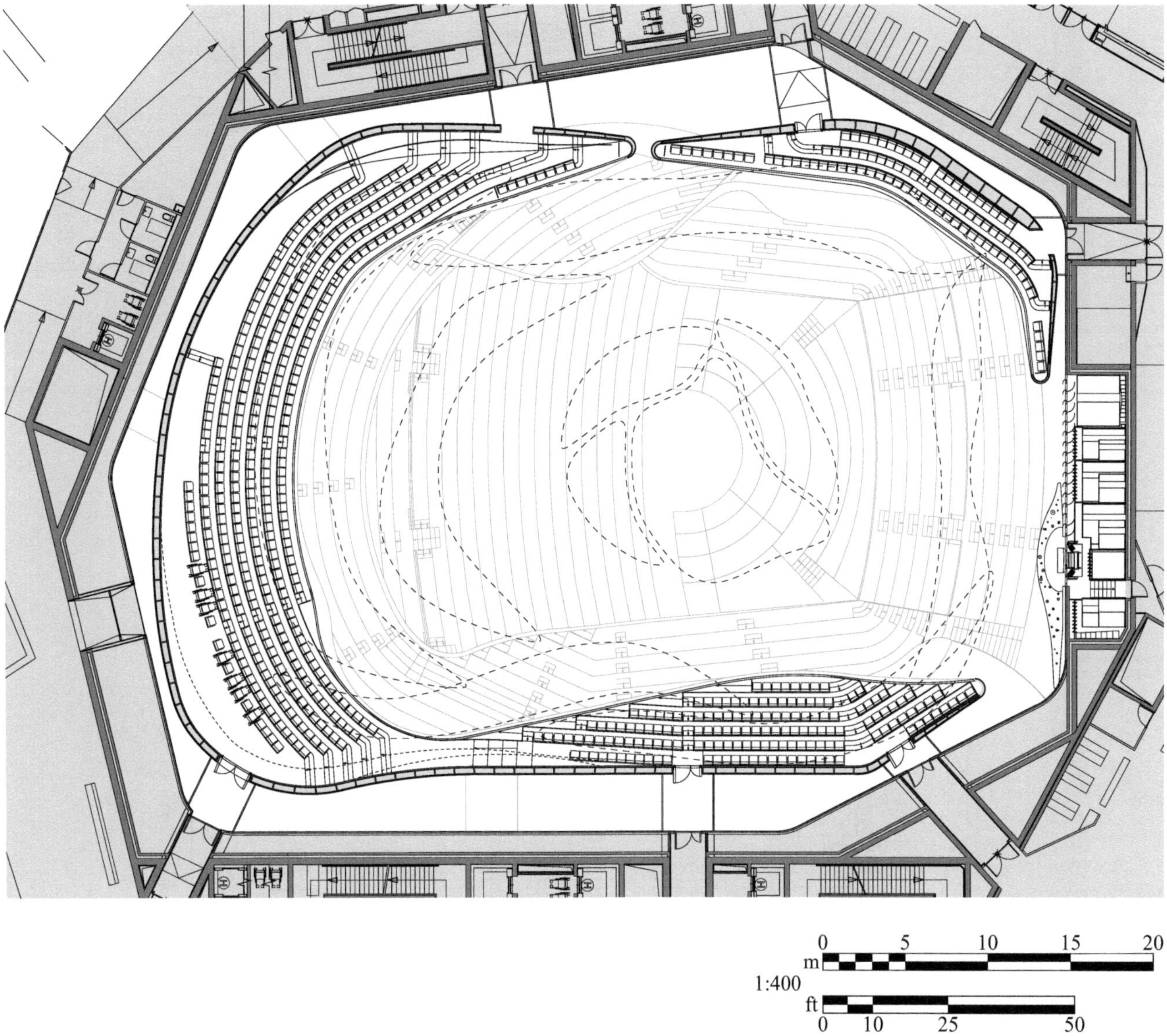

Fig. 23.5 Upper balcony plan
©2020 Jean Nouvel / Artists Rights Society (ARS), New York / ADAGP, Paris

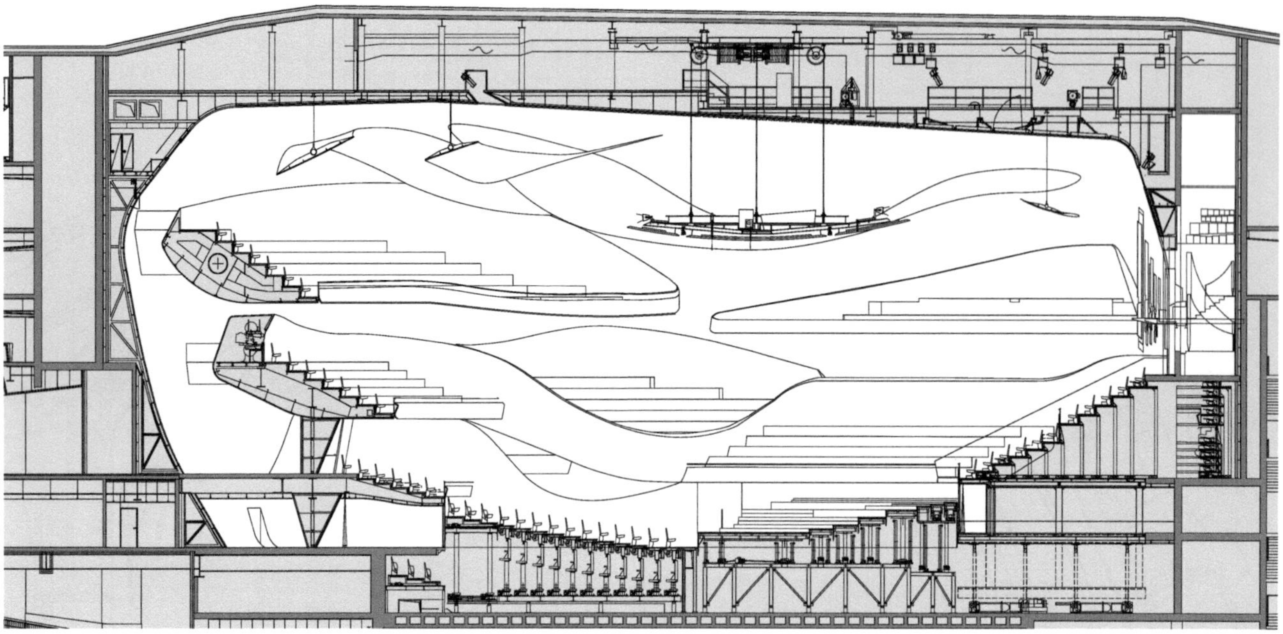

Fig. 23.6 Longitudinal section

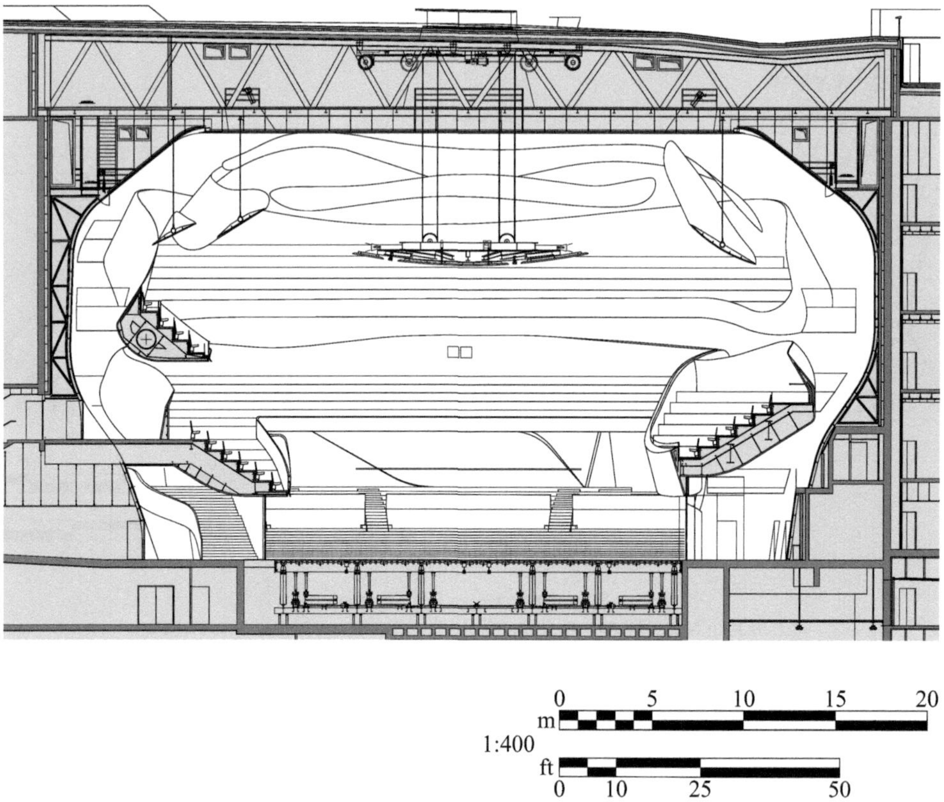

Fig. 23.7 Cross section

24. Musco Center for the Arts

Classical concert mode. For classical music concert, an orchestra shell system can be lowered from the stage fly tower. On the wall pieces, two levels of acoustic reflective panels pivot into place to provide acoustical feedbacks to the musicians. A two-piece orchestra lift can be raised to expand the usable stage area in order to accommodate a full sized orchestra, and this configuration further increases the sense of intimacy between the musicians and the audience

Y. Toyota et al., *Concert Halls by Nagata Acoustics*, https://doi.org/10.1007/978-3-030-42450-3_24

Musco Center for the Arts is located on the Chapman University campus in Orange County, California. Started in 2008, the center opened on March 19, 2016 after three and a half years of construction. In addition to a world class, 1,044-seat multipurpose hall, the center contains recording studios, nine dressing rooms, conference rooms, and many other features and amenities. Pfeiffer Partners (now Pfeiffer) was appointed as the project architect joined by Theatre Projects Consultants as the theater consultant to create a strong team well versed in the design of performing arts venues.

Building exterior

The main multipurpose hall is officially named the Julianne Argyros Orchestra Hall. From the beginning of the architectural design process, the project team prioritized the visual continuity and the closeness feeling between the stage area and the audience. The hall is hexagonal in footprint and viewer-listeners arranged on the main floor and sixteen side loges over three levels. With this configuration, all viewer-listeners are quite close to the stage. The distance from the stage front edge to the farthest seat at the top balcony is only 29 m (97 ft). In combination with the perspective view effect in the shape of hall, all seats feel very close to the stage.

Visually, the space is dominated by an intricate pattern of terracotta-colored petals which climb the walls. Acoustically, the textured, doubly curved geometry, and heavy plaster on metal lath construction are beneficial in reflecting and scattering sound. The heavy ceiling in the audience area also consists of plaster on metal lath but is essentially planar. To blend with the wall motif and hide two technical lighting bridges, acoustically transparent metal mesh elements are suspended from the ceiling.

Conceived with a wide range of programming in mind, the space needed be adaptable for acoustical concert performances, opera, Broadway-style musicals and dramas, lectures, and amplified musical performances such as jazz and pop concerts. Based on this design concept, realizing two distinct configurations—Concert Mode and Opera/Theater Mode—simultaneously became the primary challenge for the design team. To this end, the hall is equipped with a large stage including spacious wings, a fly tower, a system of high-quality stage reflectors, and an orchestra pit capable of accommodating up to seventy musicians.

Due to height limitations on the university campus and to keep enough ceiling height on stage, the building is sunken into the ground with stage level labeled as the first basement, although it is already two floors below grade. A large sloping lawn leads down to the public entrance, making the lobby inviting and full of light despite the fact that it is below street level.

View from stage. The hall is hexagonal in footprint and audience seats are arranged in the main floor and sixteen side loges over three levels. Visually, the space is dominated by an intricate pattern of terracotta-colored petals, and acoustically, the plaster petals are beneficial in reflecting and scattering sound. The ceiling in the audience area also consists of heavy plaster, and to blend with the wall motif, acoustically transparent metal mesh elements are suspended from the ceiling

Classical Concert Mode

In the stage area, an orchestra shell system can be lowered from the stage fly tower to provide early sound reflections mainly for classical music performances. When combined with the steel frame and lighting equipment, the total weight of these elements is 55,000 kg. While the weight of the movable panel system is quite heavy, previous experiences and recent technology led to a design which allows for convenient storing and deployment. The shell is broken into seven segments which are stored in the fly tower. To install the system, the upstage wall is simply lowered into position and the ceiling is lowered and rotated. The side walls hang on a single support which can both rotate and translate in order to move the segments first to the stage wings, next rotated in-line with the other scenery, and finally flown into the tower.

After the walls are secured, two levels of acoustic reflectors pivot into place with a manually operated network of ropes and pulleys. These reflectors were introduced to the shell to provide acoustic reflections to the musicians. In order to create a cohesive architectural impression, the "shelves" extend past the proscenium, underneath theatrical lighting positions, and connect to the side balconies. Swinging doors close the lighting pockets to create a continuous wall texture. Ultimately, there is seemingly no impression of a proscenium but complete architectural continuity between the stage and the audience areas clearly establishing the hall as more than a theater with a shell.

As for the stage, a two-piece orchestra lift can be raised to expand the usable area in order to accommodate a full sized orchestra. In this configuration, the orchestra protrudes beyond the proscenium, further increasing the sense of closeness and intimacy between the musicians and the audience.

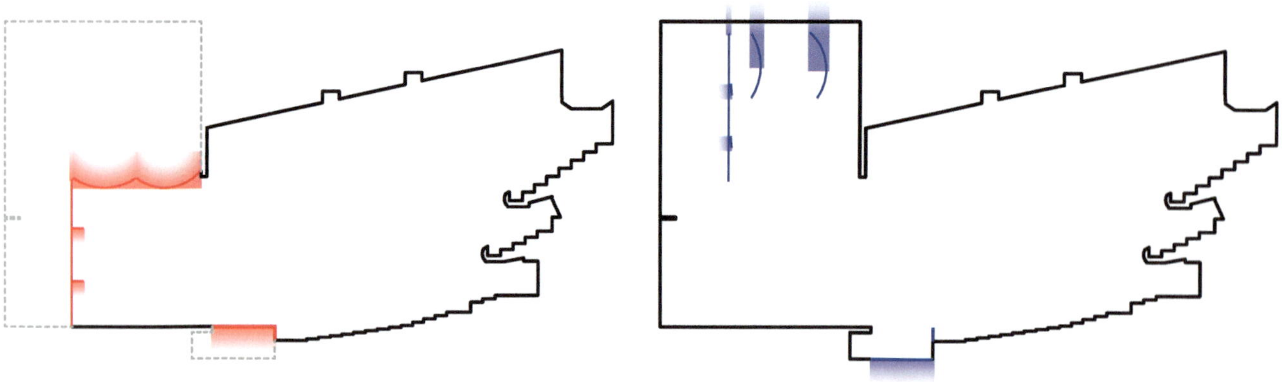

Fig. 24.1 Scope of spatial transformations

Theater Mode

For opera, musicals, and dance, the orchestra shells are nicely stored in the stage fly tower. The fixed opening of proscenium is 19 m wide and 11 m high, although these dimensions can be adjusted by the sliding acoustically transparent panels, thereby not disturbing the sound paths from the stage to the audience area. The two orchestra lifts can be lowered to create a large pit or Broadway pit.

Opera/theater mode. For opera, musicals, and dance, the orchestra shells are nicely stored in the stage fly tower, and the stage set can be installed instead. The proscenium opening dimensions can be adjusted sliding acoustically transparent panels, thereby not disturbing the sound paths from the stage to the audience area. The two orchestra lifts can be lowered to create a large pit

In order to adapt the reverberation field of the hall to the performances using electro-acoustic equipment, the hall is equipped with an extensive system of motorized acoustic banners and curtains. On the side walls, banners can be lowered between the petals. At the ceiling, curtains can be drawn horizontally along the lighting bridges. Each system can be stored individually by pushing a button making it possible to easily customize the reverberation field.

Variable acoustic banners. As one of the variable acoustic system, a motorized banner system is installed underside the balcony

The shortest reverberation condition can be set by storing the orchestra shell system, using stage curtains in the main stage area, banners on the side walls, and curtains on the ceiling. In this condition, the reverberation in the unoccupied condition can be reduced from 2.1 s to 1.4 s measured at 500 Hz.

Opening Series

At the opening gala event, Los Angeles Opera Orchestra with Plácido Domingo and the other alumni singers participated. And, many university graduates played on stage one after another, pleased audiences and made spectacular premiere concerts. On April 2, 2016, Grieg's Piano Concerto and Beethoven's Symphony No. 5 were performed by Pacific Symphony. After the opening series, Musco Center was described as "an ideal opera house, potentially the best in the West, and maybe even something more" by Mark Swed, a critic of the Los Angeles Times.

Motoo Komoda

	Musco Center for the Arts
Location	Chapman University, Orange
	California, USA
Owner	Chapman University
Construction cost	USD 82 million
Design start	May 2009
Construction start	September 2012
Construction end	February 2016
Opening date	March 19, 2016
	Julianne Argyros Orchestra Hall
Architect	Pfeiffer Partners Architects
Acoustical consultant	Nagata Acoustics (room acoustics)
	Yasuhisa Toyota, Motoo Komoda, Dave So
	Robert F. Mahoney & Associates (isolation and noise control)
	Fred Vogler (sound system)
Theater consultant	Theatre Projects Consultants
Seating capacity	1,044
Room volume	$12{,}838\,m^3$
Surface area	$5{,}607\,m^2$
Volume/seat	$12.3\,m^3$/seat
Volume/surface Area	2.3 m
Finish materials	
Ceiling, envelope, walls	Plaster on metal lath
Orchestra shell	Plywood layers with veneer
Seat manufacturer	American Seating
Stage floor	Alaskan Yellow Cedar

Table 24.1 Musco Center for the Arts—acoustical metrics at 500 Hz

	Concert configuration	Theater configuration
RT unoccupied	2.5 s	1.6 s
RT occupied	2.1 s	1.4 s
EDT	2.5 s	1.3 s
C_{80}	−6.1 dB	1.8 dB
D_{50}	16%	44%
G	22 dB	7 dB

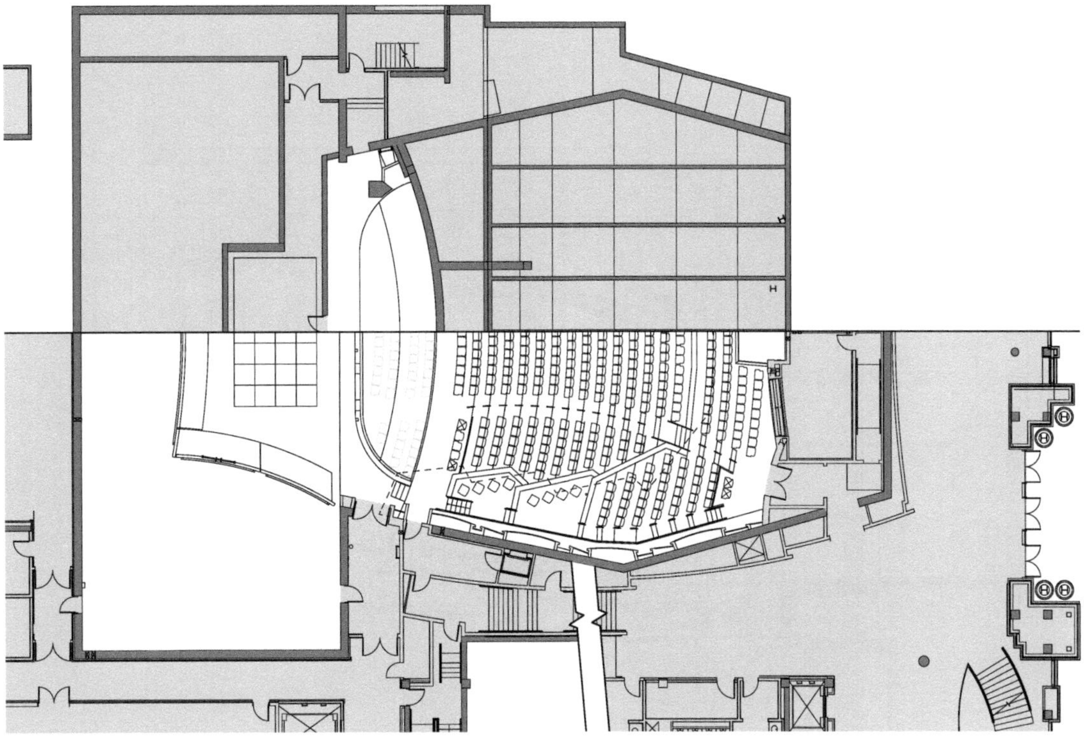

Fig. 24.2 Pit and stage levels plan
Courtesy of Pfeiffer Partners Architects

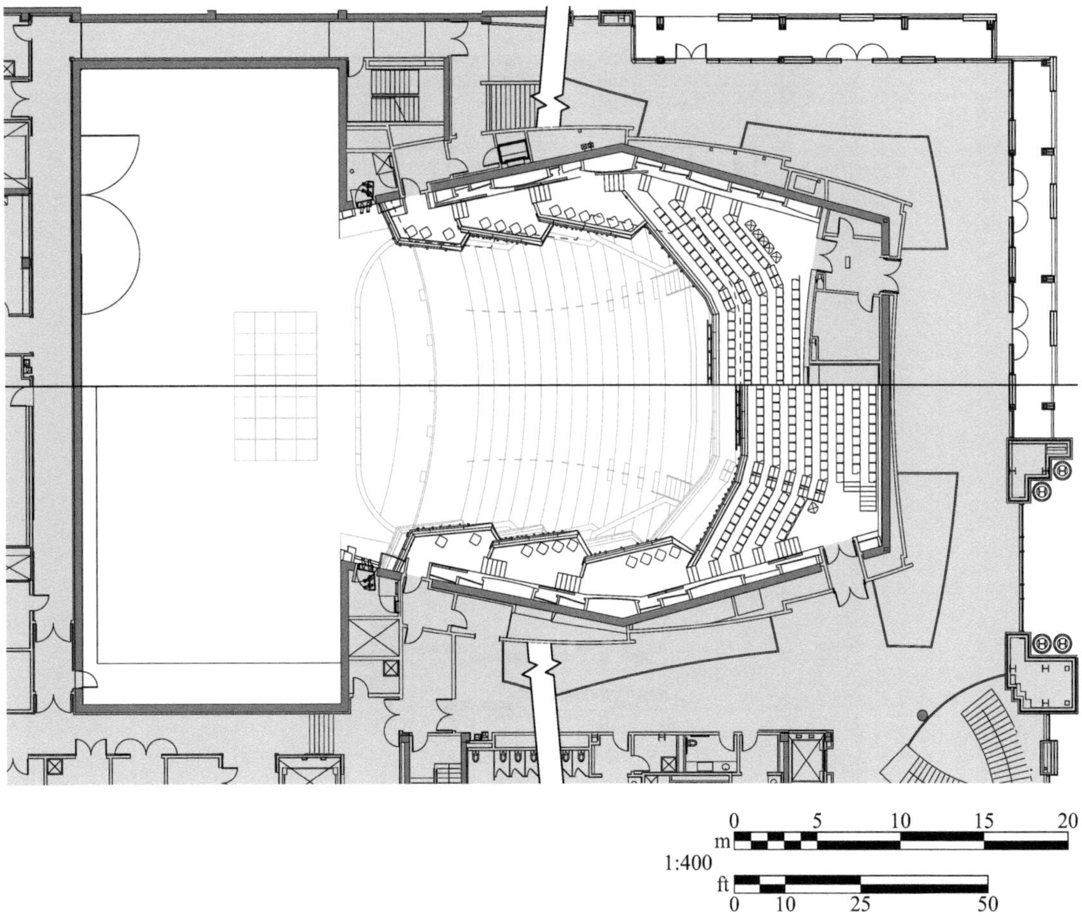

Fig. 24.3 First and second balconies plans
Courtesy of Pfeiffer Partners Architects

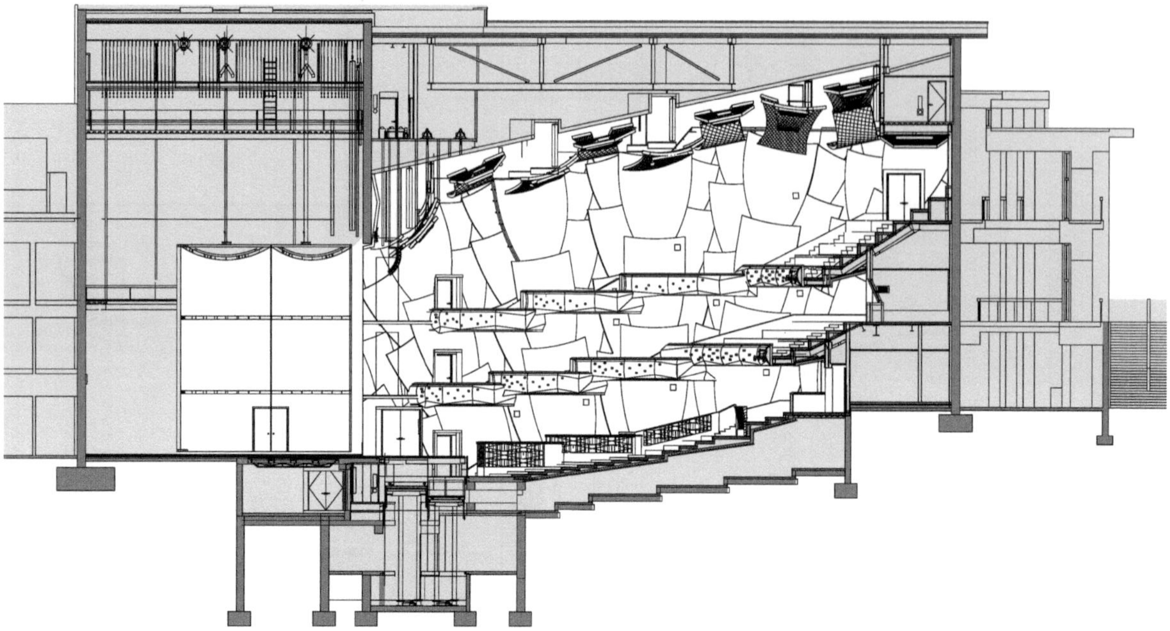

Fig. 24.4 Longitudinal section
Courtesy of Pfeiffer Partners Architects

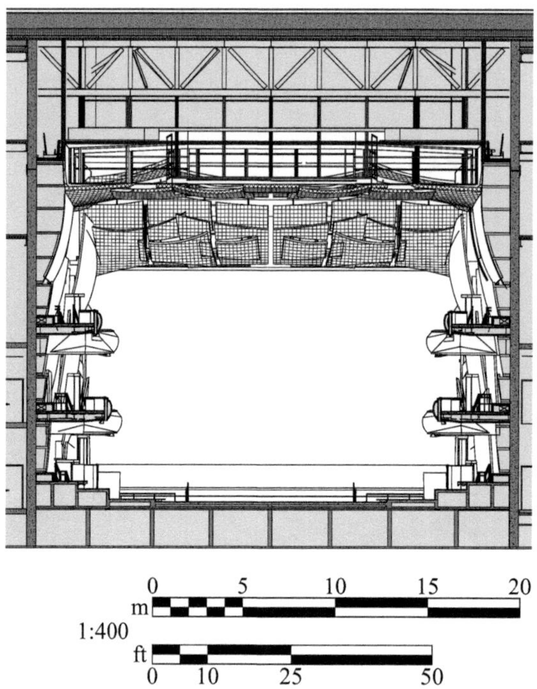

Fig. 24.5 Cross section
Courtesy of Pfeiffer Partners Architects

25. Lotte Concert Hall

View from balcony. The seat blocks surround a big stage that can accommodate a full-size orchestra. Since the room shape is almost elliptical in plan and acoustical focusing could occur, various types of shape were incorporated on the interior surfaces throughout the hall

Y. Toyota et al., *Concert Halls by Nagata Acoustics*, https://doi.org/10.1007/978-3-030-42450-3_25

In 1989, the Lotte Group of South Korea opened Lotte World Theme Park in the southeastern part of Seoul. An adjacent department store and hotel were soon to follow since Lotte World was already a famous and popular tourist spot. Next, the Lotte Group built a huge complex consisting of Lotte World Tower and shopping mall in the neighbor block of these existing facilities, including a new concert facility which opened in April 2017.

Overview of Lotte World Tower and Shopping Mall

The new complex consists of three connected buildings with a total floor area of 807,508 m^2 and a total construction cost of KRW 3.8 Trillion. The centerpiece is the "Lotte World Tower" skyscraper. At a height of 555 m, the 123-floor Lotte World Tower is the fifth tallest building in the world. Inside the building, stores, offices, condominiums, hotels and observation decks are located.

World Tower complex, with concert hall in center building

The shopping malls are separated into two buildings to the east of the tower. On the shopping floors, there are a wide variety of stores ranging from world famous luxury brand stores to supermarkets dealing with affordable items. There are food courts and entertainment spots such as movie theater, aquarium, and newly opened Lotte Concert Hall as well.

Exterior view from the garden terrace on the eighth floor

The 2,036 seat concert hall and its surrounding rooms occupy the 7th–11th floors of the center mall building. Support spaces include a rehearsal hall, a recording studio, practice and dressing rooms, an orchestral lounge, and offices and conference rooms.

Kohn Pedersen Fox Associates (KPF) in the United States was responsible for the overall building design of this complex. The shopping malls are crowded with shoppers, and visitors are constantly flowing throughout the facility, especially since there is a direct connection to the city's metro system.

The First Vineyard Style Concert Hall in South Korea

The Lotte Concert Hall is the first large-scale vineyard style concert hall in South Korea, designed for the main use of classical music concerts. The total construction area of the hall is 13,223 m^2, and the architectural design was handled by Designcamp Moonpark (DMP) of South Korea.

Since there is a subway under the shopping mall and a certain level of noise is also generated by the commercial activities inside the mall, the entire concert hall is supported by vibration damping rubber pads so that the hall will not be affected by the noise and vibration.

The interior of Lotte Concert Hall is unified in the color scheme of white and natural wood. Large arch shapes were incorporated on the walls and were designed to mirror the shape of audience blocks. Although the overall atmosphere is calm, a pipe organ by Rieger Orgelbau (68 stops, 4958 pipes) installed in front of the stage and a deep red colored audience seats give the hall a gorgeous impression.

View from stage. The interior of Lotte Concert Hall is unified in the color scheme of white and natural wood. The main audience area are divided by the terrace walls which provide early sound reflections effectively

View from main level seating. A motorized stage riser system was installed on the stage and a large-scale and heavy acoustic ensemble reflector panel was fixed to the ceiling above the stage. A pipe organ was installed at facade

Room Acoustic Design of Lotte Concert Hall

The Lotte Concert Hall is an arena-type hall where the seat blocks surround a big stage that can accommodate a full-size orchestra. A heavy ensemble reflector was fixed to the ceiling above the stage, and an electric portable stage riser system was installed in the stage.

Since the shape of the hall is almost elliptical in plan and acoustical focusing could occur, very careful attentions were paid to the three-dimensional study of the hall shape. In addition to the studies using the CAD model, a 1:10 acoustic scale model of the hall was made and the acoustic tests were carried out in the period from October 2012 to February 2013.

Furthermore, in order to prevent concentration of sound, various types of shape were incorporated on the interior surfaces throughout the hall. Relatively large scale cylinder shapes were arranged on the ceiling, and the shallow grooves were made to the surface. On the other hand, relatively small scale cylinder shapes were arranged on the white wall on the upper part of the wall, and the wood parts of the arch shaped walls and the terrace walls of the audience seats have random spaced vertical ribs.

The room air volume of the hall became 32,600 m^3, and the room volume per seat reached 16 m^3. Looking at only these numbers, the hall is spacious, however, the distance from the front end of the stage to the top balcony seat is only 33 m and feels very close to stage with the visual help of bowl shape.

Surface texture. The terrace walls of the audience seats and the wood parts of the arch shaped walls and have random spaced vertical ribs. Relatively small scale cylinder shapes were arranged on the white wall on the upper part of the wall

Tuning Period of Hall Acoustics

The period from the completion of the concert hall construction at the end of 2015 to the grand opening in August 2016 was spent for the operational test which carries out various events called pre-concerts. After experiencing some concerts and rehearsals, and fine tuning the acoustics, final acoustical measurement were carried out.

In the unoccupied condition, the reverberation time was 2.9 s, and the estimated value in the occupied condition was 2.7 s at mid frequency of 500 Hz. The reverberation time of nearly three seconds is certainly long, however, the reverberant impression was not excessive, and sufficient clarity was achieved to allow conversation without stress between the stage and the audience area. The clarity was realized and confirmed during several concerts, both classical and cross-over with amplification.

Variable acoustics. For pop concert events using sound system, motorized sound absorbing banners can be deployed underside of the side balcony seat. And, by carpets and sound absorbing baffles specially manufactured for portable operations on the stage, the reverberant sound can be suppressed

For the concert events using sound system equipments, vertically deployed sound absorbing curtains stored in the lower ceiling of the side balcony seat and carpets and sound absorbing baffles specially manufactured for portable operations on the stage, the reverberant sound can be suppressed. By appropriately arranging these sound absorbing materials and loudspeakers, it is possible to cope well with pop and jazz concerts in the hall.

At the grand opening on August 19, 2016, Seoul Philharmonic Orchestra conducted by Chung Myung-whun presented Beethoven's *Leonore* Overture No. 3, the world premiere of Chin Unsuk's *Le Chant des Enfants des Étoiles*, Saint-Saëns's Symphony No. 3 "*Organ*", and three encore tunes. The enthusiasm of the audience who were longing for the completion of the new concert hall was very impressive.

Motoo Komoda

	Lotte World Tower
Location	Seoul, South Korea
Owner	Lotte Corporation
Architect	Kohn Pedersen Fox
	Lotte Concert Hall
Architect	Designcamp Moon Park
Acoustical consultant	Nagata Acoustics
	Yasuhisa Toyota, Motoo Komoda, Kayo Kimotsuki Kallas, Dave So, Erik Bergal
Theater consultant	Fisher Dachs Associates
Design start	September 2011
Construction end	December 2015
Opening date	August 19, 2016
Building size	13,200 m^2
Seating capacity	2,036
Room volume	32,600 m^3
Surface area	10,440 m^2
Volume/seat	16.0 m^3/seat
Volume/surface area	3.1 m
Finish material	
Ceiling + canopy	Plaster
Wall	Wood ribs with veener
Stage floor	Alaskan Yellow Cedar
Seat manufacturer	Kotobuki Seating
Organ builder	Rieger Orgelbau
Scale model	1:10

Table 25.1 Lotte Concert Hall—acoustical metrics at 500 Hz

RT unoccupied	2.9 s
RT occupied	2.7 s
EDT	3.2 s

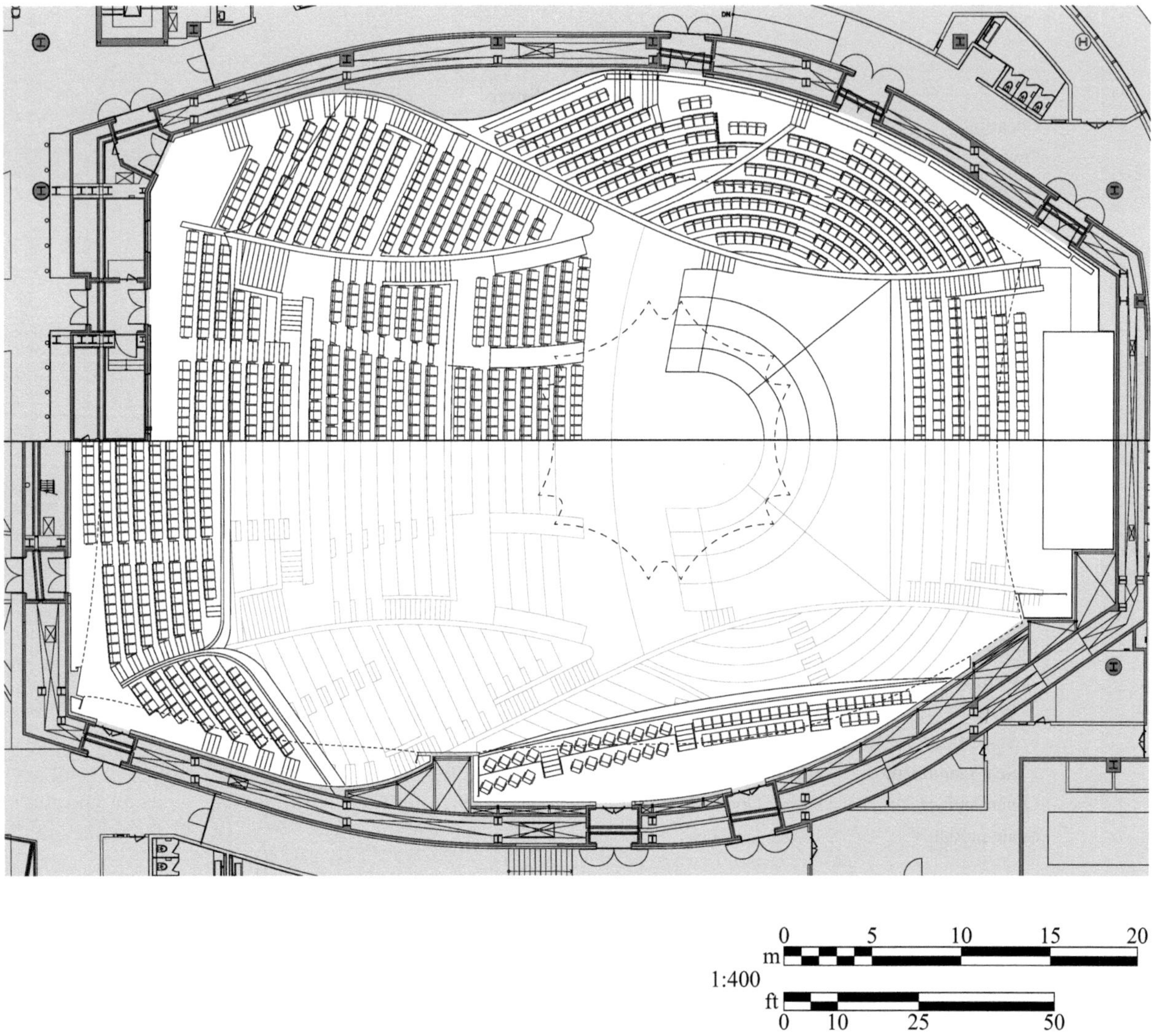

Fig. 25.1 Main level and balcony plan
Courtesy of Designcamp Moon Park

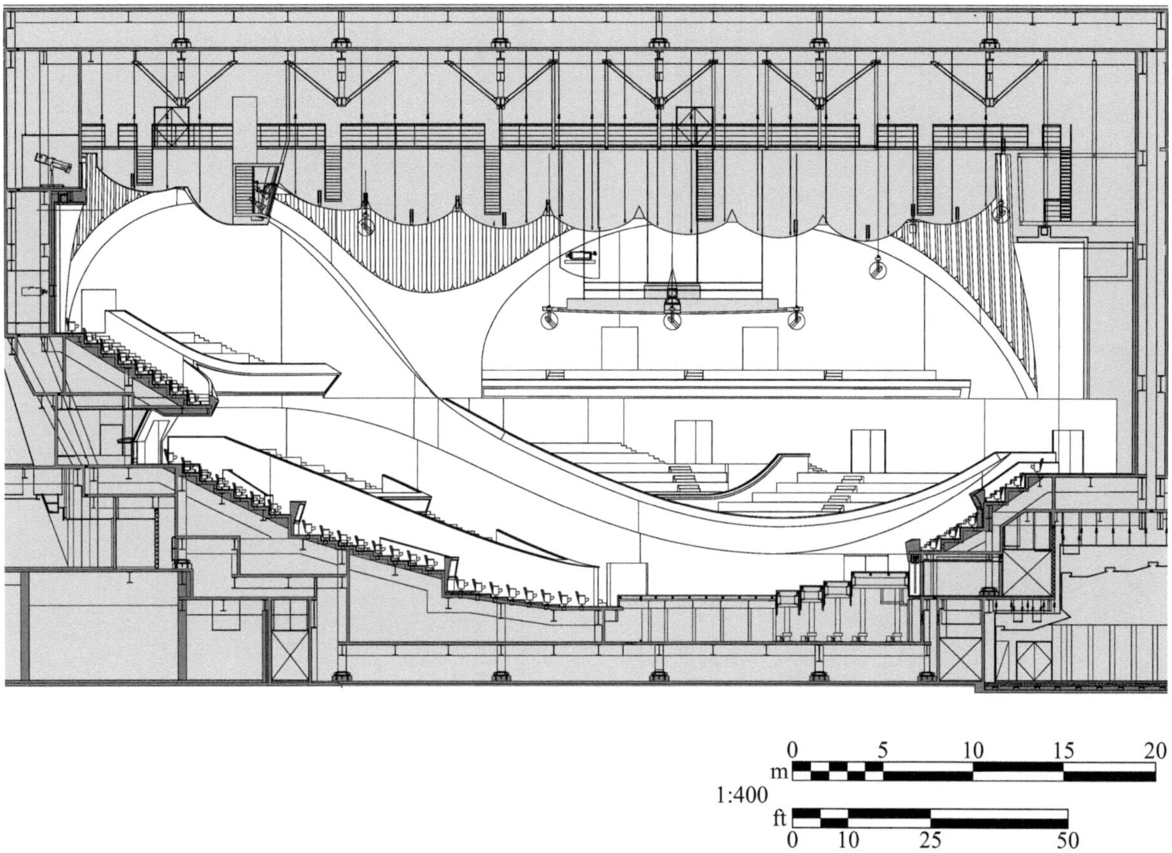

Fig. 25.2 Section
Courtesy of Designcamp Moon Park

26. Elbphilharmonie Hamburg

View from house left. Within the limitations of a constrained footprint, the hall developed vertically from a central stage at the bottom, with terraced seating tiers and balconies. Steep rakes ensure unobstructed direct sound and sightlines

Y. Toyota et al., *Concert Halls by Nagata Acoustics*, https://doi.org/10.1007/978-3-030-42450-3_26

With roots dating back to the early 2000s as a mainly private initiative, Elbphilharmonie soon transformed into a major cultural and architectural ambition for the City of Hamburg. The building was finally inaugurated in January 11, 2017 with a concert by freshly renamed resident ensemble, NDR Elbphilharmonie Orchester, to great public and critical acclaim.

The Winding Road to Completion

In early 2001, developer and architect Alexander Gerard and his wife, art historian Jana Marko, formulated the idea of developing a new cultural building and a world-class concert venue in a then-abandoned warehouse in the harbor. The project was to be privately financed by sponsorships as well as residential apartments and a hotel.

Rather unusually for a project of this importance and magnitude in Europe, the choice of the design architect did not follow a competitive process. Instead, the mutual trust and understanding between developers and the designers, nurtured by their prior relationship, provided the grounds and fuel of the collaboration. Swiss architects Jacques Herzog and Pierre de Meuron were commissioned in 2003 to develop concepts for the building and concert hall. Nagata Acoustics was appointed as acoustician for the project in 2004, along with the design team, and design formally began in early 2005. By the time the foundation stone was laid in 2007 the project and its development had been claimed by the City of Hamburg.

After 9 years of construction, unfortunately tainted by delays, suspensions, and arguments, works finalized in mid-2016 to put an end to controversies.

The Building

Elbphilharmonie building stands more than 100 m over the Elbe River, at the western tip of the HafenCity district, south of the city center. At its base, a warehouse built in the 1960s, Kaispeicher A, was entirely gutted to painstakingly retain only its landmark red brick facade. Inside, new piling work was added and a parking garage erected. This eight-story footing is crowned with a sculptural new glazed construction which appears as if lifted from the mostly blind and solid base and adds 17 floors to the edifice. While this glass ship strictly follows the trapezoidal footprint set by the original warehouse, its profile is reminiscent of large and sharp waves, and each of the four faces is scattered with smaller ripples expressed with bent glass or dotted inlays. This transparent vessel houses the main functions of the facility, comprising a five-star hotel, private apartments, the 2,100-seat "Grosser Saal," and 550-seat "Kleiner Saal."

Visitors enter the building from the widest side of the building on the east and climb the 80 m long "Tube" escalator to the eighth floor "Plaza," a large and open public space nested between the old and new parts of the construction 37 m above ground. From there, a large staircase takes concert goers to the main foyer which spirals tightly around Grosser Saal. Unusually tall doors lead to the comparatively narrow sound locks that channel audience into the hall.

The Design of the Hall

The concept of the Grosser Saal borrows from Herzog & de Meuron experience in designing large stadiums, and draws inspiration from acclaimed venues as different as Berlin Philharmonie and Teatro Alla Scala in Milan. From early on in the design process, the goal was to kindle a deep sense of communion and sharing, and to bathe listeners in music, while creating a truly unique identity for the hall.

The stage sits at the bottom and center of a large bowl approximately 50 m in diameter which peaks 25 m above the stage level. The first row of audience is a 60 cm drop from the front of the stage. The wide frontal seating tier is terraced to allow for unobstructed sightlines and create essential sound reflecting walls. It connects with a ring of seating around the stage. To the back, guardrails almost fade out and seating appears as if directly extending the stage when orchestra risers are in their highest position.

By contrast to the almost-symmetrical layout at the bottom of the bowl, the upper levels smoothly rise in a triple helix up to the tent-like ceiling, populating the walls all around the stage with seats. But rather than continuously ascending, each helix regularly slows its climb to create compact blocks of seats, never more than five or six rows deep. From any viewing angle, it appears as if seating tiers are organized vertically in four distinct layers, which are all connected and can all be reached from anywhere in the hall. The highest rows, close to 17 m above stage level, dramatically feel as if literally overhanging the stage.

With this vertiginous layout, compact seating blocks and steep viewing angles all around the hall, the result is an exceptional proximity of all listeners to the performers, with no more than 30 m from any seat to the stage, and a wonderful sense of intimacy and sharing throughout the audience.

Located slightly off-center stage left, the 69-stop organ sits directly behind multiple levels of balconies. Its presence is made visible by some of its largest metal pipes acting as a discreet facade, while most of the instrument remains hidden behind intricate acoustically transparent partitions continuing the geometrical patterns of the walls. In spite of its modest appearance, its unique location and layout make it an integral part of the design of the hall, in close proximity to the audience which can even physically touch it.

Ensemble reflector. Suspended directly under the apex of the ceiling and above the stage, the circular reflector is crowned by acoustically transparent fabric concealing structure, theatrical equipment, and an antiphonal division for the main pipe organ

Above the stage, a large circular reflector, 15 m in diameter is suspended to serve sound reflections to the stage and surrounding audience. It is crowned by a horn-like structure concealed with acoustically transparent fabric, making it appear as if dropping from the apex of the ceiling above. In addition to theatrical equipment, this structure also hides an antiphonal division of the organ with four stops, the sound of which becomes almost impossible to locate, adding a sense of mystery and mystique to the already unique instrument.

Audience is seated in individual armchairs carefully designed for comfort, durability, and acoustics. They comprise a molded high-density polyurethane internal frame, steel tubes, and elastic belts upholstered with molded cushions and durable wool and nylon fabric. The back of the backrest is finished with a curved plywood board with oak veneer. A full week of acoustical lab testing was necessary to adjust final details of assembly and upholstery, to ensure that the overall absorptive power of the chair would be minimum and consistent between occupied and empty situations. Sofa-like chairs and matching pillows were added to precisely fit in the rounded corners of seating rows.

"What Type of Wood Is This?"

As one would naturally expect in a concert hall of this nature, wood is certainly not absent from the interior surfaces, from the light oak of the audience floor wrapping over guardrails and outlining the flow of balconies around the hall, to the bright pine of the stage floor. The surfaces of the hall are nonetheless dominated by the mineral tones of the walls and ceiling, which are comprised of over 10,000 panels, approximately 80 kg each, made up of ground gypsum mixed with recycled paper. Panels are directly affixed to steel frames and joints are individually filled with soft airtight silicone running an aggregated total of approximately 15 km.

Surface texture and acoustically transparent organ grille. Walls and ceiling are realized with fiber-reinforced gypsum panels, each carved with a unique three-dimensional pattern, with depths ranging from a few millimeters to 9 cm. The motif is continued to create acoustically transparent partitions in front of the organ and the upstage wall

Each panel exhibits a unique three-dimensional texture pattern reminiscent of the profile of the building's roof. The pattern and its variations were designed by the project architect based on fundamental acoustical requirements on dimensions and locations. It then developed through a lengthy process of discussions and reviews before the final design was confirmed. The design was also directly informed by the testing of a physical 1:10 scale model of the hall and its surfaces. Specifically, model testing helped determine the necessary depths and their precise locations to eliminate detrimental long path echoes. Final depths range from a few tens of millimeters, providing acoustical diffusion to soften sound reflections, to up to 9 cm. Additionally, the overall visual design was unified by applying shallower irregularities of less than 10 mm in depth on surfaces where acoustical design studies did not require a specific treatment. The sea-shell motif is also continued on the acoustically transparent partitions concealing the organ, and surrounding the stage level.

Variable acoustics banners deployed. Retractable acoustical banners are distributed around the room. They consist of two layers of acoustical drapery which are lifted up from the floor

Distributed around the hall, acoustical banners can be deployed in front of the walls to adjust the reverberance of the hall, primarily for programs relying on recorded sound or amplification. Rather than dropping from the balcony soffits, they are stowed under the flooring, and the two rolls of drapery are lifted up by scissors mechanisms.

The Little Sister

Located to the east, only two levels below the main stage of Grosser Saal and their structures a mere few meters apart, Kleiner Saal was designed for chamber music and recitals in natural acoustics, as well as more flexible uses and programs including sound amplifications.

While the lower level features a simple rectangular footprint, 15 m in width and 30 m in length, the envelope at the gallery level tapers in on the western wall to make way for the structure and shape of Grosser Saal. The visual symmetry of the design was maintained however by shaping this technical balcony which narrows towards the back of the hall.

Contrasting with the mineral aspect of the Grosser Saal and its sharp surface irregularities, the visual finish in the hall is dominated by wood and rounded surface textures. At the far end of the hall, 17 telescopic seating rows can transform the hall from a raked seating to a fully flat floor. The stage is equipped with 35 mechanical risers, which can be lifted to raise the performers or create steps, but also lowered to extend the raked seating and improve sightlines. A total of 50 independent acoustical banners can be rolled in the balcony or deployed in front of the walls to cover all lower walls with absorption and adjust the reverberance to the needs of the varied possible configurations of the space.

Kleiner Saal with risers deployed. Designed as a multifunctional space, the hall can accommodate up to 550 seats. Motorized lifts can vary the stage layout or extend the rake of seating

Muting the Horns

It is not unusual for concert hall buildings to be surrounded by sources of noise and vibration such as heavy road traffic or rail lines, above or underground. Solutions to protect against these disturbances range from heavy or multiple partitions to structural separations by means of resilient bearings and expansion joints.

Standing over the river far above ground, on the edge of the city, the halls at Elbphilharmonie are not faced with such acoustic hazards typical of urban environments. Nevertheless, the unique situation of the building resulted in an even greater source of environmental noise which required careful structural design and construction. Located on a peninsula of the Elbe River, the facility is regularly brushed past by large ocean freighters and ocean liners. As they make their way out to the ocean from the nearby harbors, they often blow powerful low-pitched air horns designed to be heard kilometers away to signal their maneuvers or position.

In order to isolate the halls from these low-pitched blasts and maintain adequate levels of quietness, sound level reductions of up to 90 dB at 500 Hz and 75 dB at 125 Hz were necessary. They were achieved by nesting each of the two venues in two structurally independent concrete enclosures. The inner-most envelopes are punctually supported by metal spring boxes—over 350 for the Grosser Saal alone—designed to damp horizontal and vertical vibrations with the lowest possible natural frequency.

Inaugural Festival

The inaugural concert on January 11, 2017 by resident ensemble NDR Elbphilharmonie Orchester and its music director Thomas Hengelbrock started three full weeks of festivities. Concerts featured not only the resident ensemble but also local orchestras Hamburg Philharmonic State Orchestra and Symphoniker Hamburg, guest orchestras Chicago Symphony Orchestra and Vienna Philharmonic Orchestra, renowned soloists Mitsuko Uchida and Yo-Yo Ma, and jazz pianist Brad Mehldau. In total, over 30 concerts in Grosser Saal and Kleiner Saal celebrated the long-awaited arrival of a prominent new concert venue on the European and world stage.

Marc Quiquerez

View from upstage seating. Rising in a triple helix, the balconies are divided in compact blocks of seating which can all be reached from within the hall

Occupied hall preconcert. The surround and vertical layout populates the walls around the hall with audience, creating a spectacular sense of sharing

Kleiner Saal with risers stowed and curtains deployed. When stowed, the telescopic risers clear a fully flat and rectangular floor. Fifty acoustical banners can be lowered individually from the balcony to adjust the reverberance of the space

Building approach from the North

Elbphilharmonie Hamburg	
Location	Hamburg, Germany
Owner	HamburgMusik gGmbH
Architect	Herzog & de Meuron
Executive architect	Höhler+Partner
Construction cost	EUR 789 Million
Design start	Q1 2005
Construction start	2007
Construction end	Q3 2016
Opening date	January 11, 2017
Building size	120,400 m^2
Grosser Saal	
Users	NDR Elbphilharmonie Orchestra (resident)
	Hamburg Philharmonic State Orchestra
	Symphoniker Hamburg
Acoustical consultant	Nagata Acoustics
	Yasuhisa Toyota, Keiji Oguchi, Motoo Komoda, Daniel Beckmann, Marc Quiquerez
Theater consultant	dUCKS scéno
Seating capacity	2,098
Room volume	23,000 m^3
Surface area	8,500 m^2
Volume/seat	11.0 m^3/seat
Volume/surface area	2.7 m
Finish material	
Ceiling and walls	Milled high-density fiber-reinforced gypsum panels
Audience floor	Wood on high-density fiber-reinforced gypsum board
Seat manufacturer	Poltrona Frau
Stage floor	Oregon Pine
Organ builder	Johannes Klais Orgelbau Bonn
Model scale	1:10
Kleiner Saal	
User	Ensemble Resonanz
Acoustical consultant	Nagata Acoustics
Theater consultant	dUCKS scéno
Seating capacity	up to 550
Room volume	3,800 m^3
Surface area	1,800 m^2
Volume/surface area	2.1 m
Finish material	
Ceiling	Painted concrete, Woodfiber boards
Walls	Milled wood
Stage floor	Oregon Pine

Table 26.1 Elbphilharmonie Hamburg, Grosser Saal—acoustical metrics at 500 Hz

	Reflective condition	Curtains deployed
RT unoccupied	2.5 s	2.1 s
RT occupied	2.3 s	1.9 s
EDT	2.3 s	
C_{80}	0.1 dB	
G	6.2 dB	

Table 26.2 Elbphilharmonie Hamburg, Kleiner Saal—acoustical metrics at 500 Hz

	Risers stowed no chairs		Risers deployed with chairs	
	Reflective condition	Curtains deployed	Reflective condition	Curtains deployed
RT unoccupied	2.3 s	1.2 s	1.6 s	0.9 s
RT occupied			1.4 s	0.8 s

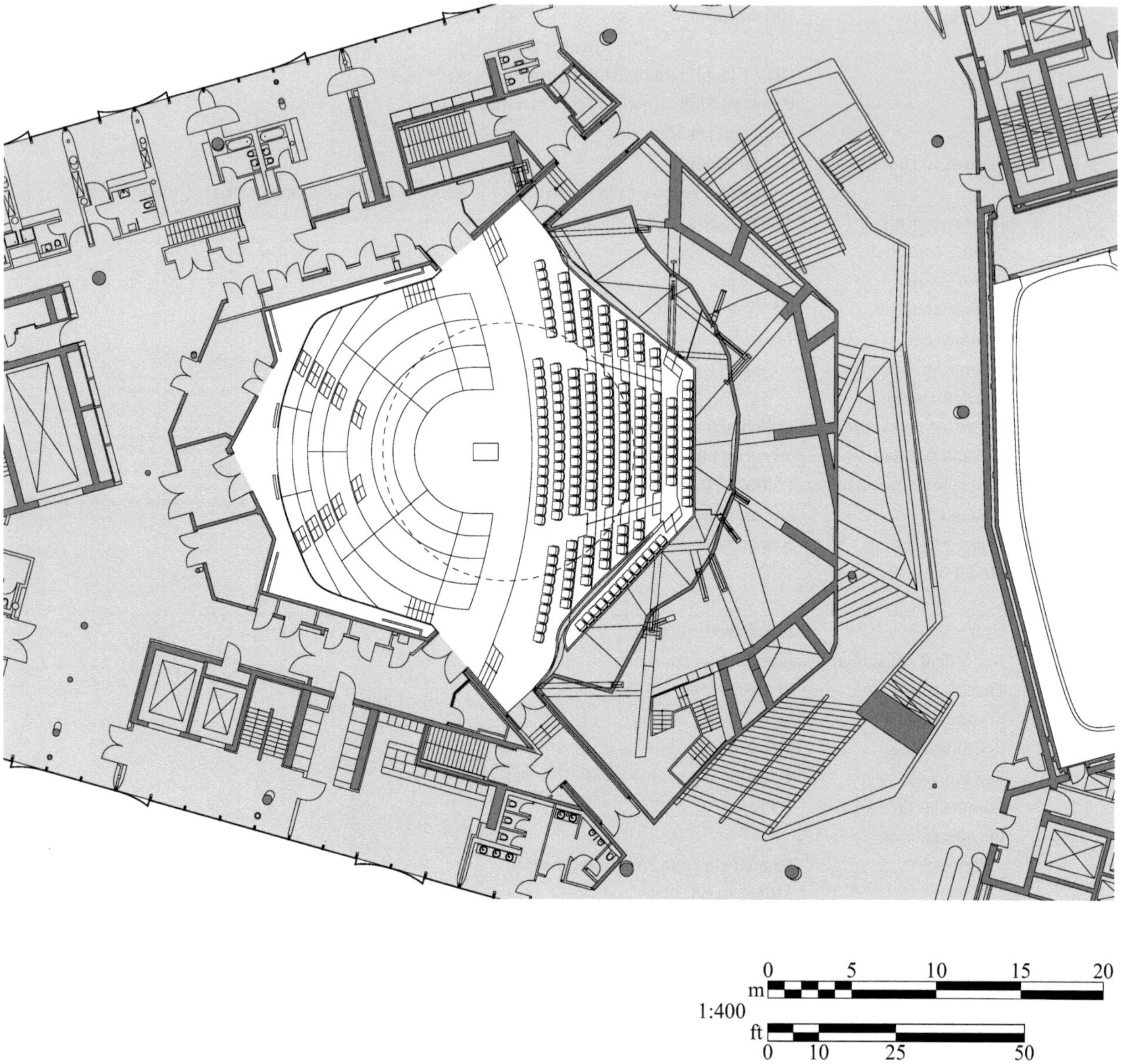

Fig. 26.1 Stage level plan
Courtesy of Herzog & de Meuron

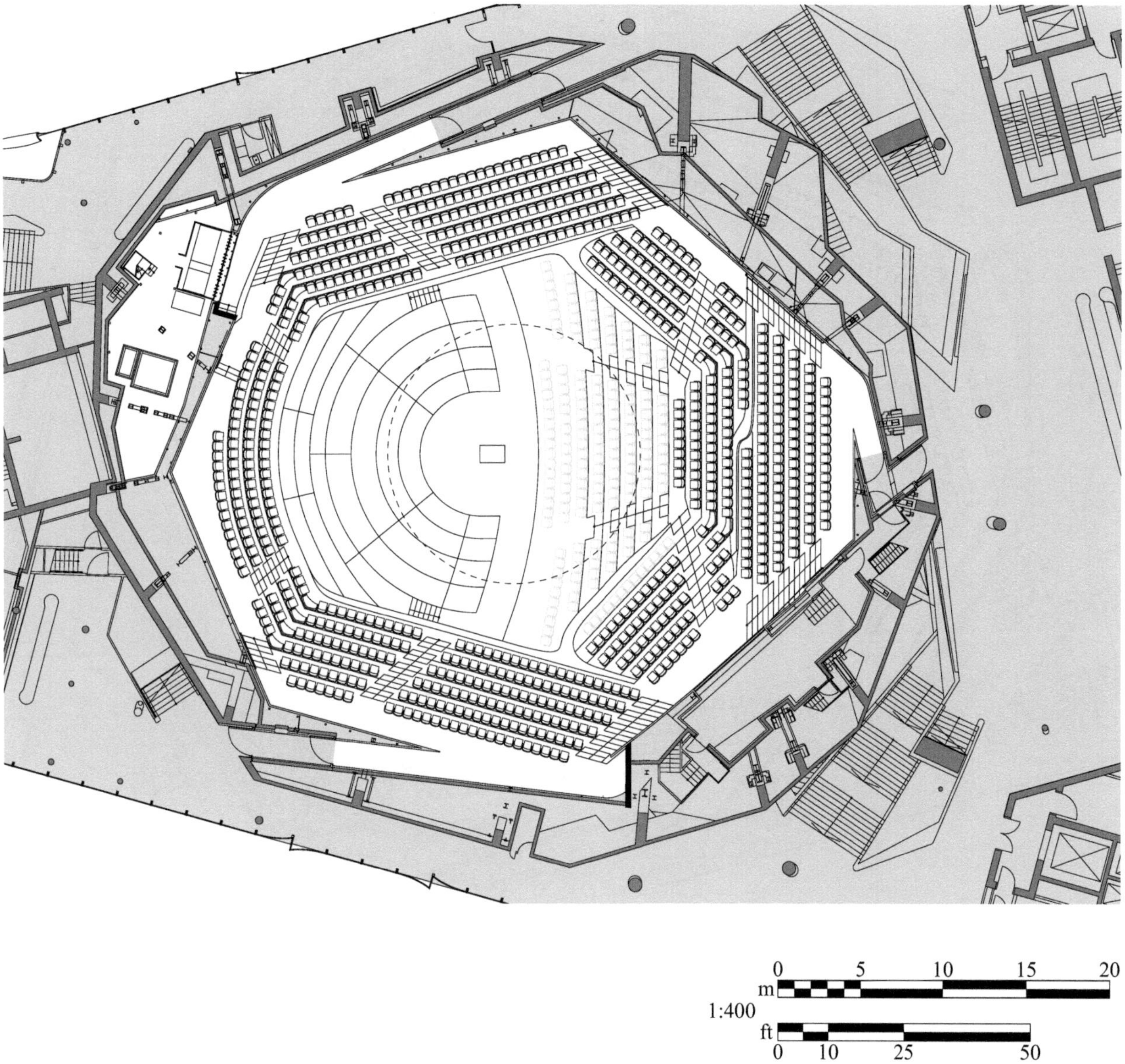

Fig. 26.2 Terrace level plan
Courtesy of Herzog & de Meuron

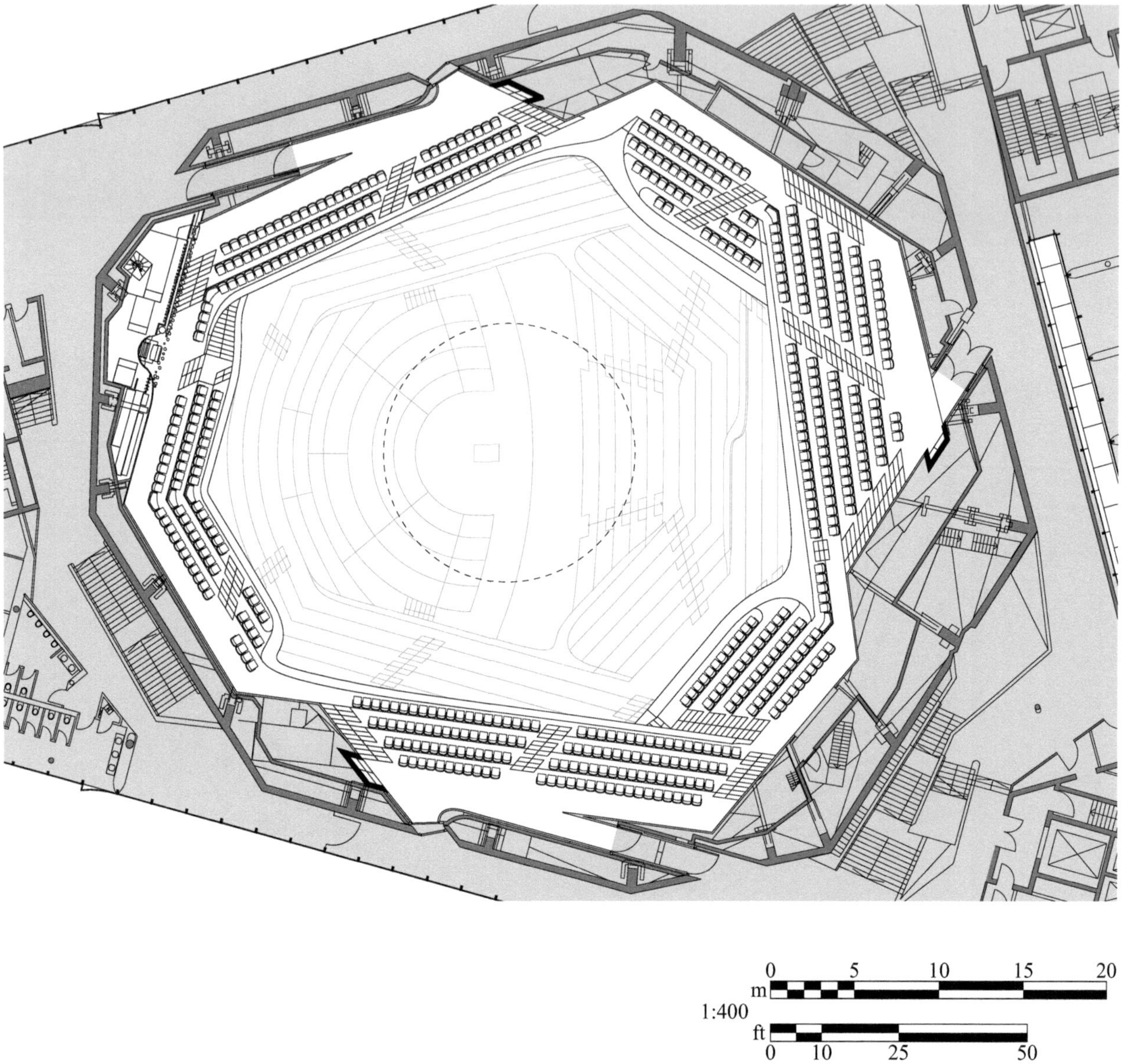

Fig. 26.3 First balcony plan
Courtesy of Herzog & de Meuron

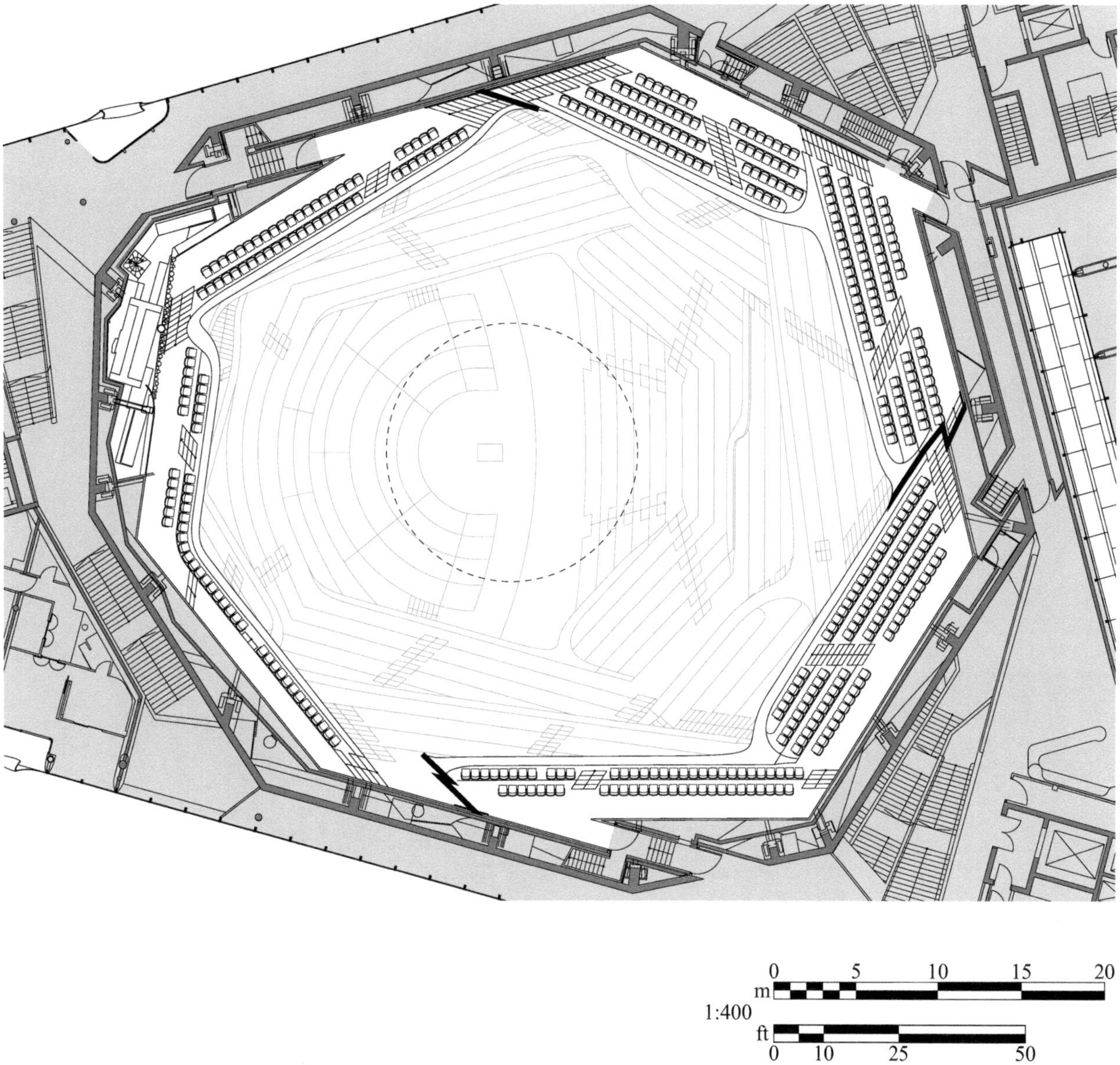

Fig. 26.4 Second balcony plan
Courtesy of Herzog & de Meuron

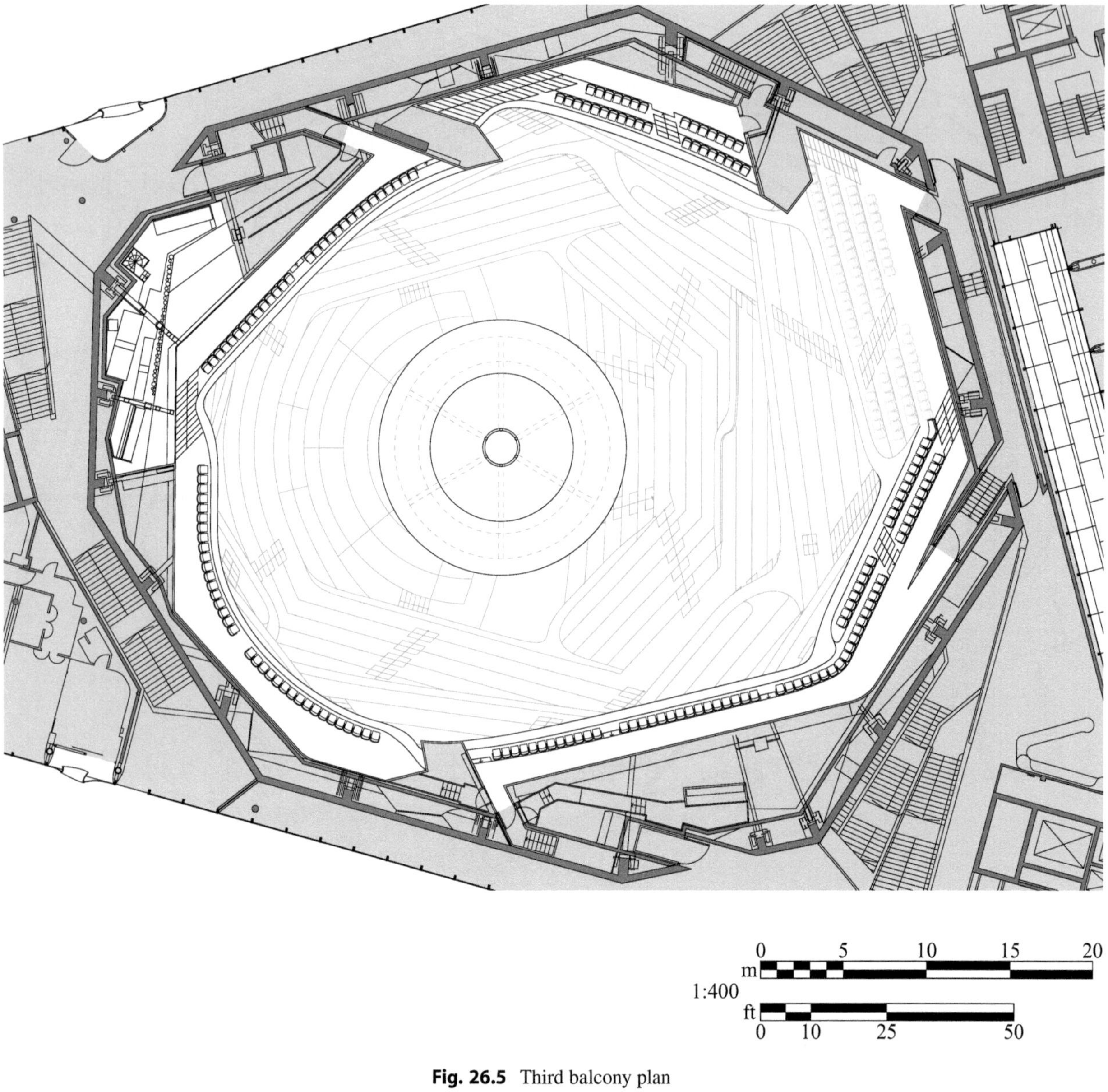

Fig. 26.5 Third balcony plan
Courtesy of Herzog & de Meuron

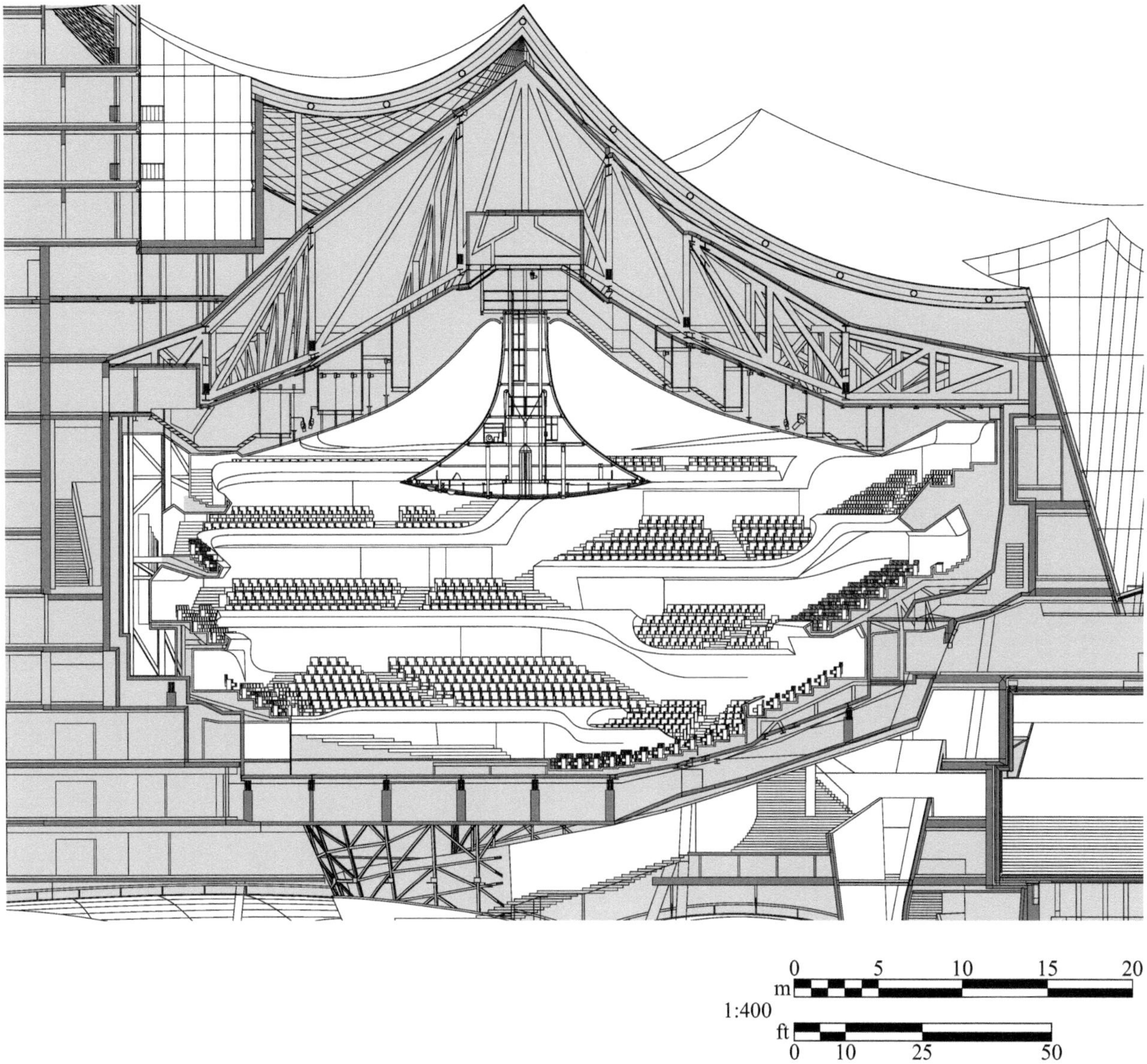

Fig. 26.6 Section
Courtesy of Herzog & de Meuron

27. Pierre Boulez Saal

View from entrance level. The stage is at the center, with three rows of chairs which can be arranged on the flat floor depending on the performance

Y. Toyota et al., *Concert Halls by Nagata Acoustics*, https://doi.org/10.1007/978-3-030-42450-3_27

The Pierre Boulez Saal is housed inside the Barenboim-Said Akademie, a new higher-education institution focusing on music and the humanities, located in the Mitte district in Berlin. Daniel Barenboim, the renowned conductor and pianist, founded the Akademie in memory of his close friend, the Palestinian humanist and philosopher Edward Said. In support of this mission, and as a seventieth birthday gift to Barenboim, the city of Berlin allowed him to turn an unused scenery warehouse adjacent to the Staatsoper Unter den Linden, where Barenboim has been the General Music Director since 1992, into the home for the Akademie. The Akademie will offer ninety students from the Middle East an "Education through Music," a combined music and humanities curriculum. As a central feature of the Akademie's facility, a small performance hall was planned. Through their mutual friend, the avant-garde composer and conductor Pierre Boulez, Barenboim met the architect Frank Gehry who pledged to design the performance space.

The building donated by the city was built in 1955, though it is composed in the Neobaroque style. The renovation formed the building into three principal sections: the four-story foyer occupying the central fifth of the building, flanked on the left by offices, classrooms and practice rooms of the Akademie, and on the right by the Pierre Boulez Saal. Since the hall is located within an existing building, the space available to the hall is a cubic volume measuring approximately 25 m square and 14 m tall. Windows on two sides of the simple volume flood the space with natural light. The primary shape of the hall is that of an oval within a square, the oval being the result of the proverbial napkin sketch made during the first meeting between Gehry and Barenboim. The oval form serves to create a strong sense of community and intimacy within the audience by minimizing the distance from all audience to the stage in the center of the room: no seat is more than 14 m from the central performance area.

The Pierre Boulez Saal was designed as a highly flexible space, suited to serve many different programs. There does not seem to be one primary usage amongst the many events which take place in the hall: rehearsals and recitals by the students of the Akademie, performances for the resident Pierre Boulez Ensemble, rehearsals and performances by both the West-East Divan Orchestra and the Staatskapelle Berlin, and smaller chamber music concerts. In order to accommodate this wide variety of ensemble sizes, Gehry took the radical step of placing the stage in the center of the hall, surrounded on all sides by up to 682 audience seats which are divided between the lower bowl and a floating balcony. Depending on the size of the ensemble, up to four rows of seats may be placed directly on the stage floor, for a string quartet or violin and piano solo performance. The next four rows of seats rise from the stage level on oval-shaped retractable risers. With all the seats removed, the entire stage is enlarged to more than 300 m^2 for a full orchestra rehearsal. The retractable seats surrounding the stage are divided into four segments, allowing the hall to be set in a more conventional configuration with the performers at one side of the room and the majority of the audience facing them from one direction. In addition to the flexible seats, there is one row of fixed bench seats seats at the entry level. All of these flexible elements contribute to the creation of one version of Boulez's "Salle modulable."

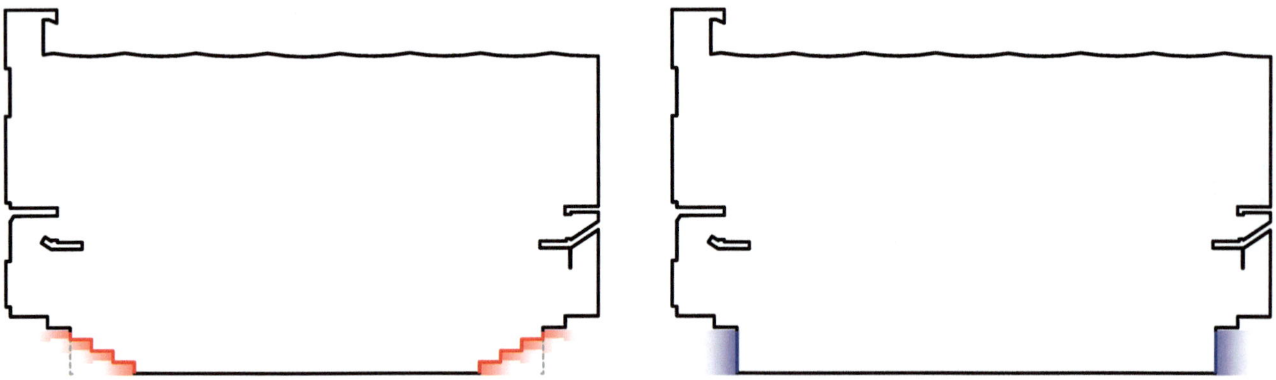

Fig. 27.1 Scope of spatial transformations

Above the main floor is a balcony with two rows and 221 seats, apparently floating in the middle of the room. The balcony is another oval-shaped ring, rotated slightly against the axes of the oval seating risers of the main floor. The first row of the balcony is about 6 m above the stage level, and since the distance to the stage is relatively short, the second row of the balcony is 1.4 m above the first row in order to have good sightlines to the stage. The dramatic structure of the balcony, supported only from a few discrete points near the centers of the walls, is further enhanced by the gentle up-and-down motion of the floor: as one traverses the path of the balcony, the elevation of the floor rises and falls by 80 cm.

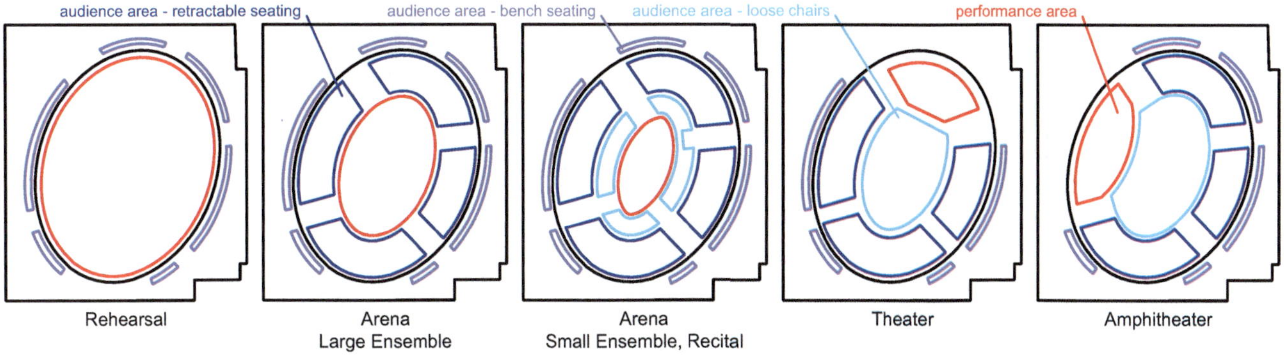

Fig. 27.2 Stage and audience configurations

Balcony floats within the volume. The balcony is not connected to the wall at the corners, in order to make use of the entire volume of the space. Sound can pass through and around the balconies to give a sensation of reverberance

The acoustical design of the hall is focused on providing a space for full-orchestra rehearsal, while also serving as a chamber music performance venue. The existing building allowed a 14 m ceiling height, which is sufficient to accommodate a full-sized orchestra for rehearsal. Therefore, the stage was placed at the lowest point in the hall. In order to provide useful early reflections back to the stage and to the audience seated on the main level, the rear edge of the underside of the floating balcony is projected downwards with glass by 1.2 m, the so-called fin, above and behind the last row of the lower audience.

Glass fins. Glass fins attached to the balcony floor reflect sound back to the performers, as well as the audience. The fins are convex in order to break the focusing caused by the overall convex shape, and to diffuse harsh reflections caused by the hard surface of glass

The fins, which cover roughly two-thirds the length of the balcony, are semi-cylindrical, exposing a strongly convex shape to the center of the room in order to control the sound focusing of the otherwise strongly concave balcony shape. The balcony structure was designed to be as acoustically transparent as possible. This is achieved by using a concrete ring beam structure, into which large holes measuring 54 cm wide and 107 cm tall have been cut. These holes are covered by an acoustically transparent fabric, visible on the rear side of the balcony from below, as well as behind the seats of the lower balcony row. All the handrails for the balcony are made of thin pipes, in order to have minimal acoustic impact in the space. The horizontal floor surfaces of the balcony remain the only areas which were not made acoustically transparent. The spaces between the balcony and the corners of the room are also left open, vertical space contributing to the drama of the balcony and the overall space. This has the effect of allowing the entire volume of the space to contribute to the acoustics. After the completion of construction, the reverberation time was measured at 1.9 s (unoccupied, at 500Hz), an unusually long reverberation time for a hall with only 682 seats.

The Pierre Boulez Saal opened to great fanfare in March 2017. The first concert was performed by the Boulez Ensemble, and featured a very wide variety in instrumentation, showing all the possibilities for the use of the hall. The concert began and ended with works by Pierre Boulez, with ensembles ranging from solo clarinet to a 15-piece chamber orchestra. It is a unique experience to join an audience which completely surrounds and embraces the performers at the center. Visual and acoustic intimacy of the highest caliber is achieved in this room, since the viewer-listener can focus her attention not only on the performers on stage, but also on a large proportion her fellow audience members at the same time. By placing the stage at the center of the room, the dynamic architecture of the space manages to recede subtly into the background of the viewer-listener's consciousness, and at the beginning of the performance her focus shifts to the performers and fellow viewer-listeners. Together with the closeness to the performers enabled by the flexible audience and stage size, the shift of focus enhances the unique experience of attending a live performance, especially when the viewer-listener is seated on the stage floor within arm's reach of the performer. Once the performance has concluded, the audience moving through the space activates the architecture which gives another dimension to the visual pleasure of the space.

Daniel Beckmann

View from balcony. The floor of the balcony is not level; it rises and falls by 1m as the ring is traversed

	Barenboim-Said Akademie
Location	Berlin, Germany
Owner	Barenboim-Said Akademie
Architect	Richard Paulick (original building)
Construction cost	EUR 33.7 million
Design start	November, 2012
Construction start	May, 2014
Construction end	December, 2016
Building size	6,500 m^2
	Pierre Boulez Saal
User	Barenboim-Said Akademie
	Boulez Ensemble
Architect	Gehry Partners, LLP
Executive architect	rw + architekten
Acoustical consultant	Nagata Acoustics (room acoustics)
	Yasuhisa Toyota, Motoo Komoda, Daniel Beckmann
	Müller-BBM (building acoustics)
Theater consultant	Ingenieurbüro Schaller
Opening date	March 4, 2017
Seating capacity	682
Room volume	7,600 m^3
Surface area	3,500 m^2
Volume/seat	11.2 m^3/seat
Volume/surface area	2.16 m
Finish material	
Ceiling	Douglas fir veneer troughs with concrete fill
Walls	Douglas fir veneer on cement board
Audience floor	White Oak
Stage floor	Alaskan Yellow Cedar
Seat manufacturer	Seda

Table 27.1 Pierre Boulez Saal—acoustical metrics at 500 Hz

RT unoccupied	1.9 s
RT occupied	1.7 s
EDT	1.7 s
C_{80}	2.1 dB
D_{50}	48%

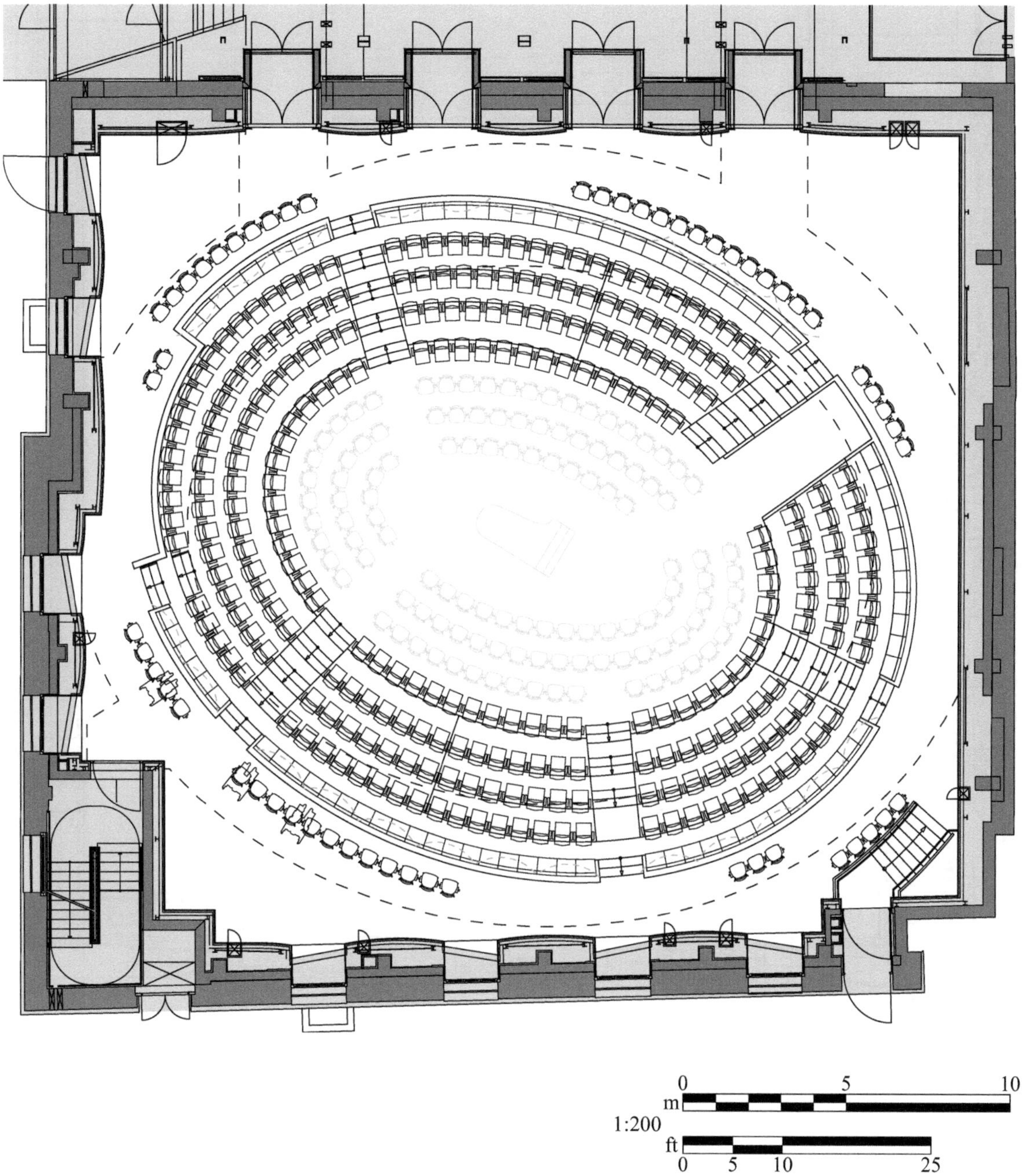

Fig. 27.3 Main level plan
Courtesy of Gehry Partners, LLP

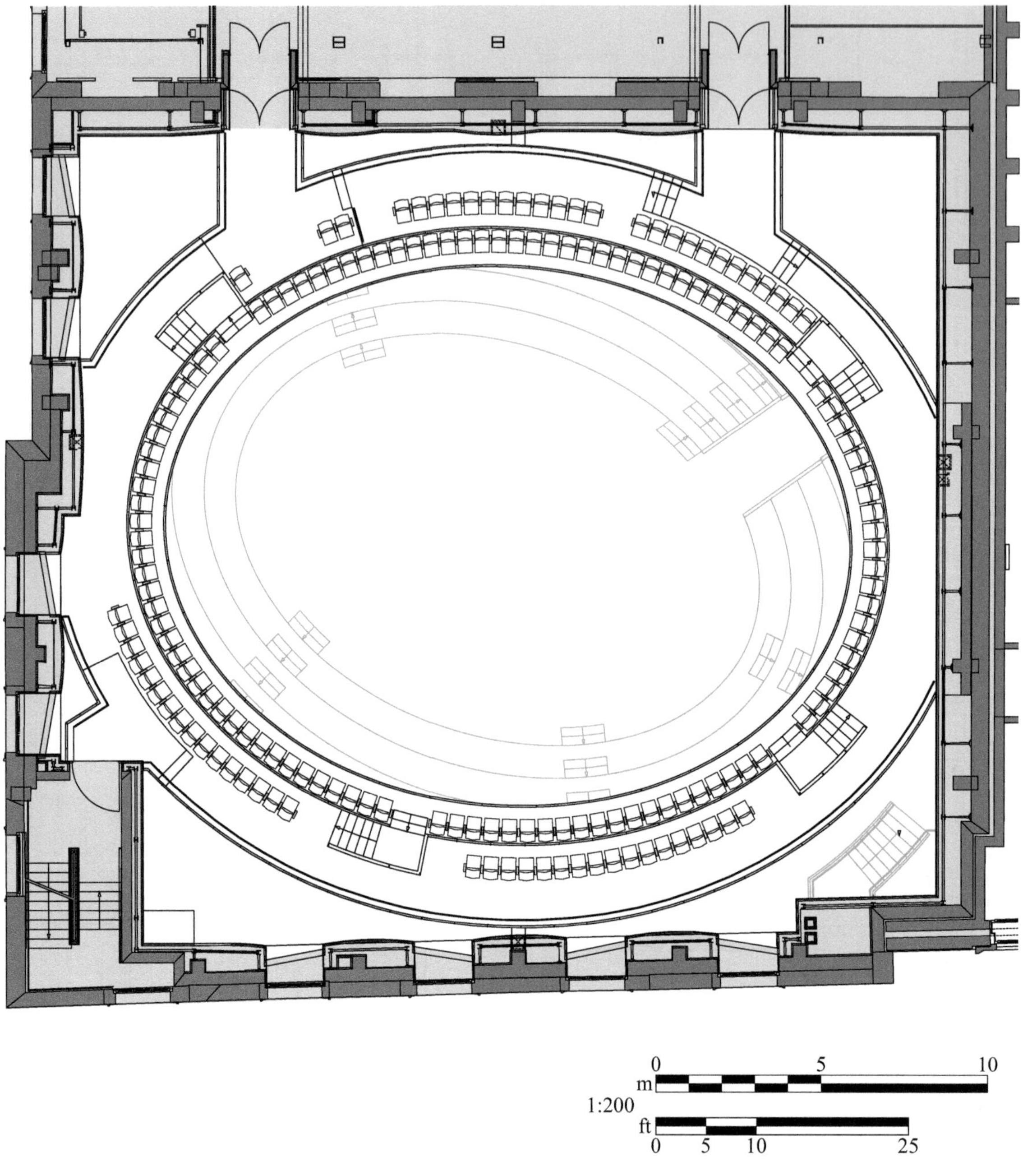

Fig. 27.4 Balcony plan
Courtesy of Gehry Partners, LLP

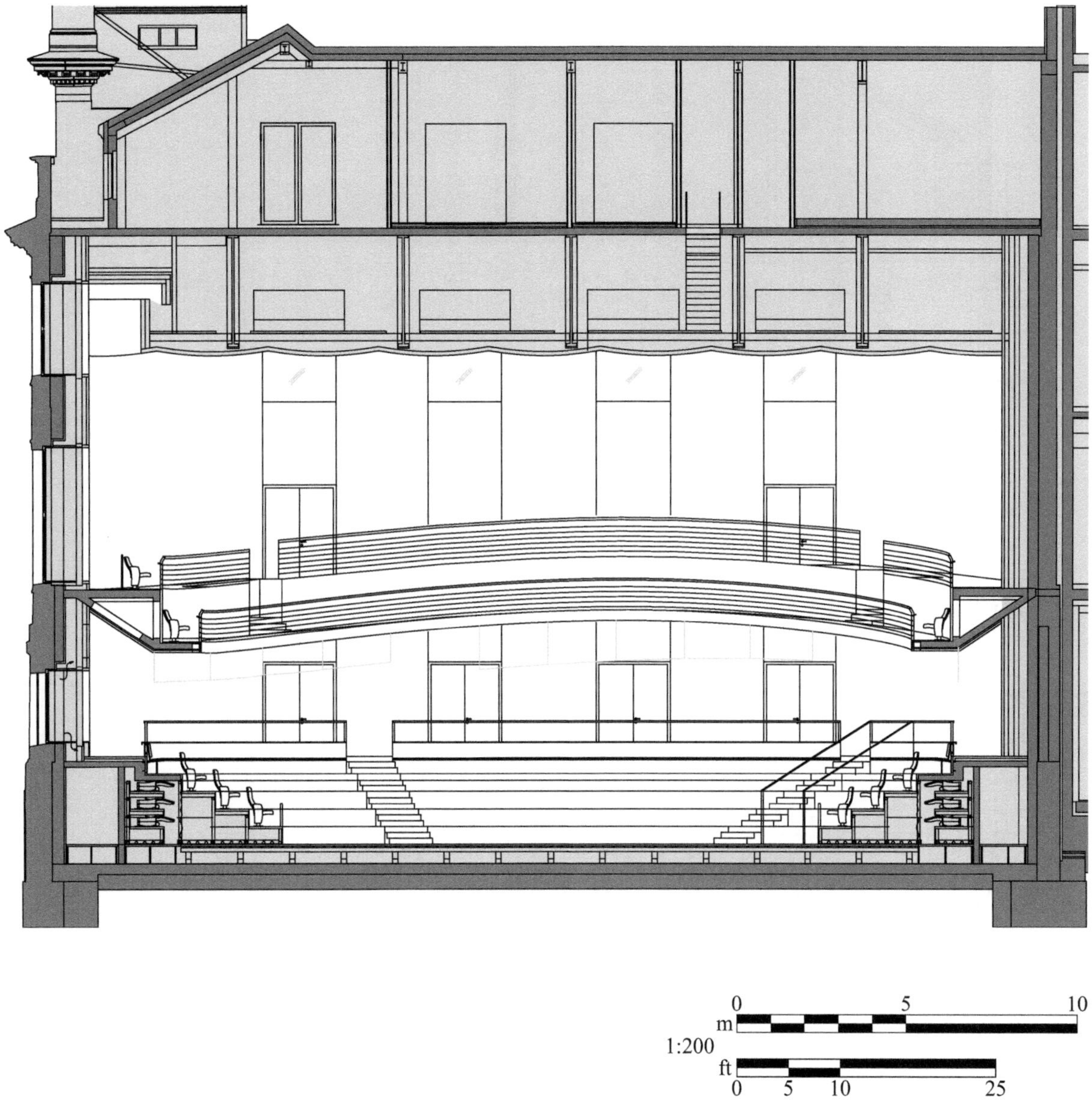

Fig. 27.5 Longitudinal section
Courtesy of Gehry Partners, LLP

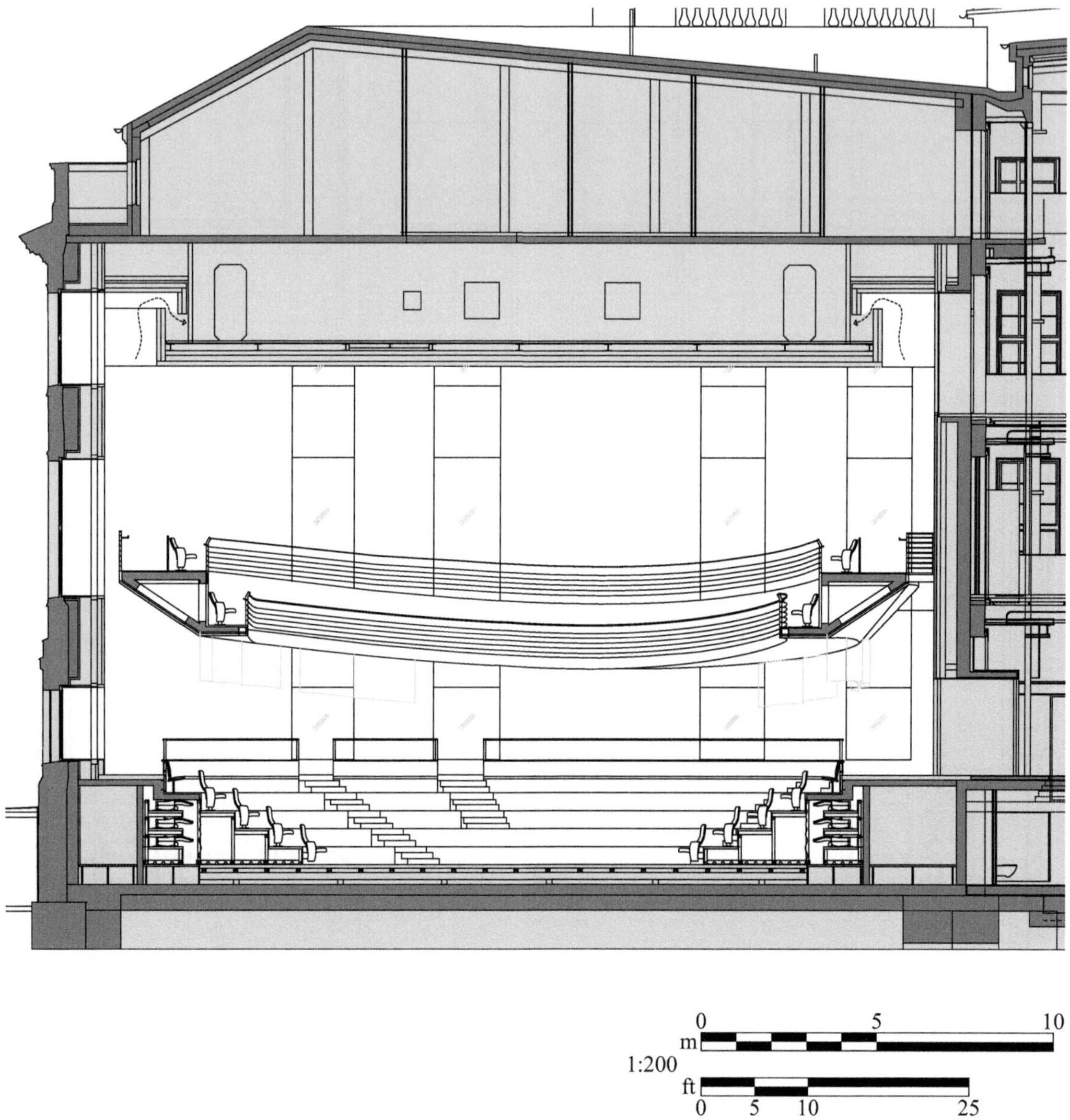

Fig. 27.6 Cross section
Courtesy of Gehry Partners, LLP

28. La Seine Musicale

View from the balcony. The large stage is equipped with mechanical risers for musicians and singers alike. The first four rows of seats are installed on a platform which can be lowered to reveal an orchestra pit. The acoustical ceiling is concealed by an acoustically transparent array of hexagonal frames fitted with sections of lacquered paper tubes

Y. Toyota et al., *Concert Halls by Nagata Acoustics*, https://doi.org/10.1007/978-3-030-42450-3_28

La Seine Musicale is located on Île Seguin island in the city of Boulogne-Billancourt, in the western outskirt of Paris, France. The building opened on April 22, 2017 after two and a half years of construction. Designed by Japanese, Pritzker Prize laureate Shigeru Ban and his Paris-based partner Jean de Gastines, the project was developed through a Private Financing Initiative initiated by the local government of Hauts-de-Seine region.

La Seine Musicale is the first new permanent construction to land on the island, and occupies the downstream third. True to the its initial name of a "city of music," the building houses two major performance venues and a host of music rooms of various sizes and purposes for three distinct users. Overall, it amounts to 36,500 m^2 of built surface area.

While the building silhouette is an obvious nod to the surrounding waters with its naval references, it also draws from the industrial history of the island, with its strict exposed concrete base. From the late 1920s and 1992, the island was fully occupied by French car manufacturer Renault whose factory spanned the entire site with assembly lines across multiple levels. The factory was demolished a decade later, but the unique piece of land stayed at the center of debates and arguments about the revitalization of the area and preservation of its heritage.

Sitting atop the base at the bow of the ship-like construction, the Auditorium is raised three levels above the ground floor and enclosed in an egg-shaped structure of glass and hexagonal wood framing. The transparent structure offers dramatic views to the surround river and hills from foyers and backstage, and acts as a beacon to the building. To the south-east, it is partially shaded by a large, spectacular sail of solar panels which can smoothly follow the movement of the sun.

Audience lobby around the concert hall, overlooking the river

The Auditorium

Primarily designed for natural acoustics and orchestral performances, the Auditorium can seat up to 1,152 people. The design goal established by the client and design team was to create a layout emphasizing intimacy and proximity between audience and stage, while favoring frontal seating for staged performances or end stage or thrust stage uses. Therefore, the conductor's podium sits close to the center of the hall and the most distant seat is located only 27 m from the edge of the stage, but over two-thirds of the seats are located in front.

The Auditorium is ovoid in footprint, but the perimeter walls are divided into convex segments to prevent focusing and evenly distribute sound reflections. The central front seating section is divided into three tiers and bounded by walls oriented to provide early sound reflections. The stage and central section are surrounded by a ring of terrace seating in continuity with the rear of the main section. A single level of audience balcony wraps around the hall, and an additional level of technical balcony flanks the stage on both sides.

The acoustical ceiling height averages at 14.5 m above the level of the stage in a continuous undulating long section. Above the stage, the wave is locally altered: smaller scale undulations were introduced to prevent focusing and flutter on stage. The shaping of this particular area, as well as the overall room shape, was studied by means of 3D computer simulation as well as measurements and aural evaluations in a physical 1:20 scale model built for this purpose.

The main ceiling consists of factory-made staff molds filled on site with projection plaster, which results in a massive surface of over 100 kg/m^2. While this heavy acoustical ceiling is perfectly smooth, the project architect designed a spectacular suspended structure underneath. This visual netting consists of over 900 hexagonal wooden frames fitted with sections of lacquered paper tubes of four different sizes to create four distinct modules. With each module used approximately 230 times across the ceiling, an astonishing 24,000 paper tube sections were used to create this striking honeycomb-like structure. Lit from below by projectors embedded into the balcony front, the structure casts interlaced shadows onto the immaculately white surface above. This structure was designed to be mainly acoustically transparent but also to provide scattering for high-frequency sound reflections.

Wall textures and audience chairs. The design of the chair was inspired by paper tube constructions, and exhibit cylindrical cushions and formed back panels. Extensive studies ensured their structural durability, acoustical performances and high-level of comfort

The paper tube construction dear to Shigeru Ban also served as the design foundation for the chairs, with their surprising cylindrical cushions upholstered in red velvet. Careful studies during design development ensured their structural durability, acoustical performances, and high-level of comfort in spite of their unusual shape and assembly.

Interior walls mainly consist of fiber-reinforced gypsum panels lined with a wooden assembly of beech plywood and medium-density fiberboards forming wavy horizontal strips laid out like parquet boards. Alternating with different offsets throughout the hall, they create changing patterns of irregularities on the surfaces. Similar waves are intertwined and perforated on the rear walls behind the frontal audience to create acoustically transparent grids. They conceal sound absorbing mineral wool used to prevent long path echoes.

Stage and Orchestra Pit

Measuring 21 m in width by 15.5 m in depth, the stage can accommodate large orchestra ensembles of more than one hundred musicians. It is equipped with 29 mechanical risers for orchestra and choir which offer a variety of stepped arrangements for the performers and can create a continuity with the seating block behind. In front of the stage, the first four rows of seating constitute a movable wagon and rest on a lift platform which can make way to a 130 m^2 sunken orchestra pit. The flooring of the stage and orchestra pit consists of an Alaskan Yellow Cedar decking over pine sleepers and beams, creating a soft resonating structure.

Adaptability and Acoustical Variability

Behind the stage, a retractable acoustically transparent screen hides a space where acoustical curtains can be deployed. The walls on each side, though finished like the walls around the audience, are divided into independent panels which can slide outward or flip around a vertical axis to expose a sound absorbing face. This allows for even more flexibility of the stage layout and acoustical fine tuning of the ensemble. On the sides of the hall and behind the stage, below the main audience balcony, acoustical curtains can be deployed or stored in dedicated closets behind the walls.

Stage lighting is embedded in the hexagon ceiling, and motorized modules can lower them down to stage level for maintenance. Rigging elements or microphones can be hung through penetrations in the main ceiling which align with the patterns of the suspended hexagons and paper tubes. Three retractable light bridges are hidden in the ceiling, and can be lowered when required by the program, bringing with them a segment of the acoustical ceiling and paper tube grid. They are also each equipped with an acoustical curtain, thus directly participating in the acoustical variability of the space.

Fig. 28.1 Scope of spatial transformations

Inaugural Concert

On April 22, 2017, the opening concert in the Auditorium was presented by the resident ensemble of La Seine Musicale, Insula Orchestra, led by music director and founder Laurence Equilbey. They were accompanied by soprano Sandrine Piau, mezzo soprano Anaïk Morel, tenor Stanislas de Barbeyrac, baritone Florian Sempey, pianist Bertrand Chamayou, and Accentus choir, and performed pieces by Mozart, Weber, and Beethoven.

Marc Quiquerez

View from upstage seating. The central seating tier is divided into three blocks and bounded by titled walls which also delineate compact seating terraces to the sides. While audience is seated all around the stage to emphasize intimacy and proximity, the seating layout is mostly frontal in order to accommodate staged or semi-staged programs

Convex walls and varied surface textures. Perimeter walls are divided into convex segments to prevent focusing from the mostly ovoid footprint of the hall. Walls are lined with wavy horizontal strips in alternating offsets to create changing patterns of irregularities. Undulating wood battens similarly finish the balcony soffits

Building approach across the bridge over the Seine river

La Seine Musicale	
Location	Boulogne-Billancourt, France
Owner	Conseil Départemental des Hauts-de-Seine
Architects	Shigeru Ban Architects Europe
	Jean de Gastines Architectes
Construction cost	EUR 170 million
Design start	January 2012
Construction start	July 2014
Construction end	January 2017
Opening date	April 22, 2017
Building size	36,500 m^2
Auditorium	
User	Insula Orchestra
Acoustical consultant	Nagata Acoustics (room acoustics)
	Yasuhisa Toyota, Motoo Komoda, Marc Quiquerez
	Lamoureux Acoustics (building acoustics)
Theater consultant	dUCKS scéno
Seating capacity	1,152
Room volume	13,800 m^3
Surface area	5,100 m^2
Volume/seat	12.0 m^3/seat
Volume/surface area	2.7 m
Finish material	
Main ceiling	Plaster
Suspended ceiling	Medium-density fiberboard frames, lacquered paper tubes
Walls	Fiber-reinforced gypsum boards, medium-density fiberboard and beech plywood
Audience floor	Oak decking on concrete
Stage floor	Alaskan Yellow Cedar
Seat manufacturer	Concept D
Model scale	1:20

Table 28.1 La Seine Musicale—acoustical metrics at 500 Hz

	Reflective condition	Curtains deployed
RT unoccupied	1.8 s	1.5 s
RT occupied	1.7 s	1.4 s

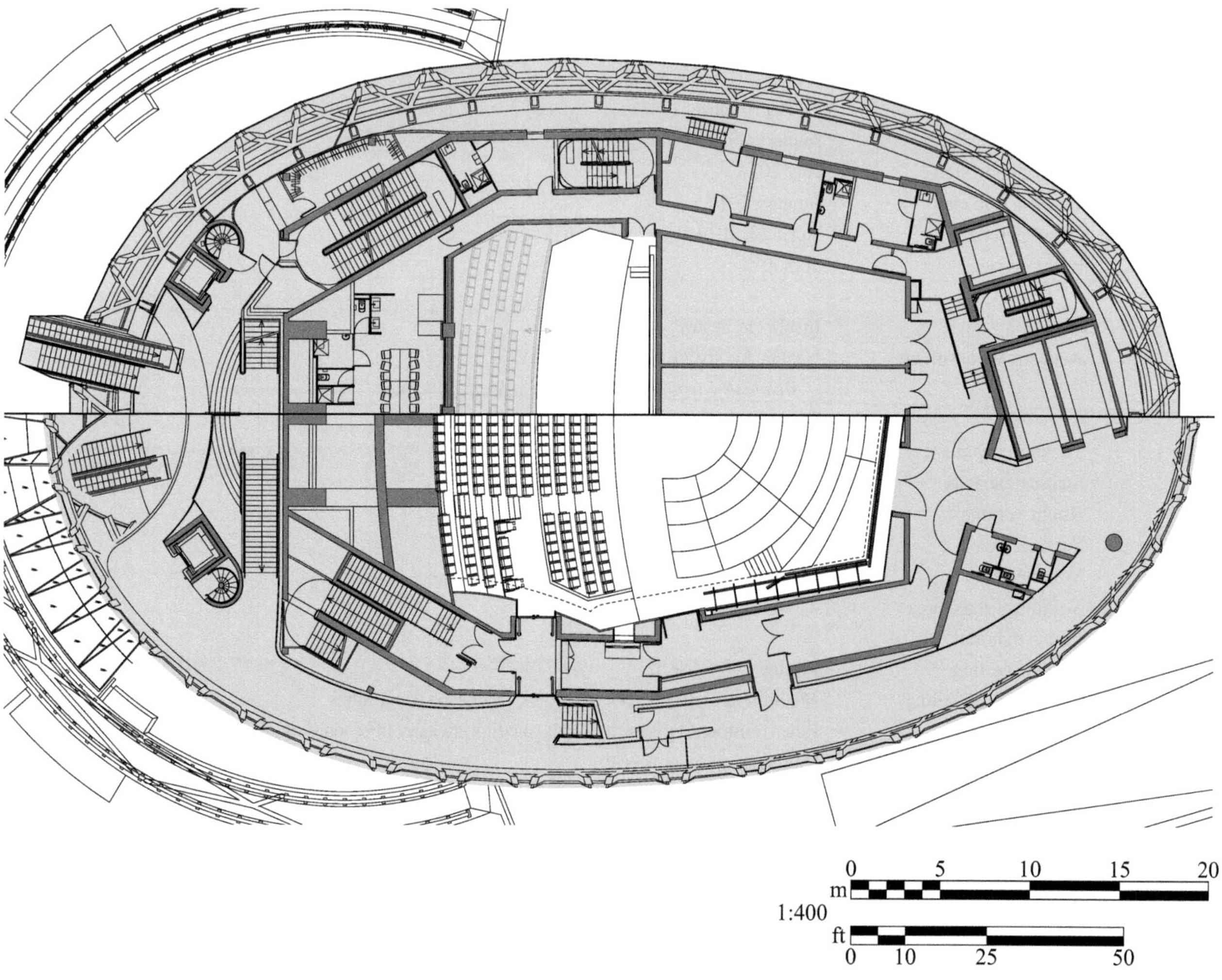

Fig. 28.2 Pit and stage level plans
Courtesy of Shigeru Ban Architects

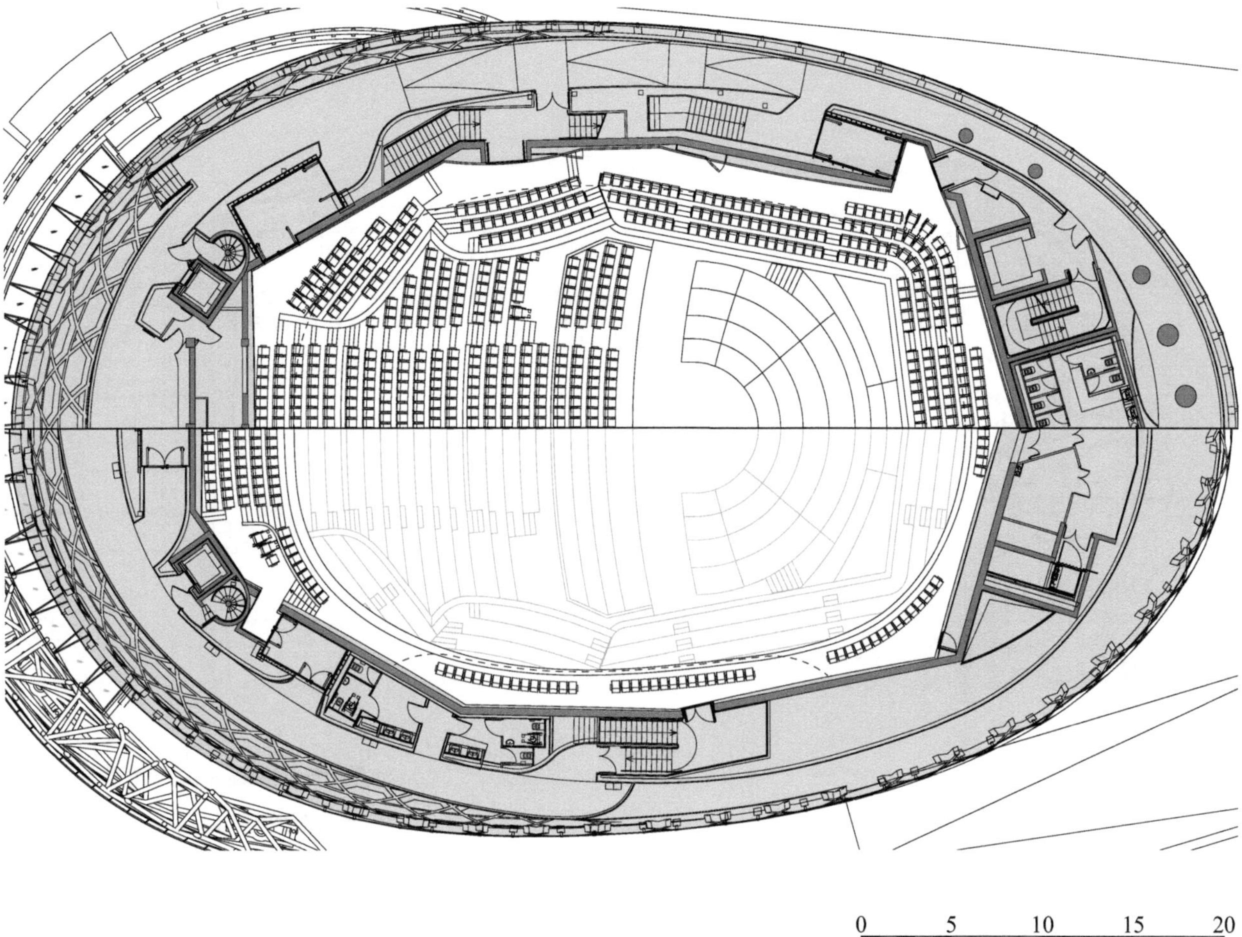

Fig. 28.3 Terrace level and balcony plans
Courtesy of Shigeru Ban Architects

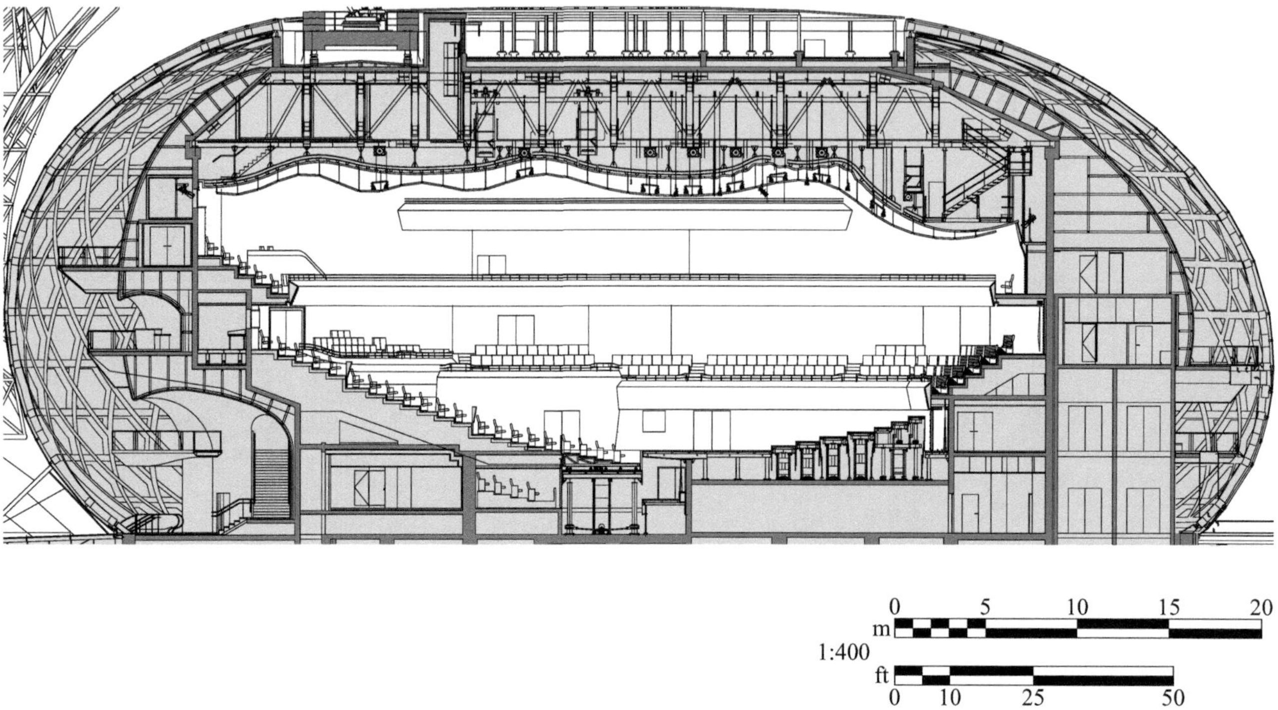

Fig. 28.4 Section
Courtesy of Shigeru Ban Architects Europe

29. Repino Hall

View from 2nd balcony. In the rather compact footprint, the audience layout developed vertically with two narrow levels of balcony to the sides and rear

Y. Toyota et al., *Concert Halls by Nagata Acoustics*, https://doi.org/10.1007/978-3-030-42450-3_29

Repino is a small locality of the city of Saint Petersburg (Russia), approximately 30 km north-west of the center. Located in the Karelian Isthmus along the coast of the Gulf of Finland, it is mainly known as a tourist and resort area, with many cottages scattered along the coastline and in the dense woods.

In 2010, Valery Gergiev, renowned conductor and Artistic Director of Mariinsky Theatre, shared with Nagata Acoustics his ambition to build a chamber music hall on a plot of land he had recently acquired in Repino. Initial plans of a private "dacha" quickly transformed into artistic and educational ambitions: creating a guest house to welcome talented young musicians in residence, and invite world-class performers visiting St Petersburg to meet and exchange. A hall which could accommodate rehearsals and workshops as well as performances was a natural extension to fulfil these ambitions. It was not, however, until early 2013 that the project development was firmly put in motion, picking up from prior unfinished construction works on site. French architect Xavier Fabre–who had previously worked with Gergiev and Nagata Acoustics on Mariinsky Concert Hall–was appointed as principal architect for the project, and work on the design of the hall finally began.

The hall is inscribed in a rectangular footprint 15 m in width and 20 m in length, comparable to the dimensions of a philharmonic stage or a large orchestra rehearsal room. As the design was taking shape, the seating capacity gradually increased, and continued to do so up until the final moments of the construction. But as the plot was naturally limited, the hall could not grow horizontally. Quite naturally, the room shape therefore developed vertically, raising the building roof and interior ceiling height and introducing balconies.

From the entrance doors at the rear of the main floor, cascading steps lead down to the main floor. The staggered platforms each offers a depth of approximately two rows. Rather than a clearly demarcated stage area, a large flat floor is reserved at the center of the room with an approximate surface area of 150 m^2. This allows for varying the positions of performers or audience and the size of ensemble, either in a traditional frontal arrangement or in varieties of surround layouts. A shallow podium at the back of the room extends up and down to the sides and raises audience or performers. It also houses a storage chamber for a grand piano. The steps and podiums on the main floor were also shaped to allow for movable podium to be later built and positioned indifferently on the front, back, or side to create various configurations.

View from upstage. The orchestra can be seated in a multitude of arrangements, utilizing the upstage and side podiums or directly onto the main floor

At house right, open stairs lead to a first balcony which ascends in a helix as it wraps around the room. A second balcony level, approximately 10 m above the main floor, provides additional seating to the sides and rear. This upper gallery is detached from the sides walls so as not to create an excess of sound reflections back to the performers while maintaining the benefits of the large unified volume.

Ceiling and skylight. The sloped ceiling panels climb to create a large volume. At the apex, a skylight introduces diffuse zenithal light

Above this upper gallery, the four faces of the ceiling surface taper up towards a skylight approximately 17 m above the main level, creating a large volume where the rich reverberance can build up. Viewed from the outside, this shape dramatically emphasizes the gentle roof slopes of the adjacent buildings, giving the concert hall extension its prominence and identity while blending with the overall language of the surrounding architecture.

The space is filled with natural light from the walls and ceiling, and offers views to the surrounding nature. Behind the stage, a tall bay window elegantly frames a preserved cedar tree just outside the building. Small windows were carved into the side walls, and protrude out of the facade like observation decks. Zenithal diffuse light from the ceiling creates an almost ethereal feeling, extending the visual impression of the ceiling beyond its physical boundaries.

The entire hall is built exclusively out of wood, using traditional building techniques. This presented a major challenge to the designers and builders to overcome limitations of local building regulations. This nevertheless gives the hall its unique character and welcoming familiarity. The surfaces of the walls and ceiling are scattered with randomly arranged joists, creating a vibrant visual texture as well as effective scattering of sound reflections. Along with the massive solid surfaces, this participates in the rich and warm acoustics of the space, ensuring that the surfaces reflect low frequency energy while softening high frequency reflections.

Rather than individual chairs, Fabre and Gergiev chose to use movable wooden pews for the seating on all three levels, in keeping with the overall finishes of the hall. Acoustical curtains can be deployed in front of the walls of the upper levels to adjust the absorption and reverberance in the hall. On the upper rear wall, these curtains are concealed by an acoustically transparent grid, and can be deployed to prevent delayed reflections from reaching the stage while remaining invisible in the space.

The hall was finally inaugurated on May 31, 2017, after 2 years on construction.

Marc Quiquerez

View from 1st balcony. Sitting under the tall window framing a preserved cedar tree outside, a shallow podium at the back of the hall extends to both sides to raise performers or audience

Balcony soffit and surface textures. Wall and ceiling surfaces feature random arrangements of wood shingles and battens to create sound-scattering irregularities. While the first balcony soffits serve useful early reflections to the audience and performers, the second balcony is detached from the walls, as if floated in the volume

	Repino Hall
Location	Repino, Russia
Owner	Valery Gergiev
Architect	Fabre/Speller
Construction cost	EUR 1.4 million
Design start	January 2013
Construction start	2016
Construction end	Q2 2017
Opening date	May 31, 2017
	Concert Hall
Acoustical consultants	Nagata Acoustics (room acoustics)
	Yasuhisa Toyota, Marc Quiquerez
Seating capacity	300
Room volume	3,400 m^3
Surface area	1,600 m^2
Volume/seat	11.3 m^3/seat
Volume/surface area	2.1 m
Finish material	
Ceiling	Siberian pine
Wall	Siberian pine

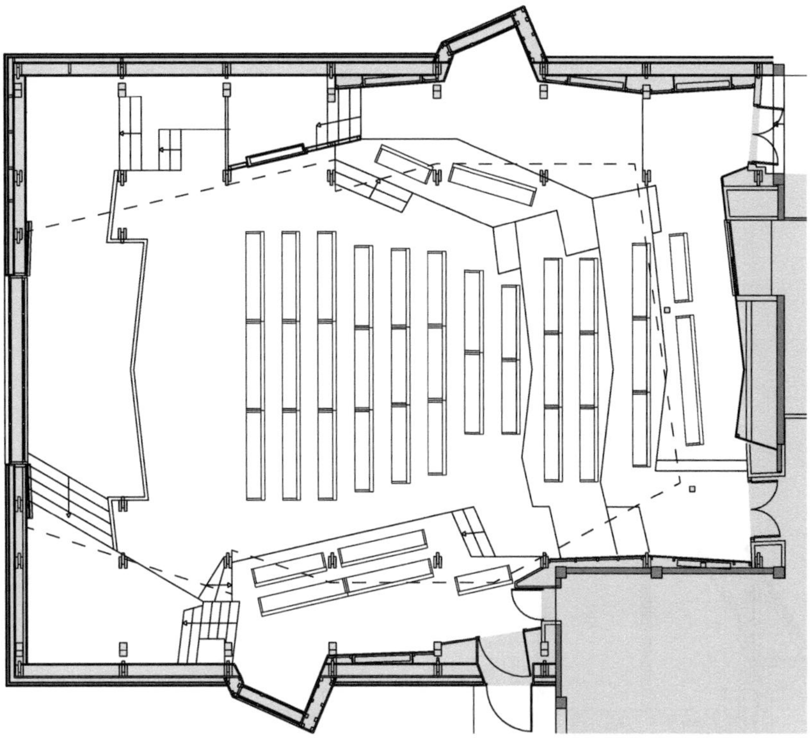

Fig. 29.1 Main floor plan
Courtesy of Fabre/Speller Architectes

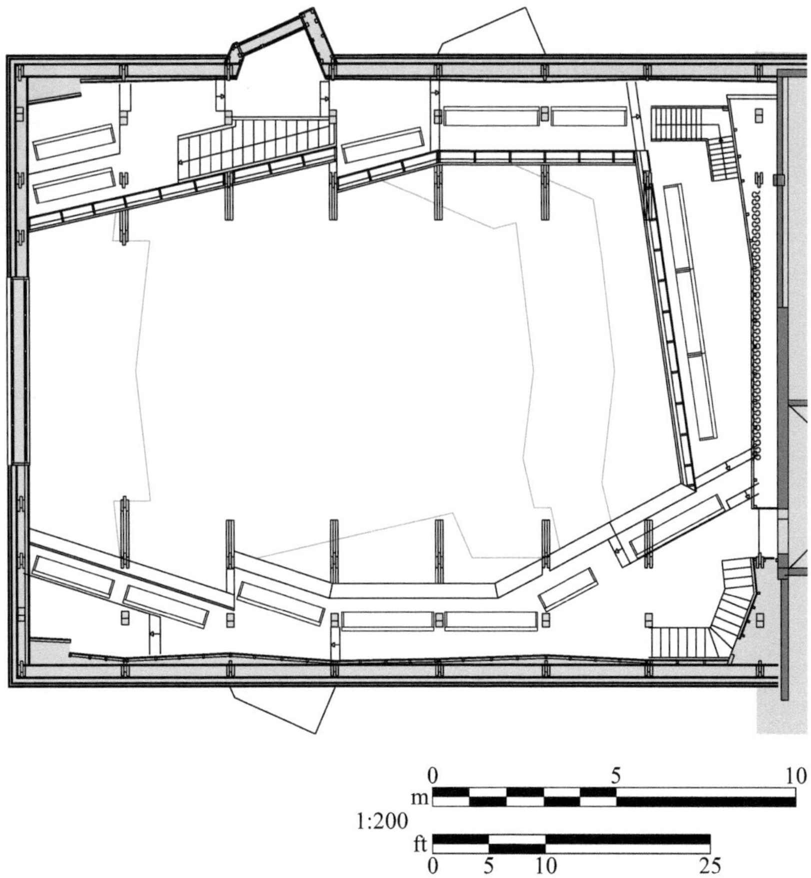

Fig. 29.2 Balcony plan
Courtesy of Fabre/Speller Architectes

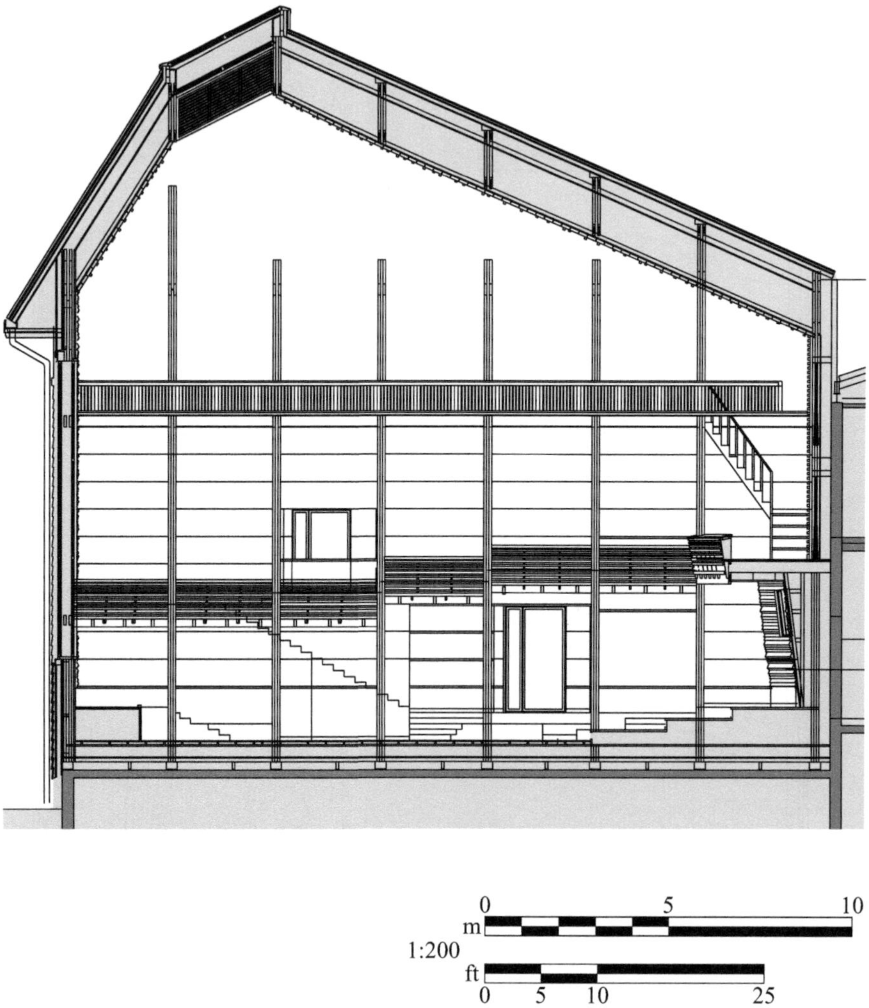

Fig. 29.3 Section
Courtesy of Fabre/Speller Architectes

30. Jinji Lake Concert Hall

View from side terrace. Much of the footprint of the hall is occupied by the stage such that a full-size orchestra can rehearse. A removable handrail separates the upstage seating from stage, for use with a chorus or to providing a uniquely intimate audience experience

Y. Toyota et al., *Concert Halls by Nagata Acoustics*, https://doi.org/10.1007/978-3-030-42450-3_30

While less famous in the West than neighboring Shanghai, the city of Suzhou was officially founded more than 2500 years ago. The Suzhou Industrial Park has been in development since the mid-1990s with the vision to construct a modern suburb of Suzhou encouraging high-tech development. The area includes high rise residential and commercial buildings, restaurants, shops, and cultural facilities including the Suzhou Culture and Arts Centre.

The Centre opened in 2007 and contains a large multipurpose hall, cineplex, cafe, and smaller classroom and exhibition spaces. The Suzhou Ballet Theatre uses the Centre as their primary rehearsal and performance location. Views from the central courtyard look out over the expansive Jinji Lake and the instantly recognizable "pants" skyscraper at Dongfangzhimen.

Within the Centre, a former exposition space was emptied to accommodate a new venue and rehearsal rooms to house the newly formed Suzhou Symphony Orchestra. The orchestra is made up of young, international musicians and began rehearsing only approximately a year before the opening of the hall. Originally, the orchestra had planned a rehearsal hall to prepare for concerts in the existing multipurpose theater as well as domestic and international tours. Following our suggestion, they reevaluated the space and cost needed to create a rehearsal hall and chose to include audience seats for concert performances. Since only interior construction was necessary, the project could be completed on an accelerated schedule, but was strictly limited by the available space.

The original architectural concept from Tongji Architectural Design was drawn from the traditional silk from the Suzhou region. Wide swaths of smooth, white surfaces were designed to wrap around the audience defining an oval footprint and evoking silk, while geometric bricks would adorn the remaining surfaces.

Since elliptical shapes could easily create problematic acoustic focusing, the silken, reflective surfaces were changed to acoustically transparent metal mesh. This allows the acoustical boundary of the hall to be made up of convex segments in plan without breaking the visual impression of enveloping concavity. Balcony fronts and interior dividing walls consist of interlocking steel spirals, each about 1 cm in diameter. Careful lighting creates gossamer surfaces which look alternately solid or semi-transparent depending on the viewing angle and distance. The reflective walls behind are textured with a pattern that alludes to both traditional fans and the intricate brickwork found at historical monuments in Suzhou. These panels of glass fiber reinforced gypsum were manufactured off site then installed on a steel subframe before hand-plastering of the joints, making for quick construction.

Wall surface texture. The walls are made of prefabricated glass fiber reinforced gypsum, designed with an overlapping textured pattern to scatter acoustic reflections. The pattern references traditional hand fans produced in the area

Since heavy concrete construction was infeasible, a free-standing steel structure was constructed to improve the sound isolation of the interior gypsum skin. Site conditions in the Cultural Centre were generally good, with little outside noise and plenty of distance to the subway, liberating the project from elaborate vibration isolation schemes.

In order to maximize the ceiling height and room air volume, the structural beams supporting the ceiling are located within the acoustical volume, hidden from view by an acoustically transparent, woven steel mesh that surrounds the suspended ensemble reflector. The reflector follows the convex profile of the structural beams and is constructed of the same glass fiber reinforced gypsum as the rest of the ceiling and interior walls. Instead of fan shaped geometric shaping, the ceiling is textured with pyramidal bumps of randomly varied height.

Despite the limited footprint available, a full-sized stage with mechanized risers was necessary to accommodate a variety of orchestral rehearsals. An additional riser adjacent to the upstage wall allows the stage to flow seamlessly into the chorus seating area divided into three terraces. Even when a removable railing separates these 152 seats from stage during concerts without chorus, they feel immersed in the orchestra.

The remainder of the 509 seats are spread over three levels with the farthest seat only 15 m from the stage. The upstage half of the second balcony is not used for audience seating but as a technical gallery. It acts as a soffit for stage reflections and provides space for stage lighting that would be impossible to fit above the ceiling. The underside of the soffit is slightly tilted so as to be perpendicular to the fan pattern on the wall as opposed to simply parallel to the floor. A system of mechanized curtains can lower from beneath the technical balcony as well as along the upstage wall to reduce the reverberation time when necessary.

Variable acoustics. Acoustically absorbing curtains descend from the underside of the technical balcony

Adjacent to the main rehearsal room is a rehearsal wing including three medium ensemble rooms, eight individual practice rooms, several musician lounges, and office spaces for the management team.

Erik Bergal

View from control room. Acoustically transparent metal mesh is used for the balcony fronts to prevent focusing from concave surfaces. Metal mesh around the ensemble reflector hides structural beams without shrinking the acoustic volume

Suzhou Cultural Centre. The concert hall is one space of a large arts complex in the technology park of Suzhou. Since the Centre was already existing, only interior fit out was necessary, leading to a remarkably quick design and construction totalling only 18 months

	Jinji Lake Concert Hall
Location	Suzhou, China
Owner	Suzhou Culture and Expo Centre
User	Suzhou Symphony Orchestra
Architect	Tongji Architectural Design
Design start	March 2016
Construction start	March 2017
Construction end	September 2017
Opening date	September 30, 2017
Acoustical consultant	Nagata Acoustics
	Yasuhisa Toyota, Motoo Komoda, Erik Bergal
Seating capacity	509
Room volume	9,600 m^3
Surface area	3,740 m^2
Volume/seat	18.9 m^3/seat
Volume/surface area	2.6 m
Finish material	
Ceiling + wall	Glass fiber reinforced gypsum
Transparent screen	Steel mesh
Stage floor	Alaskan Yellow Cedar
Seat manufacturer	Quinette Great Wall

Table 30.1 Jinji Lake Concert Hall—acoustical metrics at 500 Hz

RT unoccupied	2.0 s
RT occupied	1.9 s

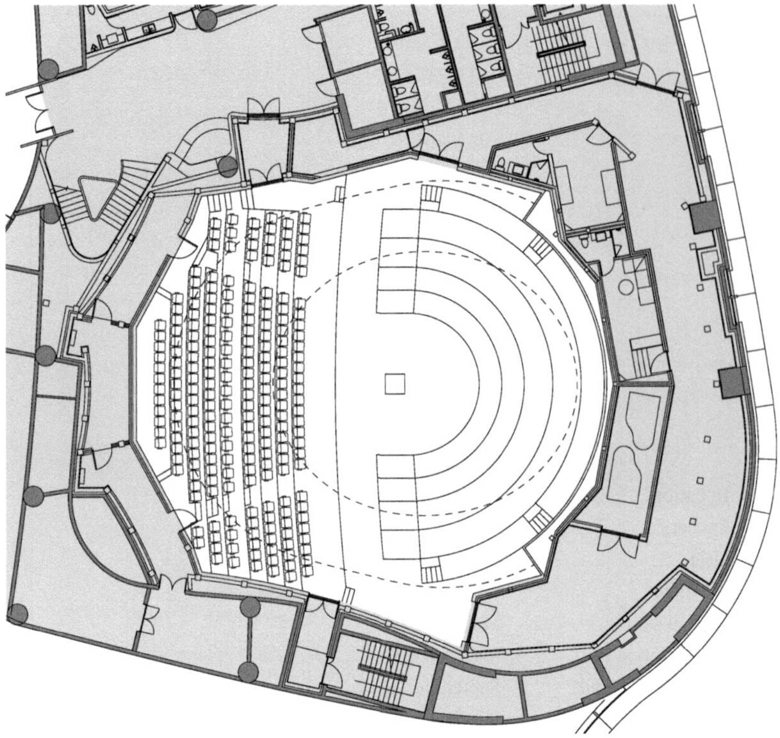

Fig. 30.1 Stage level plan
Courtesy of Tongji Architectural Design

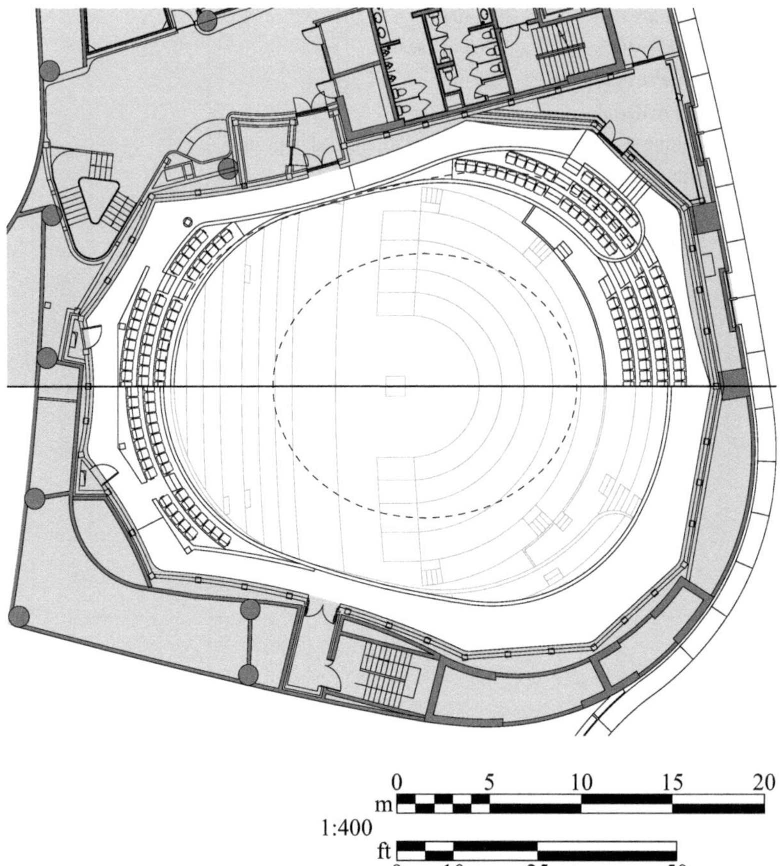

Fig. 30.2 1st and 2nd Balconies plan
Courtesy of Tongji Architectural Design

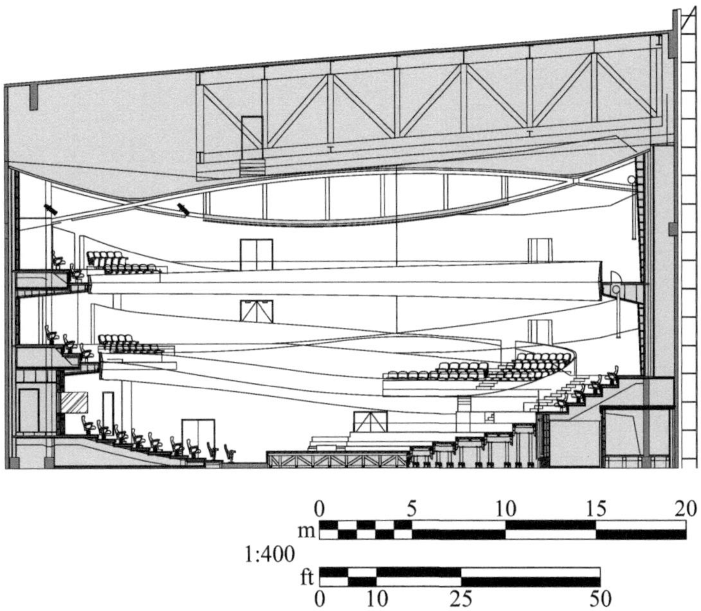

Fig. 30.3 Section
Courtesy of Tongji Architectural Design

31. Zaryadye Concert Hall

View from central seating tier. The central seating tier is divided into three blocks by tilted walls and can seat close to half of the capacity of the hall. Each row can be lowered, and chairs retracted, to create a flat floor

Y. Toyota et al., *Concert Halls by Nagata Acoustics*, https://doi.org/10.1007/978-3-030-42450-3_31

In September 2017, the new Zaryadye Park in central Moscow was opened to the public. It immediately became a favorite of locals and tourists alike, but a portion of the eastern side remained fenced off for a full year. Under the glazed canopy and outdoor amphitheater, the brand new Zaryadye Concert Hall was still under construction. It was finally inaugurated on September 8, 2018, opening a new chapter of the long history of Zaryadye district.

Project Background

Stretching over 400 m along the left bank of Moscow River, just east of Bolshoy Moskvoretsky Bridge and the Kremlin complex, Zaryadye district humbly started as a trading settlement in the twelfth century. At the height of the Soviet era, it housed the once largest hotel complex in the world, which eventually closed to be demolished in 2006. Initial plans to develop a vast commercial complex on the site, including performing arts venues, were soon relinquished. In 2013, the city of Moscow launched an international competition to transform the abandoned lot into an urban park and cultural district, reserving a section of the plot for the development of a brand new symphony hall. A consortium led by New York based firm Diller Scofidio + Renfro was awarded first prize later that year and, in early 2015, Moscow-based architects TPO Reserve were appointed as designers for the concert hall building.

View from the park's cantilever footbridge

Zaryadye Concert Hall complex features two concert spaces. The Large Hall is a philharmonic hall designed mainly for performances in natural acoustics and suited for large ensembles. The adjacent Small Hall, seating up to 400 people, is a more flexible space, designed for recital or chamber music as well as performances or events with sound amplifications. It can also double as a rehearsal hall for a full-size orchestra.

The Design of the Large Hall

Nagata Acoustics started work with the design team in the summer of 2015. By that time, a concept for the building and for the hall was being developed, which had already received the approval of local authorities. It was directly inspired by the curvaceous landscape of the park around and over the building. Although the team at TPO Reserve was interested in conceptualizing a new approach based on our mutual discussions, the tight project schedule—pushed not the least by construction works ready to break ground at a moment's notice—incited to abandon a blank slate approach in favor of expanding on the original concept with our specific inputs. The heritage of this earlier concept remains clearly visible in the completed hall today.

As an integral part of the park under construction, the building envelope—and most particularly its roof and load-bearing structure—needed to be ready for the scheduled opening of the park, well before the opening and completion of the hall. This not only imposed a quick pace to the early design development, but also a set of strong constraints on the height of the room and general proportions. Initial studies were therefore aimed at securing and confirming the most fundamental dimensions and shape requirements. Chief among them were the ceiling height and basic profile as well as stage dimensions.

The large stage area, 21 m in width and 16 m in depth, is designed to accommodate large orchestra ensembles. It can be extended upstage by retracting the four rows of the chorus seating section, while the orchestra can also make use of a 120 m^2 orchestra pit which is revealed by lowering two independent lifts in front of the stage. Beyond these lifts, the deep parterre area extends all the way to the rear of the hall. It is divided into three raked blocks by tilted terrace walls, but the entire area can be lowered to create an extensive flat floor for a wide range of programs and functions. From the chorus seating behind the stage, audience rows form terraces to the sides which climb towards the rear of the hall to meet as an overhanging balcony and frame the stage and parterre with a continuous ribbon-like ring. Overall, the hall accommodates 1,560 seats, with just under half of them on the main floor.

Surface textures on the walls. As the walls alternate between flowing ribbon-like white bands and darker wood-colored patches, they feature different matching shapes and geometries of surface irregularities. The textures vary in depth and density to create sound scattering patterns

Wall surfaces alternate between staggered triangular shapes with wooden finishes and smooth white outlines which extend to the ceiling surfaces, punctuated by shallow indentations. Detailed geometries and small-scale irregularities were studied and confirmed with acoustic tests on a scale model. Consisting mainly of thick plywood assemblies, with occasional addition of medium-density fiberboard, these massive surfaces contribute to developing a warm sound in the space, by effectively reflecting low frequencies while scattering higher frequencies.

Located directly behind the chorus section, the 82-stop pipe organ is the work of Manufacture d'Orgues Muhleisen (from France), in association with Daniel Kern. Its facade pipes cover the full width of the upper back wall in a dynamic asymmetrical arrangement, while the rest of the instrument, located in the center, is concealed behind tall louvers which can be controlled directly from the console by the player.

The Small Hall

The initial program and concept proposed two superimposed rehearsal studios next to the large hall, with one room doubling as a recital hall. However, this resulted in both rooms having a relatively low ceiling and limited volume. We therefore suggested to merge the two rooms into a single double height room to achieve a 14 m ceiling height, more suited to large ensembles and performances in natural acoustics. This allowed for the creation of a true performance space, thus expanding the programming opportunities of the complex.

Sitting at the intersection of curved and rectilinear structural grids at the north-west corner of the building, the footprint was subjected to strong shaping constraints, but the dimensions still allowed to secure appropriate proportions to lay out a large ensemble within the 300 m^2 surface area. The main floor is fully flat, but audience can also be seated on a balcony which surrounds the hall, with two rows in the front and a single row on the other sides.

The walls and ceiling, consisting mainly of painted plywood, are randomly faceted with panels varying in sizes to favor an even distribution of sound reflections in the space and prevent focusing and flutter. The dark gray ceiling and balcony underside contrast with the stark glossy white of the walls. Around the lower level and directly behind balcony audience, irregularities were added to the surfaces and create the visual illusion of pleated cloth wrapped around wall panels.

Retractable acoustical draperies can be deployed in front of the lower walls on the long sides to adjust the acoustics for rehearsals, and additional curtains can be deployed all around the room in front of the upper walls to adjust the reverberance of the space to the specific needs of the programs or occupancies.

First Notes and Grand Opening

Construction works in the building and the halls were rushed for an early completion in July 2018, when the complex was scheduled to host an international urban forum. Construction then quickly resumed in late July to complete the hall in time for the inaugural concert scheduled to coincide with the yearly celebration of Moscow Day.

Less than 2 weeks before the grand opening, Maestro Valery Gergiev assembled a group of musicians from Mariinsky Orchestra to perform their first rehearsal in the finished hall in preparation of the upcoming concert. The previous day, all chairs still sported their protective covers and the organ area, still an empty volume waiting for the instrument, only had scaffolds for a facade. As the first notes started filling the large volume of the hall, the musicians already sounded at ease on the new stage. Maestro Gergiev quickly stepped off the podium and the stage to stroll around the different seating sections of the hall while instructing and discussing various changes to the music pieces and the layout of musicians. Playing through various musical registers and seating arrangements, and later joined on stage by pianist Denis Matsuev, the ensemble enthusiastically explored the potentials of the hall which responded with a consistent balance of clear and precise yet warm and rich sound throughout the audience. Some recommendations for local adjustments to specific assemblies and materials were nevertheless formulated to finalize the hall and secure its fundamental acoustical qualities.

A larger Mariinsky Orchestra took the stage again with Maestro Gergiev and Mariinsky Chorus in late afternoon on September 8, 2018 for the grand opening. The concert opened with Mussorgsky's aptly titled Overture to Khovanshchina opera, "Dawn over the Moskva River," as a tribute to the new hall emerging on the river bank. An all-Russian program followed, with an impressive cast of soloists joining the orchestra on stage: vocalists A. Shagimuratova, A. Netrebko, M. Petrenko, and I. Abdrazakov, pianists D. Trifonov and D. Matsuev, violinist P. Zukerman, and trumpeter T. Martynov.

Marc Quiquerez

View from upstage seating. Side seating terraces climb towards the rear of the hall to form a balcony in a continuous ribbon-like ring around the stage and central seating tier

View from the balcony on opening night. The large stage can accommodate a full-size orchestra ensemble and choir. The first rows of seats rest on motorized platforms which can be lowered to create a large orchestra pit. The chorus seating area can also be retracted and lowered to extend the depth of the stage

Building approach and main entrance

Zaryadye Concert Hall	
Location	Moscow, Russia
Owner	City of Moscow
Architect	TPO reserve
Construction cost	RUB 7 billion
Design start	Q3 2015
Construction start	Q4 2015
Construction end	August 2018
Opening date	September 8, 2018
Building size	23,800 m^2
Large Hall	
Acoustical consultants	Nagata Acoustics (room acoustics)
	Yasuhisa Toyota, Motoo Komoda, Marc Quiquerez
	Acoustic Group (building acoustics)
Seating capacity	1,560
Room volume	23,300 m^3
Surface area	7,700 m^2
Volume/seat	14.9 m^3/seat
Volume/surface area	3.0 m
Finish material	
Ceiling	Medium-density fiberboard, plywood
Walls	Medium-density fiberboard, plywood
Audience floor	Wood over concrete or fiber-reinforced gypsum
Stage floor	Alaskan Yellow Cedar
Seat manufacturer	Ascénder
Organ builder	Manufacture d'Orgues Muhleisen
Model scale	1:10
Small Hall	
Acoustical consultants	Nagata Acoustics (room acoustics)
	Acoustic Group (building acoustics)
Seating capacity	330
Room volume	3,800 m^3
Surface area	1,800 m^2
Volume/seat	11.5 m^3/seat
Volume/surface area	2.1 m
Finish material	
Ceiling	Medium-density fiberboard, plywood
Walls	Medium-density fiberboard, plywood
Stage floor	Siberian pine

Table 31.1 Zaryadye Concert Hall—acoustical metrics at 500 Hz

RT unoccupied	2.9 s
RT occupied	2.3 s

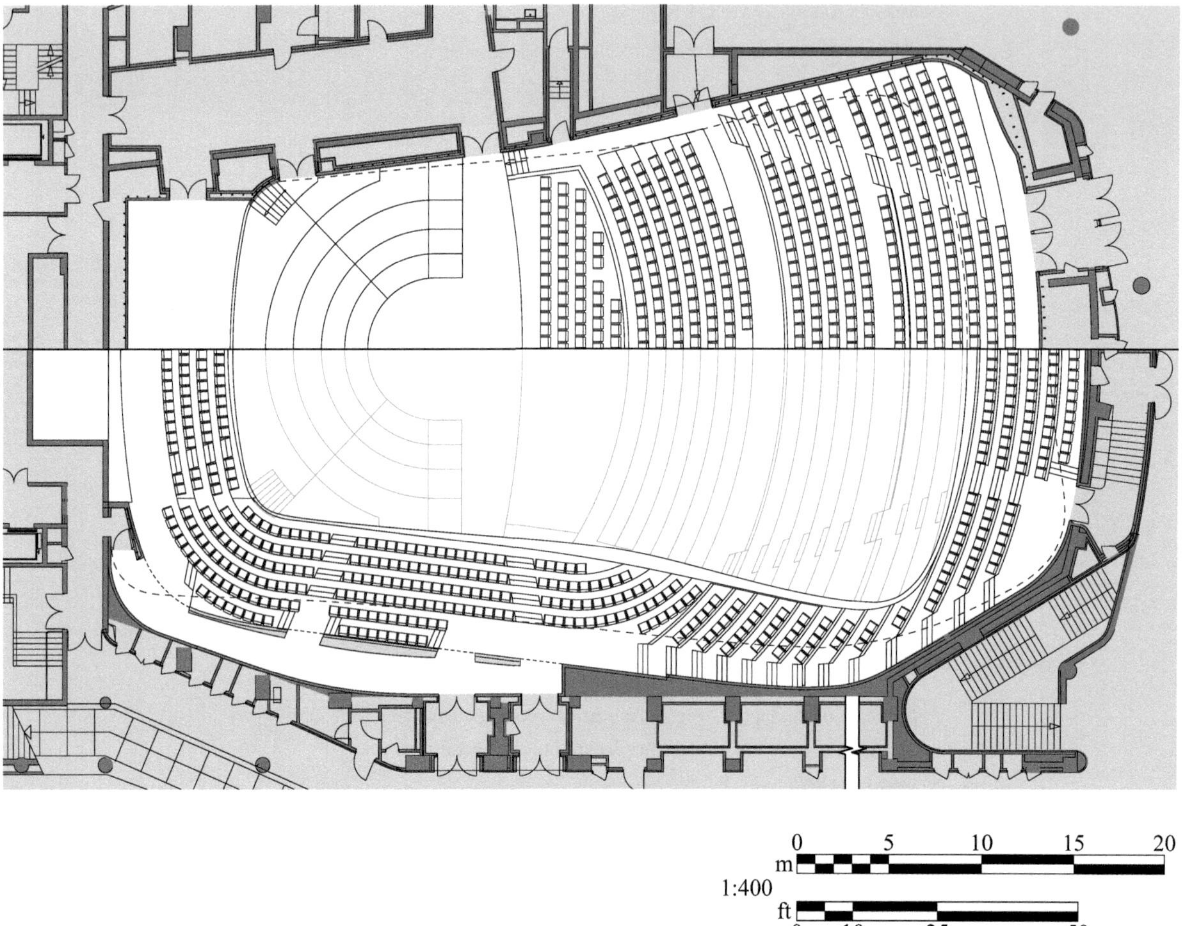

Fig. 31.1 Stage and terrace levels plan
Courtesy of TPO Reserve

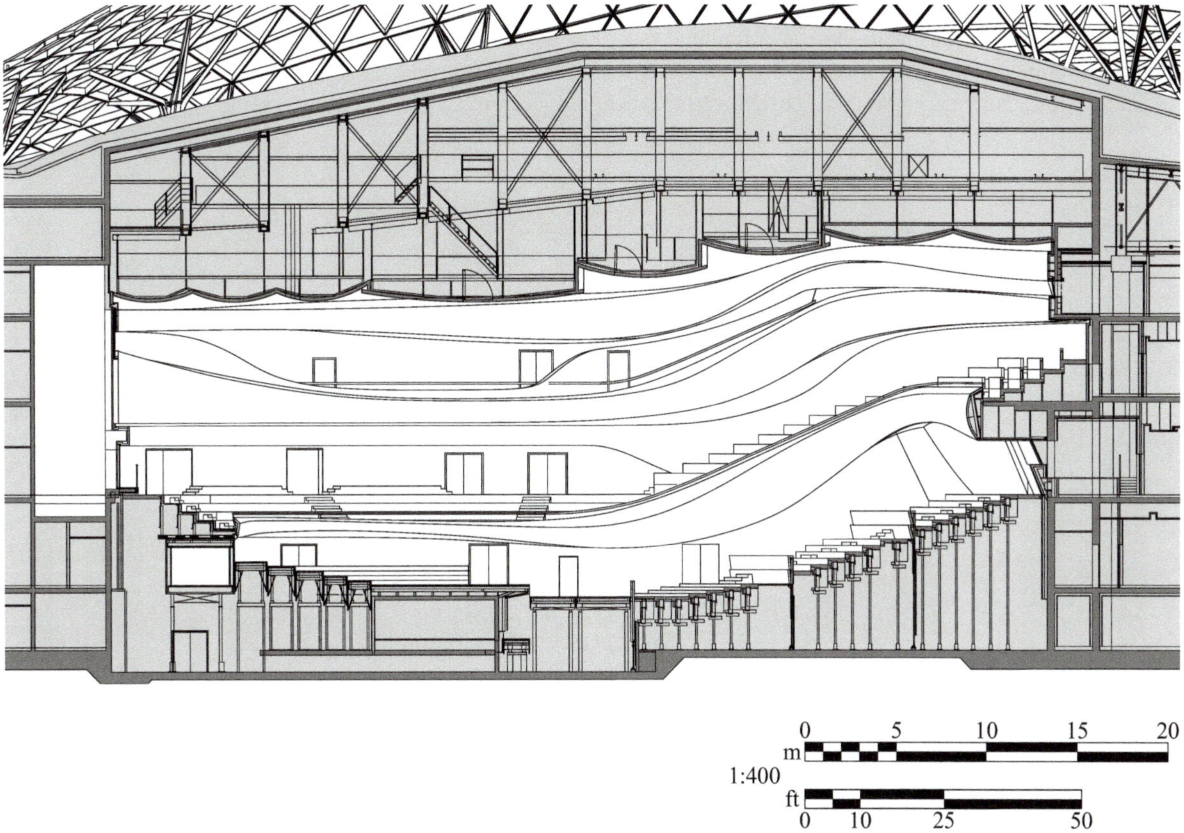

Fig. 31.2 Section
Courtesy of TPO Reserve

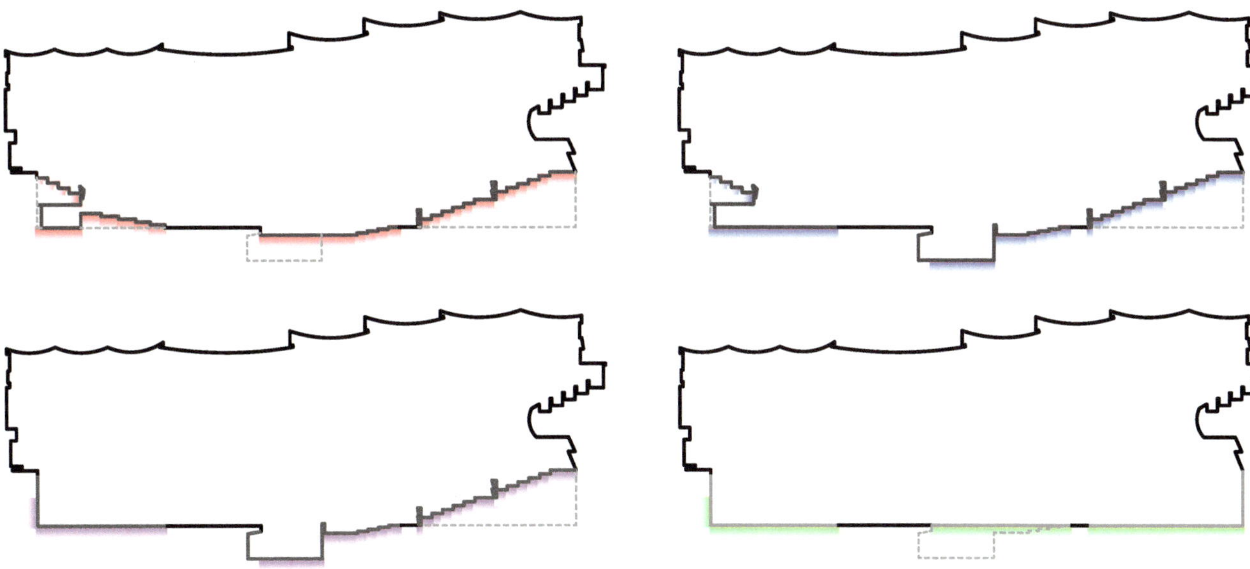

Fig. 31.3 Scope of spatial transformations

32. The Conrad Prebys Performing Arts Center

View from main floor. Most of the wood slat walls that make up the hall interior are acoustically transparent and connect the main volume with the space behind

Y. Toyota et al., *Concert Halls by Nagata Acoustics*, https://doi.org/10.1007/978-3-030-42450-3_32

The La Jolla Music Society was established in 1941 in a luxury sea-side community about a 20-min drive from downtown San Diego, California. The Society's activities have enhanced the vitality and given depth to the cultural life of the local community, transforming the organization into a close-knit and influential member of the local arts scene. One major focal point of the Society's calendar is the "SummerFest" chamber music festival, held every summer since 1986. Many famous musicians have participated in the festival, leading to an international name recognition for the organization among audiences and performers alike.

Exterior

The Conrad Prebys Performing Arts Center, known as "The Conrad," opened in the center of La Jolla Village in 2019 as a new home for the La Jolla Music Society, replacing the existing hall at a nearby museum. The new building is anchored by two performance spaces, the 513-seat "Baker-Baum Concert Hall" and the 128-seat multi-purpose space "The JAI," which are separated by a large courtyard. Rehearsal rooms and offices for the Society are also included.

Programming for The Conrad envisions a broad spectrum of events and activities that will benefit the surrounding community both culturally and economically. In addition to concerts, the facility will host expanded daytime educational programming by La Jolla Music Society, such as lectures, demonstrations, and performances for children and students. Other local arts presenters also use the new facility, and it is available for meetings and even weddings.

The design architect was Cambridge, Massachusetts-based Epstein Joslin Architects, and Joseph Wong Design Associates of San Diego was the local executive architect.

Interior Visual Design and Acoustic Volume

The Baker-Baum Concert Hall was designed with classical chamber music concerts in mind. The design concept began with the intent to make the relationship between audience and performer as intimate as possible. Instead of simply arranging the stage and audience to face one another, the audience is made as compact as possible in a horseshoe shape. In this arrangement, the gaze of each viewer-listener can be felt closer.

View from stage

View from the balcony. Instead of simply arranging the stage and audience to face one another, the audience is made as compact as possible in a horseshoe shape. In this arrangement, the gaze of each viewer-listener can be felt closer

New buildings in the La Jolla area, which includes The Conrad, must abide by a strict height limit of 9 m (30 ft.). This resulted in a final ceiling height in the hall of 8.3 m (27 ft. 4 in). Although this was a severe constraint for the acoustic design of a concert hall, we decided to explore new directions of design with the architects. To counteract the inability to raise the ceiling, the hall was made as wide as possible on the site to secure the largest possible room air volume. Most of the wood slat walls that make up the hall interior are acoustically transparent and connect the main hall volume with the space behind. On the plan, the visual border is the horseshoe shaped line, and the light-gray portion is the space behind the visual wall.

Surrounding acoustical volume. The space between the rigid wall and the grille wall allows early reflections to be returned at the appropriate timing, and the additional room volume helps develop rich sounds

Even though the interior of the hall looks like a horseshoe at first glance, acoustically it has more in common with a shoebox. The hard walls forming the outer perimeter are rough plaster on heavy concrete block, with blue lighting illuminating the entire wall to create a gentle impression. The space between the rigid wall and the grille wall allows early reflections to be returned at the appropriate timing, and the additional room volume helps develop rich sound. On the three sides of the stage, small soffit surfaces at lower positions are installed behind the grille wall, so that extremely early reflections can be obtained to support the musicians on stage.

Sound Absorption Curtains

In order to accommodate a variety of programs which use amplification, in addition to the core classical chamber music program, the upper portion of the hard wall behind the grille can be almost entirely covered with motorized, sound-absorbing curtains. Manual curtains are also available around the stage, at the lower level, to allow the acoustic condition to be finely tuned without giving any visual impact to the audience. By deploying all the curtains, the reverberation time can be reduced by 0.5 s, and there is a significant difference in the hearing sensation.

Opening Concerts and Delicate Piano Playing

The opening concert was held on April 5, 2019, after a cocktail party and ribbon-cutting ceremony. The 90-min performance featured various artists, including the current and all former music directors of the SummerFest Music Festival. The calm interior of the hall and the intimate atmosphere created by the gentle lighting supported the pleasant sound. Piano performances have sometimes saturated the hall when loud sound was created. It is a sensitive hall where the skill of the pianist is tested, especially the ability to control dynamics. Especially for string ensembles, the sound of each instrument was very clear, simultaneously with rich reverberation—the effect of making the room air volume as large as possible.

2019 was the 50th anniversary of the La Jolla Music Society, and the annual SummerFest was centered around that celebration.

Motoo Komoda

The Conrad Prebys Performing Arts Center	
Location	La Jolla California, USA
Owner	La Jolla Music Society
User	La Jolla Music Society
Architect	Epstein Joslin Architects
Executive architect	Joseph Wong Design Associates
Acoustical consultant	Nagata Acoustics (room acoustics) *Yasuhisa Toyota, Motoo Komoda, Daniel Beckmann* Robert F. Mahoney & Associates (isolation and noise control)
Theater consultant	Theatre Consultants Collaborative
Construction cost	USD 82 million
Design start	August 2014
Construction start	February 2017
Construction end	April 2019
Opening date	April 5, 2019
Building size	4,570 m^2
Baker-Baum Concert Hall	
Seating capacity	513
Room volume	5,200 m^3
Surface area	2,800 m^2
Volume/seat	10.1 m^3/seat
Volume/surface area	1.85 m
Finish material	
Ceiling	Corrugated metal/concrete
Walls	Plaster on concrete block, wooden grille
Audience floor	Hardwood
Stage floor	Alaskan Yellow Cedar
Seat manufacturer	Ducharme

Table 32.1 Baker-Baum Concert Hall—acoustical metrics at 500 Hz

	Reflective condition	Curtains deployed
RT unoccupied	1.6 s	1.1 s
RT occupied	1.5 s	1.0 s
EDT	1.6 s	
C_{80}	1.8 dB	
D_{50}	60%	

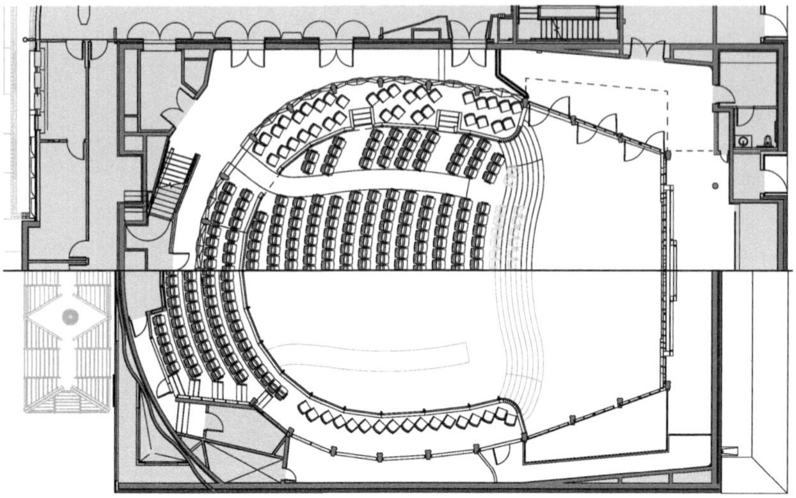

Fig. 32.1 Stage level and balcony plan
Courtesy of Epstein Joslin Architects

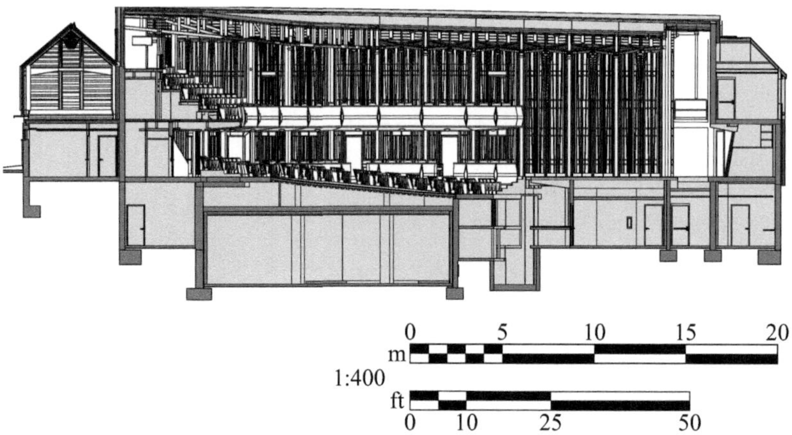

Fig. 32.2 Section
Courtesy of Epstein Joslin Architects

Part II

Topics on Acoustics and Acoustical Design

1. Vienna Musikvereinssaal as the Starting Point of Discussion

Few consensus opinions exist concerning acoustics except perhaps the preeminence of the Vienna Grosse Musikvereinssaal. Opened in 1870, the Musikvereinssaal is the classic starting point for any discussion of acoustics. Well loved since that time by musicians, composers, and audiences alike, the hall has no doubt been influential in forming the image of how spaces dedicated to the performance of classical music should be expected to look, sound, and feel.

Vienna Grosse Musikvereinssaal

Y. Toyota et al., *Concert Halls by Nagata Acoustics*, https://doi.org/10.1007/978-3-030-42450-3_33

The long rectangular room of the Goldener Saal seats 1,680 viewer-listeners on a flat main floor and a single balcony which wraps around all four sides of the room. Slightly wider than it is tall, the hall measures just under 20 m wide with a ceiling height of 16 m above the stage. In conjunction with the 36 m length, the total volume is quite large. The hall is the standard example of the "shoebox" style hall, along with Boston Symphony Hall and the Amsterdam Concertgebouw, so called because of its simple rectilinear shape. The excellent acoustics are a result of reverberance from the large room volume together with strong early, lateral reflections produced by the narrow side walls.

Understanding these early reflections is integral to understanding the Musikvereinssaal's quality and eventually to designing new concert spaces. Every room, from a concert hall to a living room, will have a distinctive thumbprint of reflections reaching the listener. Figure 1.1 shows an impulse response obtained from recording a sharp impulse such as a balloon popping and Figure 1.2 shows an echo diagram which is a typical impulse response. Each bar in the diagram represents the magnitude of an incoming reflection. First, the direct sound reaches the measurement position followed by the other reflections, one after another. The human auditory system, however, does not hear each reflection discreetly, but rather integrated together as a continuous sound. Thus, we hear the resulting decay as smoothly decreasing or, if long enough, reverberating.

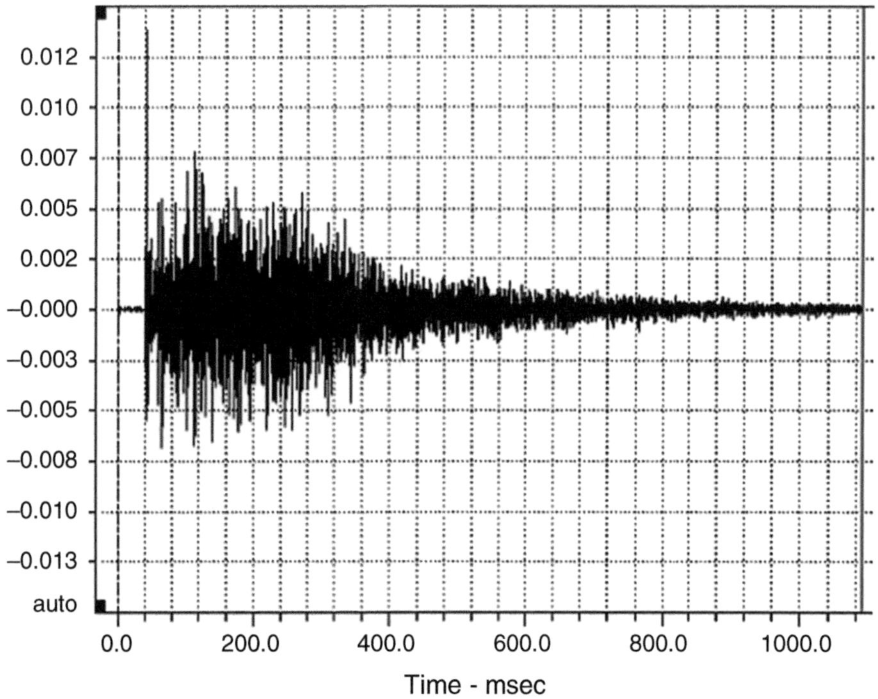

Fig. 1.1 Impulse response

Similarly, the reflections immediately following the direct sound are heard as integrated with the direct sound. These reflections which characterize the impression of the direct sound are called "early reflections." There is no clear definition of the border between early reflections and the reverberant sound, but it is generally considered to be before one hundred milliseconds after the arrival of the direct sound to the viewer-listener. Since they color the impression of the direct sound, these are the reflections which are the most important for the room acoustics of concert halls. A sound, for example, a violin, will sound rich and present in a hall with good acoustics, while the same violin will sound thin and weak in a hall with poor acoustics. The extreme case is analogous to a violin played in an open field with no reflections whatsoever: only the direct sound would be heard and the violin would not sound attractive at all.

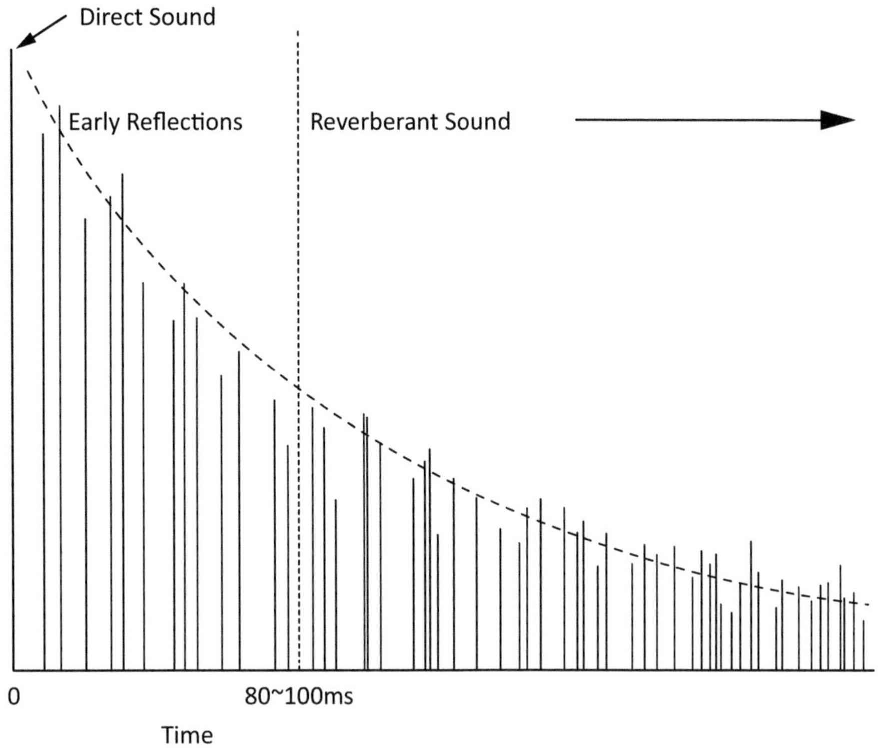

Fig. 1.2 Typical echo diagram

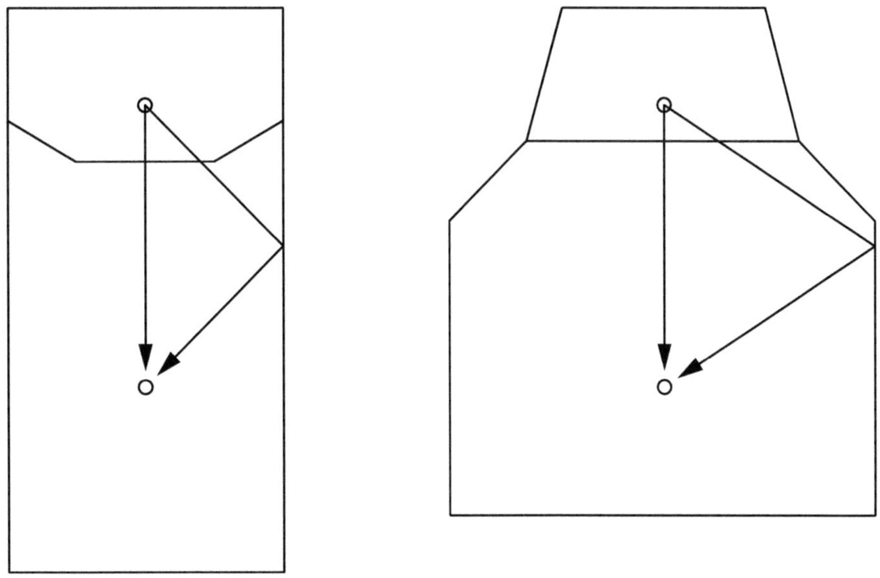

Fig. 1.3 The shoebox typology relies on narrow side walls

It is intuitive that the rich early reflections are coming from the large side walls in a rectangular-shaped—or shoebox—concert hall such as the Vienna Grosse Musikvereinssaal. This typology is as popular now as it has been historically for its reliable acoustics, but is limited by the dependance on the narrow width to maintain an appropriate delay for reflections from the side walls. This is problematic for modern concert hall projects, where there is demand for more and more audience seats.

It would be impossible to match the seating density of the Musikvereinssaal or other historical halls. Comfortable seat width and material—to say nothing of modern safety regulations—mean that significantly more space is required to fit the same number of seats. To meet modern expectations, 1,100–1,200 seats could fit in the footprint of the Musikvereinssaal. Similarly, a design exercise once attempted to fit modern-sized seats into the floor plan of Amsterdam Concertgebouw, another of the great, historical halls. Although wider than Vienna Musikvereinssaal, only 1,200–1,300 seats would fit, a substantial reduction from the roughly 1,700 seats it currently holds. Architects and acousticians needed to find a way to reach an economically viable seat count without compromising the acoustical quality.

Increasing the basic dimensions of a shoebox-style hall could take several forms:

What if we increase the width?
This would lose the advantageous early reflections which are the strength of the shoebox layout.
What if we extend the length?
Seats would be too far from the stage, suffering both in acoustical quality and in visual intimacy.
What if we add balconies?
Two balconies would be the practical limit, since both sightlines and distance become more challenging with each additional level.
What if we increase the depth of the balconies?
Seats underneath the balconies already have notoriously challenging acoustics.

A more radical departure from the shoebox typology was needed.

2. Berlin Philharmonie and the Birth of the Vineyard-Style Concept

With the advent of acoustical science by Wallace Clement Sabine in the late 1890s, it became possible to measure and plan interior materials to control reverberation, and acousticians began to make more quantitatively informed designs. However, the narrow, rectangular footprint of the shoebox layout remained the foundation on which objective design and planning was developed. Nearly a century after the opening of the Vienna Musikvereinssaal, architect Hans Scharoun's Berlin Philharmonie finally set a new precedent in 1963.

View from side balcony

For this novel design, the audience is grouped into seating blocks or "terraces" which are divided by small walls and are arranged in a steep, bowl-shaped configuration. The terraces recall the layout of grapevines on a hillside, hence making the Philharmonie the first "vineyard-style" hall. While some historical halls such as the Amsterdam Concertgebouw feature a small number of seats beside and behind the stage, Scharoun moved the stage away from the end of the room and almost to the center. The number of audience members around the stage was drastically increased, making the average distance to the 2,250 seats much shorter than would be possible in a shoebox typology. Spiritually, the shift broke the dichotomy of the audience–orchestra dynamic; the public now shares the same space with the performers, who are placed at the center of attention.

It is worth noting that this monumental shift was not driven by acoustics, but by architectural, almost philosophical, design. Dr. Lothar Cremer was selected to be the acoustician for the new hall. By all accounts, Cremer was skeptical that a

Y. Toyota et al., *Concert Halls by Nagata Acoustics*, https://doi.org/10.1007/978-3-030-42450-3_34

View from upstage balcony

good acoustical result could be achieved, but resolved to attempt to work towards Scharoun's vision. By the time of design in the late fifties, the importance of early reflections was already understood. Instead of rectangular walls and a flat ceiling, the Philharmonie has a seven-sided footprint with an off-center organ, and the ceiling has a tent-like shape which peaks over the stage. Reflectors hanging above the orchestra, terrace walls, and balcony fronts provide the acoustic support formerly entrusted to the narrow walls of the shoebox concept. Within each terrace, the seating rake is much steeper than in previous halls resulting in better sightlines and better transmission of direct sound.

The architectural and acoustical success of the Berlin Philharmonie opened new possibilities for architectural creativity and flexibility. No longer constrained by the vertical walls, restricted width, and lateral balconies of the successful shoebox layout, designers could place the audience as they pleased and articulate both walls and ceiling. Two-dimensional drawings became inadequate to accurately represent the hall design, and acoustical studies were limited only to simple geometric studies. Early examples relied heavily on the use of physical models as concert hall design became more complex, and became the only method to achieve a quality result for high profile projects. Cremer used a huge 1:9-scale model to be able to make as detailed measurements as possible during the design of Berlin Philharmonie. It would take another thirty years for personal computer to achieve the computing power necessary to make repeated simulations feasible.

3. Surround Seating Paradigm

Relinquishing the rectilinear layout coupled with the tendency of each project and language to choose their own name for seating areas has led to confusing lexicon. For the purposes of this text, the seating areas in front of the stage behind the conductor will be called the "front." The seating behind the stage facing the conductor will be referred to as "upstage" seating, with the area immediately upstage called "chorus" seating. The "rear" of the hall will refer to the seats farthest from the stage but still facing the stage in the historical orientation.

This variety in seating area lexicon exemplifies the fundamental shift introduced by the surround seating paradigm. By pulling the stage and performers towards the center of the room, a variety of new seating placements is created. This brings a new-found freedom and flexibility in laying out the audience in the hall and connecting viewer-listeners and performers. Naturally, this break with century-old traditions and expectations is not without its questions and concerns.

During the design of a new hall, there is always concern over the quality of the upstage seating. It is not realistic to expect these seats to have the exact same acoustics as front seating, but this does not imply that the acoustical quality upstage is poor. Indeed, in performances where the upstage area is filled with the chorus, there is no doubt that the singers can easily hear the musicians on stage even if the left-right or front-back orientation of the orchestra is flipped.

The individual instruments in the orchestra each have their own unique directivity, but when they are grouped together, they tend to act as a plane source, radiating sound primarily upwards. Each string section, playing in unison, combines the directivity of its individual instruments, resulting in the generally upwards directivity of a plane source. Woodwind and percussion instruments are omni-directional radiators even as solo instruments. The brass instruments are the exception, since they radiate the majority of their sound directly out of the bell mouth (facing forwards for trumpet and trombone, upwards for the tuba, and backwards for the horn). If one is seated at the side or upstage of the orchestra, the effect of sitting outside of the primary directivity of the brass instruments is compensated by the reduced distance to those instruments, and the balance is preserved, even if it is not identical to the seats in front of the orchestra. When all of the instruments are placed together on the stage, the individual directivities of the instruments become harder to identify, and the orchestra as a whole takes on an omnidirectional character.

The biggest difficulty is for the case of vocal soloists. When a singer is facing away, there is a marked loss in lyric intelligibility as well as a slight to the audience who have come to hear a soloist and then cannot see their face. In a large hall such as the Hamburg Elbphilharmonie, the singers should be placed upstage so that the number of viewer-listeners behind the singers could be reduced. Similarly, the piano is challenging due to its lid which gives it a strong directivity, both visually and acoustically. Simply removing the lid can transform the piano into an omnidirectional instrument, though there can be resistance from performers for whom the acoustic condition is changed.

Given the simple choice between a front seat and an upstage seat, I would choose a front seat as long as the distance to stage was comparable, a key benefit of surround seating. But only halls with surround seating truly offer this choice, including not only direction but also distance. The close proximity of the upstage and side seats to the orchestra is a key benefit of surround seating. It translates into an average distance to the stage that is much smaller than if the same number of seats were instead relegated to the rear of the hall. Shortening the distance to stage greatly improves the visual and acoustical intimacy for the audience and—considering a concert as a shared moment between the musicians and the audience—results in a notable improvement. The upstage seats also have the unique opportunity to see the face of the conductor which adds to the intimate experience of the concert. This similarly holds true for seats to the sides of the stage. These key aspects can be factored into the choice of a seat, depending on the personal preferences of the patron.

Most tellingly, the concerns about upstage or side seating are mainly confined to the programming or design phases and seem to evaporate after actual listening experiences in concerts.

Y. Toyota et al., *Concert Halls by Nagata Acoustics*, https://doi.org/10.1007/978-3-030-42450-3_35

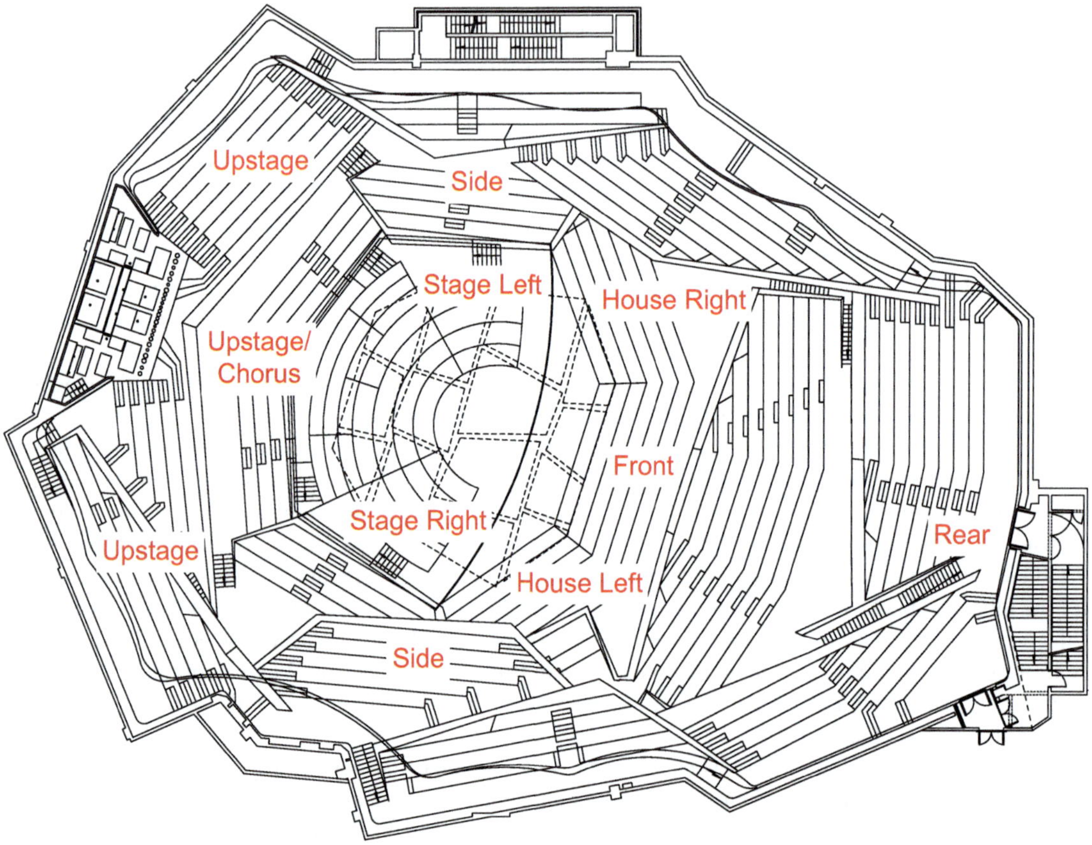

Fig. 3.1 Seating name convention

4. Foundational Projects of Nagata Acoustics

To understand our current design principles, it is useful to understand the timeline of our early work.

Tokyo Bunka Kaikan

Tokyo Bunka Kaikan (Tokyo Metropolitan Festival Hall) in Ueno Park, Tokyo opened in 1961, essentially coincident with the Berlin Philharmonie's opening. NHK Technical Research Laboratories provided acoustical consulting services, where Dr. Nagata was working at the time. The Architectural Acoustics Research Group was headed by Professor Yasuo Makita, known for his academic texts on room acoustics. Hexagonal in footprint, Bunka Kaikan seats 2,303 over a large main floor and four vertigo-inducing balconies. The inclusion of both a pit and movable orchestra shell firmly classifies the hall as multifunctional, allowing it to host a wide array of programs for classical music including opera and ballet as well as orchestral concerts.

Suntory Hall

The design of Suntory Hall drew heavily on the model of Berlin Philharmonie. Completed in 1986, the hall was designed without the use of computer modeling. All acoustical analyses were performed manually in two dimensions or in a physical scale model. It was here that our foundational understanding of the time structure of early reflections was developed. By comparing the acoustical results in various seating positions in the completed hall with these modeling studies, we continued to identify the design elements which contributed to the success of Suntory Hall.

Touring orchestras generally agreed that the acoustics in Suntory were good and were happy with the condition on stage. However, when Suntory Hall first opened, the Tokyo orchestras found the stage acoustics to be troubling. The situation is quite different from the halls—especially compared to Bunka Kaikan—in which they were accustomed to playing, and the shock of a new space tainted their impression. After a few years, many musicians felt that the stage acoustics had improved. In fact, nothing had been changed in the hall; it was only the musicians which had grown accustomed to the new hall.

Walt Disney and Sapporo Concert Halls

A donation from Lillian Disney kicked off the design and construction of a new concert hall for the Los Angeles Philharmonic in 1987 to be named after her late husband, film pioneer Walt Disney. Frank Gehry was appointed as the lead architect and design began in 1989. Shortly thereafter, in 1992, a concert hall of similar size was planned in Sapporo, Japan. By 1993, the Los Angeles design by architect Frank Gehry had developed to the point of a large-scale acoustical model test. Naturally assuming they would open first due to their head start of several years, Gehry and his team welcomed the Sapporo Concert Hall design team to their office to inspect the scale model. It is undeniable that some design inspiration and acoustical ideas were brought back to Sapporo from these meetings.

Y. Toyota et al., *Concert Halls by Nagata Acoustics*, https://doi.org/10.1007/978-3-030-42450-3_36

Work on Walt Disney Concert Hall came to a halt in 1994 due to budget issues. In the meantime, Sapporo Concert Hall opened in 1997. The successful acoustics in Sapporo were a good omen for Los Angeles and can almost be seen as a full-sized model. Several design elements were subsequently introduced back into Disney once work re-started in 1998, after the success realized in Sapporo. The two halls can be thought of as siblings considering their concurrent timelines and shared design elements, and this cross-pollination undoubtedly resulted in better halls than had they been built in isolation.

5. Reverberation Time, Other Metrics, and Underlying Goals

There has been much effort to measure and correlate quantitative acoustical parameters in existing halls to find some objective quantification of quality. The seminal texts of Leo Beranek and Vilhelm Lassen Jordan focused considerable effort on aggregating acoustical metrics for the purpose of characterizing the perfect concert hall. Their studies focus on reverberation time (RT), early decay time (EDT), clarity (C_{80}), strength (G), room response (RR), lateral fraction (LF), and interaural cross correlation (IACC), among others. During the design of Suntory Hall, we also measured more than ten successful halls in Japan to try to quantify an ideal design using RT, EDT, C_{80}, and RR. Ultimately, we found insufficient correlation with the subjective impression and the endeavor was refocused.

On Reverberation Time

We are most often asked to evaluate or prescribe a reverberation time for ongoing projects. Reverberation time is convenient as a metric since it is easy to understand and broadly used. It is defined as the time in seconds for a sound in an enclosed space to decay and become inaudible. It was established initially as a subjective evaluation, as it was originally measured by Wallace Clement Sabine using organ pipes as simple tone generators, the human ear, and a stopwatch. An objective definition was later formalized, as the time for a sound level to decay by 60 dB after it has ceased.

It is important to be mindful of the limitation of reverberation time. Since these computations assume a linear decay, there can be considerable variability in the decay curves even if the final value is the same. For this reason, comparing two halls with similar reverberation times does not imply that they provide similar acoustical experiences. Furthermore, the reported values are often averaged over several seating positions. As an absolute measure, reverberation time gives no indication of the room shape, room volume, number of seats, or program requirements further complicating the task of comparing different venues or creating design targets for new spaces.

A useful analogue is the alcohol content in whisky: easy to measure and simple to understand, but without nuance. A distiller has some target range in order to call the final product whisky, but infinitely many variables in the distilling process can eventually yield the same percentage, but vastly different products. The quality of ingredients and skill of the distiller are what may differentiate top-shelf from bottom-shelf whisky, but this discrepancy is not captured by looking only at the alcohol content. Similarly, reverberation time may be an interesting statistic of a completed space and there is even a generally accepted range, but targeting a specific value during design is no guarantee of acoustical quality. Similarly, the ISO standards for EDT, G, and LF have emphasized these metrics as objective goals, but we must wonder about of their importance as design tools. It is no coincidence that this analogy was formulated during the design of Suntory Hall, owned by the distillery of the same name.

While it is useful to calculate reverberation time in order to make some specific design decisions about interior materials, there is no rigid target. Instead of choosing a metric as a target, we aim to achieve a balance between clarity, richness, and intimacy.

Y. Toyota et al., *Concert Halls by Nagata Acoustics*, https://doi.org/10.1007/978-3-030-42450-3_37

Clarity, Richness, and Intimacy

The role of the concert space must be to faithfully transmit the music from stage to the audience with complete transparency. Each note and nuance is important and cannot be lost en route; simultaneous notes sounded by different instruments across the stage, or rapid successions played by a single section should be clearly identifiable. In the age of easily accessible recordings, it is now common to hear perfectly mixed, perfectly clear music from headphones only millimeters away from our ears. A concert hall is expected to match the clarity of recordings to offer a fulfilling experience. There is no single nor simple metric which can encapsulate clarity, although there is general consensus that early reflections are needed to create a sound which would, otherwise, be dull and lifeless.

Richness is the sense of sound energy enveloping the listener and filling the space, most closely related to reverberation. Increasing this reverberant sound energy in the lower frequency bands also improves the sense of "warmth," which is desirable. While an overly reverberant sound field can become muddy, the qualities of richness and clarity are by no means mutually independent. Instead, with the proper balance, the two can flourish in tandem.

Intimacy is the most difficult to quantify, as the acoustical intimacy is closely related to the sense of visual or architectural intimacy. The opportunity to commune with the musicians and other audience members remains the biggest benefit of live concerts. Reducing the distance to the stage and arranging the audience such that they surround the orchestra and face each other improves the intimacy. This may be the strongest point of the surround style layout: the unification of the audience and performers.

6. Computer Simulation

Once the goals of richness, clarity, and intimacy are established, we must find a way to reliably achieve them in a wide variety of designs. The advent of computer modeling allowed for an explosion in the diversity and complexity of architectural designs that need to be studied for acoustical feasibility. Since we inhabit a three-dimensional space, it is vital that room acoustical studies must also be conducted in three dimensions. This is especially necessary for vineyard style halls where the layout is rarely regular or even rectilinear, thus making simple sections futile. To prevent discrepancies in moving from 2D to 3D, architects must be prepared to provide a virtual model of the hall shape. Using these architectural models directly enables quick iteration as the shape of the hall interior evolves.

Our proprietary simulation software began development in the mid- to late-1980s. Based on the 1968 paper of Krokstad, Strøm, and Sørsdal, the ray tracing method was chosen over the image method. In our early simulations, highly simplified models allowed study of the general hall dimensions and required several hours of manually entering geometry followed by overnight computation on workstation computers. Historically successful halls were simulated with the software to create a foundational understanding of how to read the output.

By the mid-1990s, the market-wide availability of 32-bit computing allowed for these simulations to be carried out on desktop personal computers with reasonable computation times, and for increasingly complex geometries. Notably, the underlying acoustical premise of calculating time structure and spatial distribution of early reflections is robust enough that the analysis has changed very little in the past 30 years.

Modern computers enable us to run these models in minutes or less using the original 3D geometry provided by the architects. This quick prototyping allows us to test architectural innovations for viability, internally identify and solve problems, then iterate back and forth with the architect to sculpt the design. This flexibility frees us from rigidly defining major aspects of the room shape before the architect has begun. Rather than looking at a specific set of parameters or ratios, the software broadly and indiscriminately simulates whatever geometry it receives.

Y. Toyota et al., *Concert Halls by Nagata Acoustics*, https://doi.org/10.1007/978-3-030-42450-3_38

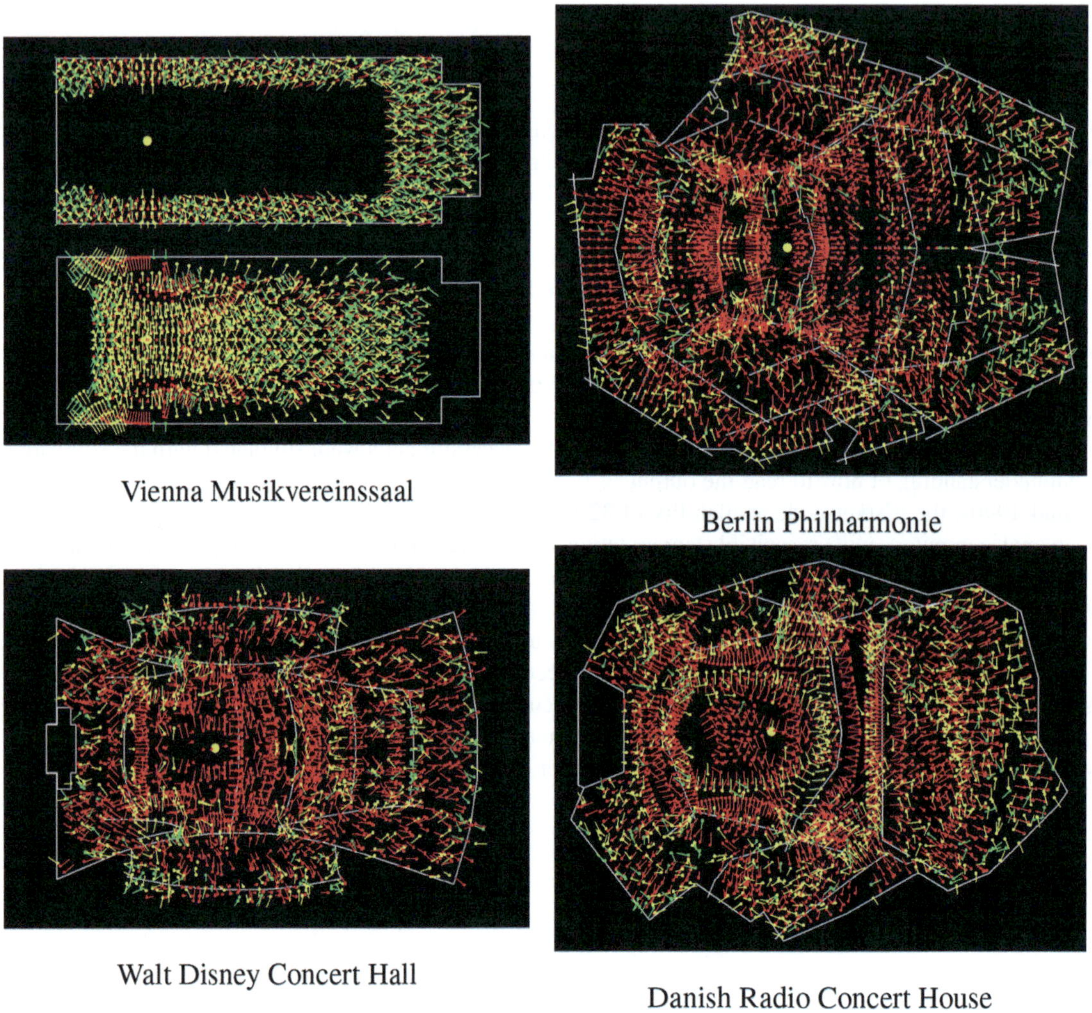

Fig. 6.1 Simulation output for the [0,100) ms time window

7. Scale Model Testing

Physical scale model testing is complementary to the computer simulations. Early in design, it is necessary to quickly and easily explore a broad design space. Virtual simulations are ideal here due to their flexibility and short turn around. As the design becomes more refined, the room shape must be inspected to ensure that there are no acoustical defects. Here, the stakes are high, and physical effects such as diffraction and room modes are difficult to detect in computer simulation.

Building an acoustical scale model allows the acoustician to take measurements using physical sound waves which accurately include these effects. For large or complex projects, 1:10 is the most appropriate scale since a smaller model will be less precise. In such a model, distances are scaled down and frequencies are scaled up correspondingly by ten times. The model is flooded with nitrogen gas to flush out oxygen, creating a closer analogue to air absorption at the original frequencies, especially upwards of 5 kHz (500 Hz at 1:1 scale). Each audience member is represented by a foam doll covered in felt to properly account for the distribution of audience as both an absorptive and diffractive surface. Solid surfaces must be extremely smooth since any imperfection would represent a large divot in the actual building. The size of the model permits it to be entered in order to measure distances, relocate acoustic absorption, and investigate possible changes to geometry. The large scale also allows for recreating and assessing intricate design features.

In our first 1:10 model, Suntory Hall, a specialized tape recorder was used to slow the recording back to the full-sized scale. Unfortunately, the quality was poorer than expected, and we have used digital processing ever since. Miniature speakers play a swept signal and a binaural, dummy-head microphone is used to measure the response at various points throughout the audience before digital processing.

After processing, the binaural recordings are auditioned at 1:1 scale. Here we can evaluate whether there are any detrimental echoes, flutter echoes, or other detrimental acoustic phenomena. Largely, these are phenomena which are clearly audible, but would be difficult to capture objectively through any metric. Ultimately, our ears are our most powerful tools, and it is beneficial to avoid the assumptions which would need to be made for computational analysis. While it is also possible to measure acoustical metrics and to convolve anechoically recorded music with the impulse response, the results are not particularly useful as design tools.

It is difficult to make changes to such models due to their size and expense. For this reason, the timing of the model tests should be conducted after Schematic Design but during the middle of Design Development. Hopefully, only minor changes to the room shape or interior material would be necessary. A smaller scale model such as 1:20 or 1:24 can be used for smaller halls or to meet budget constraints. These models have reduced accuracy, but are a better alternative than no model at all.

A scale model is useful for the architectural design. Certainly, the model is instantly understandable and inspires confidence as a representation of both the architectural and acoustical design. Complex geometry and lighting design are greatly helped by a physical model which can be used for inspection and trials. In many cases, the model is also used for publicity after it has completed its technical purpose.

Y. Toyota et al., *Concert Halls by Nagata Acoustics*, https://doi.org/10.1007/978-3-030-42450-3_39

Hall	Model	Building
Suntory Hall 1:10 Scale February 1984–February 1985		
Kyoto Concert Hall 1:10 Scale August–December 1992		
Sapporo Concert Hall 1:10 Scale August 1994–February 1995		
Walt Disney Concert Hall 1:10 Scale May–October 1993		
Muza Kawasaki Symphony Hall 1:10 Scale January–March 2001		

Hall	Model	Building
Danish Radio Concert House 1:10 Scale January–April 2004		
New World Symphony 1:24 Scale December 2006–April 2007	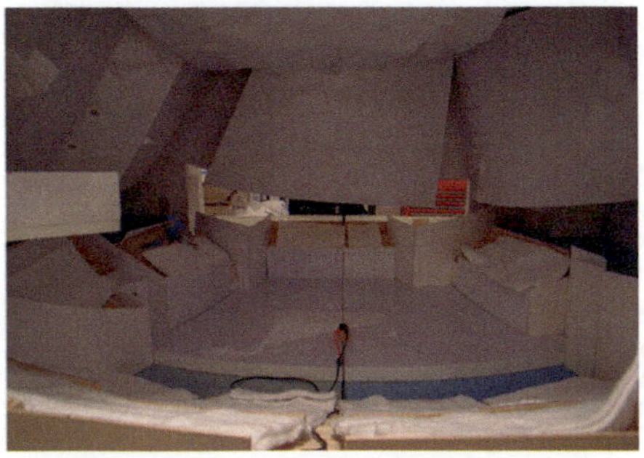	
Helsinki Music Centre 1:10 Scale October 2006–July 2007		
KCPA-Helzberg Hall 1:10 Scale January–July 2007		
Isabella Stewart Gardner Museum 1:20 Scale June–July 2008	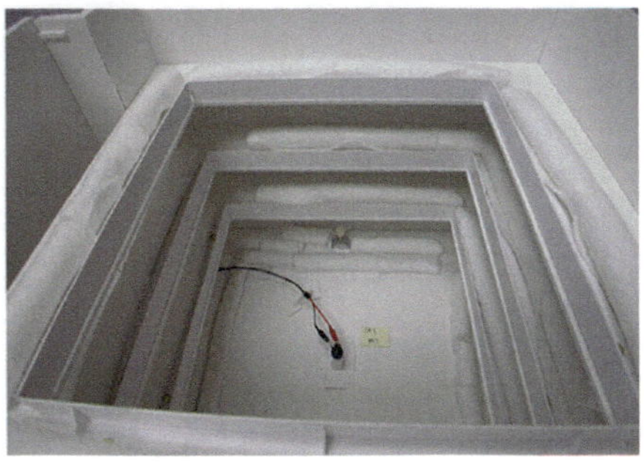	

Hall	Model	Building
Bing Concert Hall 1:24 Scale April–September 2009	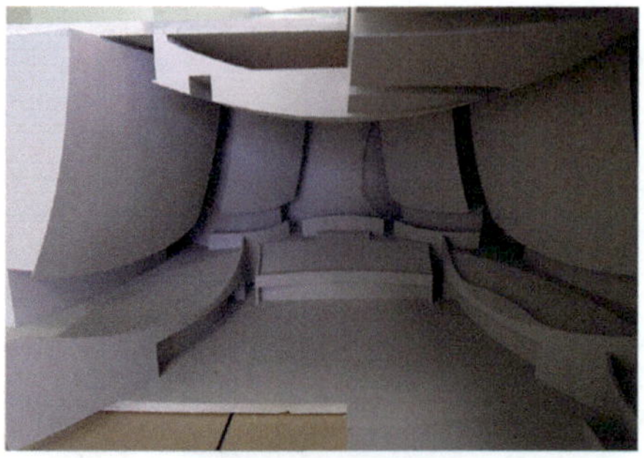	
Shanghai Symphony Hall 1:10 Scale October 2010		
NOSPR Katowice 1:10 Scale October–December 2010		
Fondation Louis Vuitton 1:20 Scale February 2009		
Radio France Concert Hall 1:10 Scale October 2006–September 2010		

Hall	Model	Building
Philharmonie de Paris 1:10 Scale November 2008–May 2009		
Lotte Concert Hall 1:10 Scale October 2012–February 2013		
Elbphilharmonie Hamburg 1:10 Scale July–October 2007		
La Seine Musicale 1:20 Scale February–June 2014	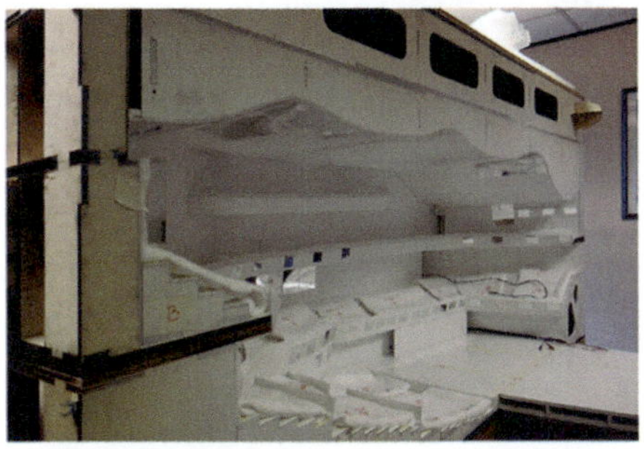	
Zaryadye Concert Hall 1:10 Scale March–September 2017		

8. The Importance of Stage Acoustics

While it is ultimately the audience who will enjoy the concert, we must pay special attention to the acoustical condition on stage. Even an astute concert-goer may never get to experience what a musician hears on stage, and therefore may not even be aware of the huge amount of design effort given to perfecting the stage acoustics. While the hall is designed to convey the music to the audience, that music is being created on the stage. With poor stage acoustics, the quality of ensemble will suffer, and even excellent hall acoustics cannot improve poor music-making.

The audience is only in the hall for a few hours at a time, rarely more than once per month. In contrast, it is the musicians who will spend the most time in a concert hall, particularly on stage, and are therefore most directly affected by difficult conditions. The performers—and especially the conductor—are in the strongest position to make their voices heard. Furthermore, there are factors such as rehearsal experience in a space which cannot be manufactured and are fostered over time.

Amsterdam Concertgebouw and Suntory Hall

The Amsterdam Concertgebouw is usually counted as one of the top three concert halls in the world along with the Vienna Musikvereinssaal and Boston Symphony Hall. However, I have heard many touring orchestras have negative comments regarding the stage acoustics. Even Bernard Haitink, the principal conductor for almost thirty years from 1961 to 1988, felt that the stage acoustics were complicated and not easy to handle. In fairness, there are some touring orchestras who give positive accounts, although this could easily be ascribed to the overall positive reputation, the flexibility and experience of touring orchestras in adapting to new spaces, or the fact that since Amsterdam is a regular stop for many tours, many musicians are already familiar with the hall.

I would assume the acoustical condition on stage is somewhat similar to Suntory Hall, which was challenging for the local orchestras to become accustomed to. The width and ceiling height of Suntory Hall is comparable to the Concertgebouw and neither have balconies around the stage that provide early supporting reflections. Without these reflections, the musicians must expend a concerted effort to hear each other and produce a clean, balanced, and beautiful sound. Counterintuitively, without some adversity on stage, an orchestra may become complacent and begin to stagnate. Relying too heavily on this feedback loop might be dangerous, however, since the effects would take time to develop and will depend on the leadership and musicians themselves.

We can divide the quality of the stage acoustics into three realms. First, there are the halls in which musicians cannot hear themselves or each other regardless of the quality of the ensemble. This situation is unacceptable, and detrimental to the ensemble quality over time. On the opposite side of the spectrum are stage acoustics which are so supportive that musicians can hear themselves and each other with little effort, offering a clear picture to the musicians, conductor, and music director and the opportunity to make fine adjustments. However, there is little obvious incentive or direction to improve their ensemble. In the middle is the condition we find in the Amsterdam Concertgebouw and Suntory Hall. The musicians can hear themselves and each other only when the ensemble is good. By leaving the quality somewhat exposed, the ensemble must continually listen and improve, and in turn, the orchestra will flourish.

Y. Toyota et al., *Concert Halls by Nagata Acoustics*, https://doi.org/10.1007/978-3-030-42450-3_40

9. Ceiling Height, Ensemble Reflectors, and Soffits

If the sound source is assumed to be omnidirectional on the stage, much of their sound energy is radiated upwards. It is then no surprise that the most important reflecting surface in a concert hall is the ceiling, especially the ceiling just above the stage.

During the design of Suntory Hall, there was no clear design direction, so once again, we borrowed from the design of the Berlin Philharmonie. The result is an ensemble reflector made up of nine plexiglass panels suspended above the stage. These panels are pentagonal with bounding dimensions of 3 × 1.6 m, and are 12 mm thick.

Within the first 6 months after opening Suntory Hall, there were several trials for the position the plexiglass reflectors above stage. Originally, the building management was cautious to lower the reflectors to avoid obscuring the views to the pipe organ. Since the impression of stage acoustics is very dependent on the performers and repertoire, it would be nearly impossible to compare between different concerts. Luckily, the English Chamber Orchestra featuring pianist Mitsuko Uchida playing and conducting a series of Mozart Piano Concerti provided the opportunity to keep both the musicians and program fixed, changing only the hall geometry. The reflectors started at the highest position and were gradually lowered as the series progressed. The lowest position was preferred and the reflectors have remained in that condition ever since, approximately 12 m above the stage floor.

During the design of Walt Disney Concert Hall, Frank Gehry wanted to avoid an ensemble reflector, and had the support of Los Angeles Philharmonic managing director Ernest Fleischmann. Gehry wanted to keep the impression of open space and even clear reflectors like Suntory or Berlin would affect the visual impression of the volume. Fleischmann was concerned that any adjustable element would be a compromise that could detract from the singular intention for the space to be used for unamplified, classical music. At this point, the challenge was that there could be no trials in the finished space; the ceiling above stage needed to be fixed during design. Our survey of ceiling heights in existing halls found that, regardless of typology, the ceiling in exceptional halls were all approximately 16 m above the stage as in Table 9.1. This became the ceiling height in Disney and values in the range of 15–16 m above stage have led to good results for both stage and audience acoustics in many of the halls presented here, large and small. This also results in a simple dichotomy in which an architect can choose a continuous ceiling like in Disney or Mariinsky Concert Hall in St. Petersburg or a lofted ceiling with an added ensemble reflector such as in Helzberg Hall in Kansas City or the Hamburg Elbphilharmonie. A heavy, monolithic structure is preferable to discrete panels floating in space so that it can reflect sound from across all frequency bands.

The ceiling above the stage—including the optional ensemble reflector—is very effective at providing reflections between musicians and to the audience seated closest to the stage. We can think of this surface as important for creating good ensemble. Musicians also need to be able to hear themselves; in other words, they need acoustic reflections that come directly back to the source.

In the same way that we can always see ourselves in a corner mirror regardless of the viewing angle, including 90° soffits around the stage can provide acoustic reflections back to each individual musician: the source and receiver are coincident and could be at any arbitrary point on stage. In the design of soffits, the distance to stage is very sensitive since they must be far enough to provide adequate coverage of the stage, but not so far as to create strong, distracting echoes. Since the reflections do not actually bounce directly at the cusp of the corner, there would also need to be some effective wall exposed which must be incorporated into the architectural design. There are a multitude of architectural directions which can satisfy this type of geometry.

Y. Toyota et al., *Concert Halls by Nagata Acoustics*, https://doi.org/10.1007/978-3-030-42450-3_41

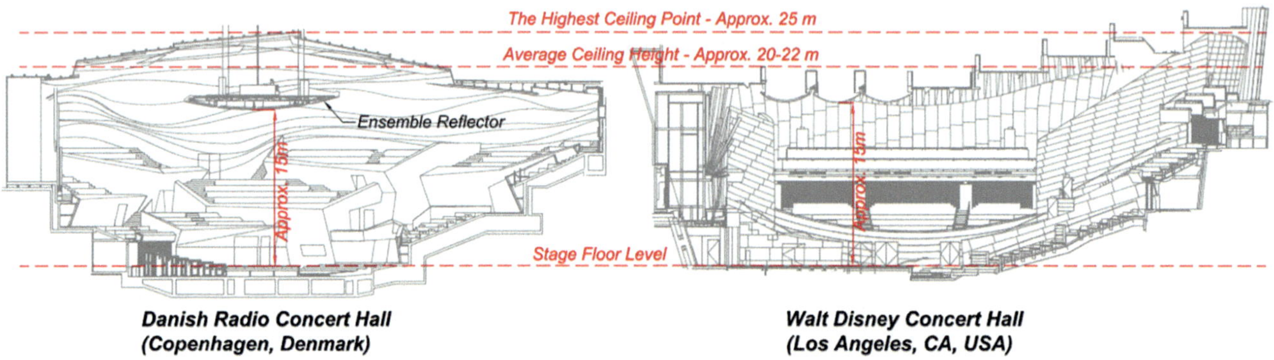

Fig. 9.1 Comparable ceiling heights with and without an ensemble reflector

Table 9.1 Ceiling height above stage

Hall	Ceiling height above stage (m)	Delay time for ceiling reflections (ms)	Suspended ensemble reflectors
Boston Symphony Hall	12.5	67	No
Vienna Grosser Musikvereinssaal	16	87	No
Amsterdam Concertgebouw	15.5	85	No
Leipzig Gewandhaus	16	87	No
Munich Philharmonie am Gasteig	19	105	No†
Tokyo Bunka Kaikan	11	61	No
Kumamoto Prefectural Theater	14	76	No
Berlin Philharmonie	19	105	Yes
Suntory Hall	17.5	96	Yes

†Added after opening

Balcony Soffit. Walt Disney Concert Hall

Ledge Soffit. Sapporo Concert Hall "Kitara"

Shelf Soffit. Musco Center for the Arts at Chapman University, Orchestra Shell

Downstand Soffit. Pierre Boulez Saal

10. Acclimation to a New Hall

I am often asked if I am nervous on opening night. While certainly exciting, we are already quite familiar with the acoustics of the hall by the time we reach opening night. For us, the most nervous moment is the first rehearsal. The hall is new to everyone: to the conductor and musicians, managers, and to us as acousticians.

The first rehearsal for Walt Disney Concert Hall was on June 30, 2003, coincidentally the 45th birthday of conductor Esa-Pekka Salonen. After a few hours of rehearsal, I had some discussion with the concertmaster, principal clarinet, and other section leaders. They were concerned with the conditions in the new hall and reported that the stage acoustics were problematic and that they found it difficult to hear themselves and each other. If this had been our first hall, I would have panicked at this point. However, this is a very normal occurrence. At the first rehearsal, the musicians may be nervous, causing them to play louder, and forcing everyone else to play louder to hear their own sound. This further deteriorates the ability for others to hear their own sound, potentially devolving into chaos. Instead, I recommended that everyone played softer and focused on listening.

Two weeks later at the second rehearsal, I met again with the section leaders who now reported that they were much more able to hear each other. They wondered what had I changed. Of course nothing had changed, there was only more time to acclimate to hearing each other in the new hall. While it is possible for conditions to change if the tail end of construction overlaps with rehearsals, the changes would be small. More likely is that the musicians are becoming more and more comfortable in the space.

I think of this process in the same way as moving in to a new house. It takes a few days to discover which way the doors open and which lights correspond to which switches. This does not mean that there is some inherent flaw in the house, only a lack of familiarity. In a concert hall, the "family" moving in can be well over a hundred people, so the effect is magnified. For this reason, we suggest that the first notes in the hall are from a soloist or small ensemble before working up to a full orchestra. The situation will improve most dramatically in the early days and weeks, but is still noticeable after two to three years depending on the relationship between the orchestra and the hall.

Even for those we now consider the best halls, there would have been some moving-in period. Vienna Musikvereinssaal and Berlin Philharmonie both have world-class orchestras, but the stage acoustics in the Musikvereinssaal are very different from the Philharmonie. Even these top-notch musicians prefer their home hall over the other, but there is no way to say which pairing of orchestra and concert hall are best. Familiarity developed through time is one of the most essential components of stage acoustics. The importance of hours of rehearsals in a concert venue cannot be overstated.

Over the course of years, many—especially musicians—may feel that the hall acoustics have changed. I believe this long-term effect is mostly tied to unreliable memory. The acclimation of the musicians to the new hall gives the effect that the hall has somehow "matured" since their last tour. For example, Daniel Barenboim reported to me that he was surprised by the improvement in the acoustics in Suntory Hall since his first visit with Orchestre de Paris soon after the opening. I have no doubt that his impression was different, but there is no indication that the room itself had physically changed.

To this end, I have no personal evidence nor have I read any research which would imply that some physical change is causing the hall to change its acoustical characteristics. Musicians report that their instruments change, and that the wood and varnish become increasingly resonant with use over the years, making it logical to assume that a hall would similarly mature as well. In some ways, a building does act like a big instrument where some components such as the stage wood may physically mature. However, the plaster surfaces and concrete structure which make up most of the reflecting surfaces in a hall have no mechanism or benefit to age.

Fortunately, the changes reported are almost always positive.

Y. Toyota et al., *Concert Halls by Nagata Acoustics*, https://doi.org/10.1007/978-3-030-42450-3_42

11. Orchestral Balance

Two experiences fairly early in my career heavily influenced my concept of the ideal orchestra balance.

Dresden Staatskapelle

On October 16, 1973, the Dresden Staatskapelle kicked off their Japan tour by visiting Tokyo Bunka Kaikan under the baton of Kurt Sanderling. Their program consisted of Wagner's Overture to *Die Meistersinger von Nürnberg* and Brahms's Symphony No. 1—both of which I was intimately familiar with from my time as an oboist in the Kyushu Institute of Design student orchestra—as well as Beethoven's Symphony No. 8. From the first note, I was shocked at the wonderful balance between strings and brass. Normally, the trumpets dominate the C-Major chord which starts the Overture, but here, the strings were emphasized. This orchestral sound where the strings form the foundation of the ensemble on which other instruments are added became my reference point moving forward.

Sapporo Symphony

One of my first projects at Nagata Acoustics was a multipurpose auditorium for the city of Chitose, about 50 km south-east of Sapporo. The hall featured a heavy, monolithic orchestra shell which made the space well suited to orchestral performances. In 1985, several years after the completion, I received a call out of the blue from one of the house technical staff saying that the Sapporo Symphony Orchestra under Hiroyuki Iwaki was recording the soundtrack to Akira Kurosawa's film *Ran* by Toru Takemitsu in the hall. He wanted to pass on his excitement that the orchestra and recording crew were amazed at the acoustics in the hall and frustrated that they could not capture the beauty on tape. Kurosawa was reportedly frustrated initially that Takemitsu had not chosen a more internationally recognized ensemble, but he was extremely happy with the final recording.

Two years later, Iwaki brought the Sapporo Symphony Orchestra to Suntory Hall when he won the Suntory Music Award. I heard the rehearsal for Brahms's Symphony No. 2 and Takemitsu's Requiem for Strings and was amazed at the quality of the ensemble. The balance and clarity was simply perfect. Like the Dresden Staatskapelle, the strings formed the foundation of the sonic pyramid, supporting the rest of the orchestra, and never overpowered by the trumpets and brass. This was the balance necessary to hear every instrument as the composer intended.

With the opening of their own dedicated hall in downtown Sapporo in 1997, there was renewed interest in the orchestra, but the balance had changed and somehow degraded. Several veteran musicians confirmed that this downturn was not only a change in my taste or some rosy, retrospective illusion. This led me to wonder how the orchestra had come to peak in the previous decade. As best I have been able to piece together, there are three reasons which led to the orchestra's wonderful sound during the eighties.

First was the appointment of Peter Schwarz as the music director in 1967. Schwarz was previously the principal cellist for the Bamberg Symphony Orchestra, but had a keen interest in becoming a conductor. During his tenure, he made Sapporo his home and focused on building a strong relationship with the orchestra. Many of the musicians in the orchestra were young and receptive, so they absorbed much of Schwarz's experience, including his roots in the rich German string sound. This caused the string section in particular to mature beyond any local orchestra.

Y. Toyota et al., *Concert Halls by Nagata Acoustics*, https://doi.org/10.1007/978-3-030-42450-3_43

Next was the dissolution of the Japan Philharmonic Orchestra in 1972. As the musicians dispersed, many strong wind players joined the Sapporo Symphony Orchestra.

The third was in 1975 when Mineo Sugiki, formerly at the Orchestre National de Lyon (France), became the principal trumpeter and effective leader of the brass section. Sugiki was conscientious of his volume and strove to achieve good balance with the strings. As their leader, none of the other brass could exceed his volume, so he effectively kept the entire section in check, never overwhelming the rest of the orchestra.

With the three main sections of the orchestra each improving under their own influences, Iwaki had the foundation to develop the orchestra when he became music director in 1975, leading to the sound that impressed Kurosawa, and later earned him the Suntory Music Award in 1987.

It is important to remember that I am not a conductor, and creating good orchestral balance is certainly a case of "easier said than done," but I have listened to a great many concerts and rehearsals and can summarize my experience as follows: ideally, all instruments in the orchestra would be heard at all times. As acousticians, we can affect the physical space of the concert, for example, boosting the level of strings through the use of a resonant stage and risers as discussed in later chapters, but ultimately, the responsibility rests with the musicians, especially the conductor and the principal musicians. Among them, the principal trumpet is extremely important.

In many orchestras, especially American ensembles, the brass section has a tendency to be excessively loud, masking the strings for the audience. From the podium, surrounded by strings, the conductor stands in a difficult position to gauge the dynamic of the brass. Compensating for this "stage effect" is challenging and requires that the conductor listen carefully to their assistants sitting in the audience during rehearsal. A compact layout on stage reduces the distance between brass and conductor, which may partly explain why orchestras which sit tightly together have better ensemble quality overall.

Additionally, the principal trumpet must have an extraordinary sense of intonation. If they drift out of tune, woodwinds and brass tend to be sharp while strings tend to be flat. Due to their prominence, if the trumpet is out of tune, some players may try to compensate while others may not, and the resulting uneven compensation will lead to a dissonant mess.

Earplugs or panels are sometimes used on stage to protect the ears of the musicians sitting directly in front of the brass section. This is quite understandable since their hearing is their livelihood. However, I strongly discourage the use of reflectors on stage. These panels are detrimental to the ensemble quality and the quality of sound overall. Earplugs would be a better solution, but the best case is for the trumpet to monitor their own volume. The trumpet can never be too soft; either there is good balance or the trumpet is too loud.

12. Orchestra Layout

The layout of the orchestra on stage has a substantial impact on the quality of the ensemble. There are two fundamental orchestra layouts: the classical and American.

In the classical layout, the violin section is split, with the 1st violins on stage right, followed by cello and double bass, viola, and finally the 2nd violins on stage left. In the American layout, however, the strings are arranged 1st violin, 2nd violin, viola, and cello and double bass moving from stage right to left. Quite often, the viola is swapped with the cello and double bass in the American layout. This layout was supposed to be beneficial to recording with clear separation of high and low registers.

In the American layout, the violin section acts as a giant plane source which can cause their sound power to be overwhelming while in the classical layout, the split first and second violins reduce their effective power. In addition, the woodwinds are better poised to hear the first violins in the classical layout as they are not blocked by the second violins.

For small to medium sized orchestras, these are clear advantages over the American style. Problems arise for large ensemble sizes, however, as the stage right becomes too crowded with the first violins, cellos, and sometimes double basses, leaving no good position for the horns and pushing the woodwinds off center.

A possible solution would be to modify the classical layout by switching the violas with the cellos and double basses. This solves the problems of crowding while still splitting the violin section. Several orchestras have tested this layout in concert, notably the Bamberg Symphony Orchestra under Jakub Hrůša, and sometimes by the Berlin Philharmonic Orchestra under Simon Rattle and the Mariinsky Orchestra under Valery Gergiev. Like with any other major change to playing conditions of the orchestra, changing the layout often is challenging for the musicians as it takes time to grow accustomed to a new layout. There will understandably be some period of transition during which the ensemble may suffer before it improves as well as negative feedback from the players themselves.

Whether in the classical, American, or alternative layouts, the woodwinds should be seated close to the conductor. They should be on the first stage riser with only two rows of strings between the flutes and the conductor. The woodwinds and strings which sit on the flat stage form the core of the orchestra and should not be moved for any changes in repertoire, from the classical ensembles of Haydn or Mozart to the large modern orchestras used by Mahler, Stravinsky, Shostakovich, etc.

Y. Toyota et al., *Concert Halls by Nagata Acoustics*, https://doi.org/10.1007/978-3-030-42450-3_44

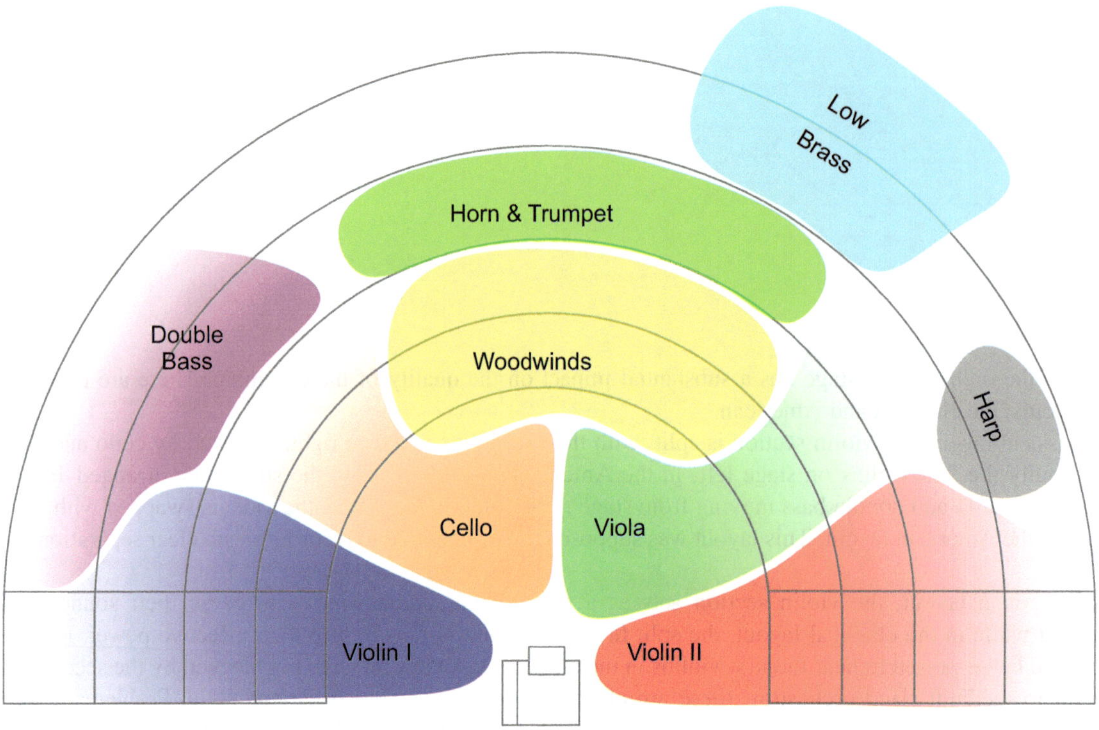

Fig. 12.1 Classical layout

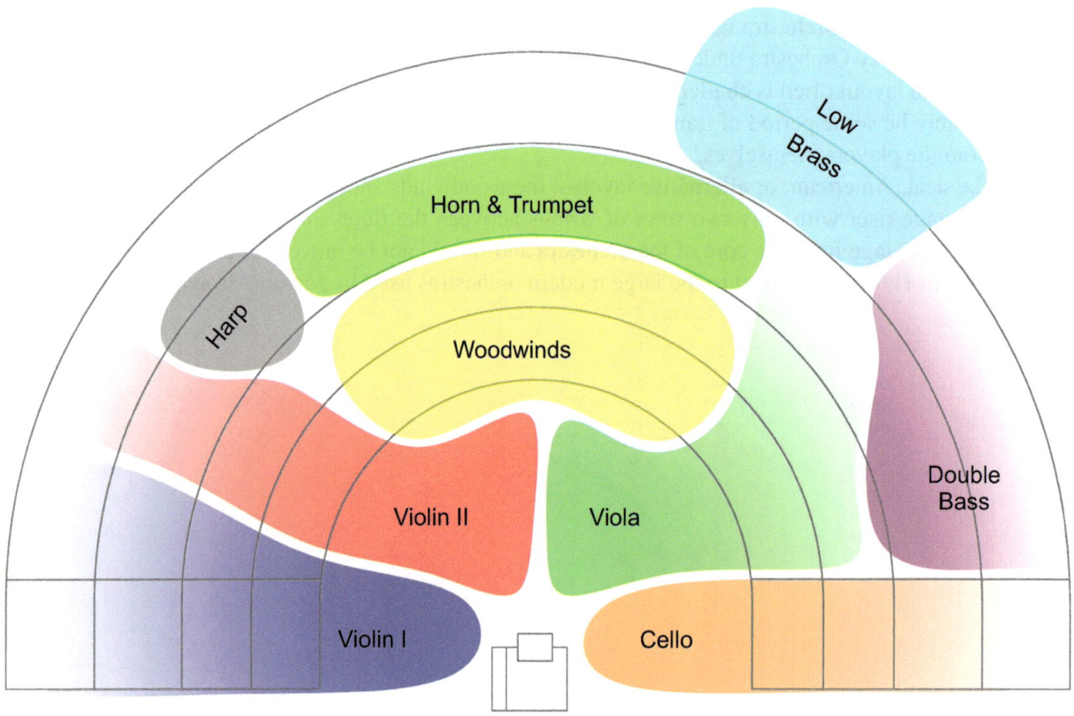

Fig. 12.2 American layout, variant 1

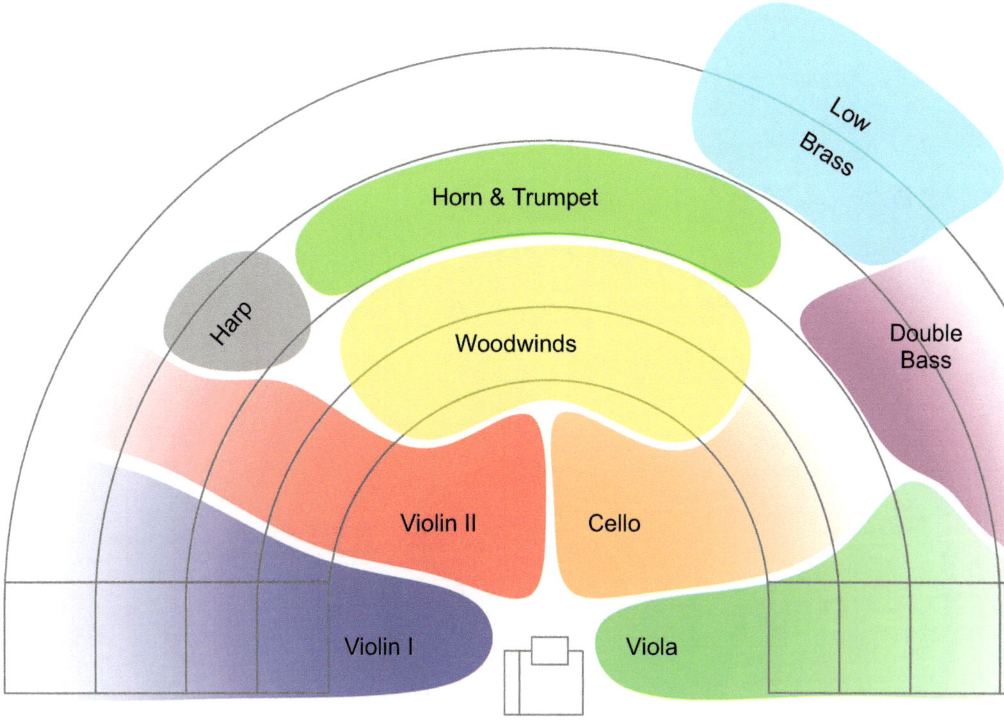

Fig. 12.3 American layout, variant 2

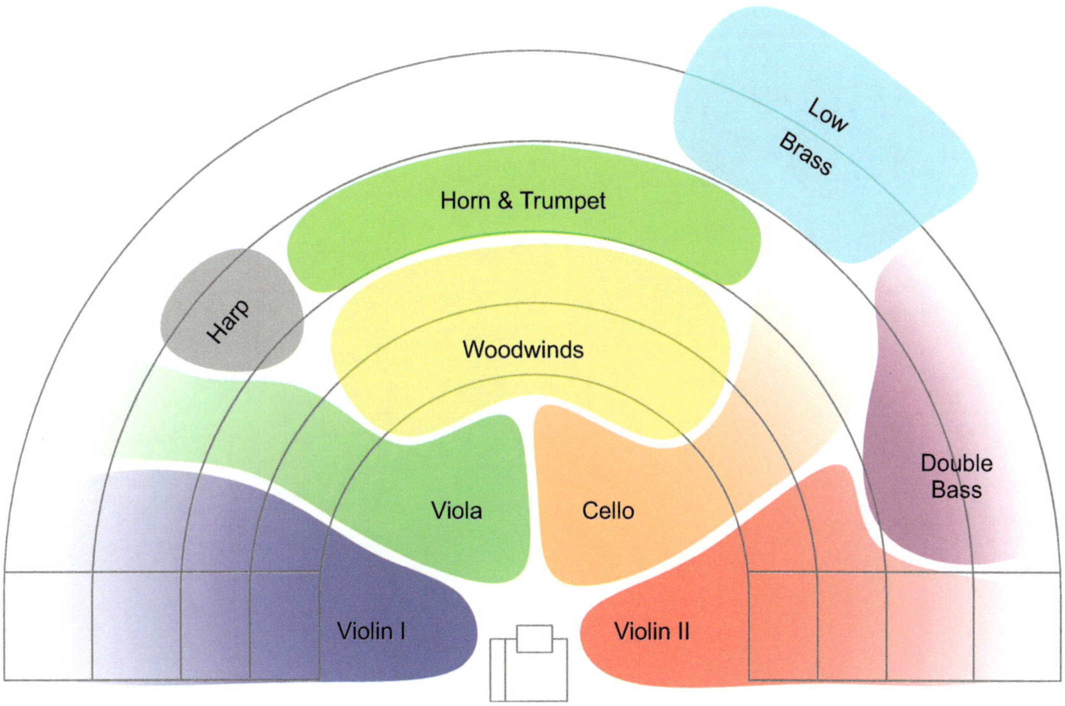

Fig. 12.4 Suggested layout

13. Excessive Sound Exposure on Stage

Placing a large number of instruments on stage will inherently result in a very loud condition on stage. The situation is very critical for a profession which depends so heavily on hearing. As musicians' groups evaluate the working conditions of being subject to this loud sound level for many hours, there is some question as to how the acoustician can improve the situation. Unfortunately, this is an issue that is mostly independent of the room acoustics. While we do put considerable effort into reflecting sound back to stage to provide acoustic support to the musicians, these do not affect the sound level on stage nearly as much as the direct sound of the instruments.

Directional brass instruments, like trumpet and trombone, and percussion are especially problematic to the musicians immediately around them, often woodwinds and sometimes strings.

Free standing, on-stage reflectors or wrap around head pieces may seem like suitable solutions to protect the musicians' ears directly in front of these instruments. While they may shield the blow of strong impulses like the hammer in Mahler's Symphony No. 6, they are not changing the overall sound saturation on stage. If anything, they reduce clarity which may even encourage the musicians to play louder.

Ear plugs are effective at greatly reducing the sound level, but interfere with the production of music. Musicians must hear themselves and each other in order to create a well-balanced ensemble. This would be impossible with ear plug which both block and distort incoming sounds.

The loudness is really a function of the composition and the choices of the conductor. They could choose to scale back the volume of every dynamic level, although one could argue that this is not the composer's wish. Organizing the stage layout to reduce the distance from the conductor to the brass may be helpful. Due to their close proximity, string sections drown out the brass, causing the conductor to incorrectly believe that the brass needs to be louder. During some rehearsals of the Swedish Radio Symphony Orchestra a few years back, Daniel Harding invited me to the podium. The orchestral balance was starkly different from what I heard in the audience. Not only were the relative levels of the various sections different but, at such close range, I heard the instruments much more like individual sound sources, rather than as a single large source.

Finally, one should not expect nor desire that an orchestral concert be as loud as a heavily amplified pop concert. As the audience, we have the responsibility to report uncomfortably loud concerts to the hall management or artists to remind them that "louder" does not mean "better," and comes as the detriment of quality and as a risk to the health of performers.

Y. Toyota et al., *Concert Halls by Nagata Acoustics*, https://doi.org/10.1007/978-3-030-42450-3_45

14. Stage Risers

While orchestra risers are present in historical halls such as in the Vienna Musikvereinssaal, our use of the deeply arched, semicircular risers used in modern halls was derived from the Berlin Philharmonie, where their original purpose was not acoustic, but visual. Herbert von Karajan produced and directed films of orchestra performances and lifted the instruments at the rear of the stage in order to better see them and to fill the frame. The incidental acoustical improvement was noticeable. In particular, sound from the strings and woodwinds is better able to project, improving their balance with the brass section. The vertical faces of the risers act as an additional reflecting surface to project sound outward towards the audience. Since the brass section sits on the last platform with no riser behind them, this vertical face is absent, further helping to boost the strings and woodwinds in comparison. While we have not conducted rigorous studies, a rise of more than 25 cm per step is appropriate.

Placing the orchestra on risers causes a noticeable improvement in the ensemble quality. The instruments are more equally emphasized, resulting in greater clarity. Since the musicians may feel more exposed, there would be more incentive to strive for excellent balance and musicality.

Automation of the risers is well worth the added expense. Time is exceptionally valuable during rehearsal, and the time to change between different repertoire must be reduced as much as possible. As a bonus, the conductor has flexibility to change the orchestra layout for spontaneous experimentation. During concerts, mechanical automation makes it possible to adjust the orchestra layout and move large instruments during intermission.

Suntory Hall

Since Maestro von Karajan was a personal friend and unofficial consultant of the client, Keizō Saji, the risers were included without question for the design of Suntory Hall. As this was our first encounter with designing risers, we had no experience in the requirements of orchestra layouts and simply copied the layout from the Berlin Philharmonie. As the six local orchestras transitioned from Tokyo Bunka Kaikan to Suntory Hall, the initial reaction was quite negative and the local orchestras did not use the risers at first. Their habit was to arrange the woodwinds in a line as opposed to an arc, so there were even a few cases where the stage was flattened and manual, straight risers were placed on stage.

In October of 1990, the Munich Philharmonic under Sergiu Celibidache performed a series of three Bruckner symphonies at Suntory Hall and made full use of the stage riser system. The result was shocking! Both the balance within the ensemble, and between low and high registers were much better than the Tokyo orchestras. The musicians reported that they were happy with Suntory Hall, especially the stage acoustics. At the time I was unaware of their frustration with the Gasteig which had opened in 1987, only a year after Suntory Hall.

In 1989, the Japan Philharmonic Orchestra increased the number of concerts in their season and was the first to move their official home to Suntory Hall. Despite complaints from the musicians, there was some experimentation with the riser settings. For the tests, they selected Benjamin Britten's *Young Person's Guide to the Orchestra* since, by design, there is variety in the combination of instruments as well as solos. However, the results were not good, in fact worse than on the flat stage. Disappointed with the tests, the orchestra moved on to rehearse Brahms's Symphony No. 1 and all were surprised to hear the clear improvement when using the risers. Using a piece which is broadly played and comfortable is important to remove extraneous variables while performing any acoustical trials.

Still, the use of risers can be a difficult adjustment for the orchestras. Tokyo's six orchestras, several of which have regular subscription concerts at Suntory Hall, now use the risers more effectively, but often experiment with different configurations. There is no single ideal riser setting or configuration; it is highly dependent on the ensemble and its leaders.

Y. Toyota et al., *Concert Halls by Nagata Acoustics*, https://doi.org/10.1007/978-3-030-42450-3_46

Kyoto Concert Hall

During the design of Kyoto Concert Hall, the Kyoto Symphony tested a stage mockup in their rehearsal hall. The mockup was built from plywood and allowed some flexibility in modifications of the layout before the construction of the final stage. Including the musicians in the design conversation must have helped them adopt the idea as partly their own. The orchestra had several years to become accustomed to the design before the hall opening in 1995.

Like the Tokyo orchestras, the Kyoto Symphony was accustomed to arranging the woodwinds in a line. During the mockup test, the musicians were convinced to arc slightly, and the final result is curved, though not a true semicircle. Since the mockup was not mechanical, the riser height had to be fixed from the beginning. Even though 25 cm was already known from Suntory Hall as the target, 15 cm was chosen to help the orchestra adjust. After the hall was completed, the orchestra conducted tests to pick the default conditions for the stage and chose to stay with a 15 cm riser step height. There was simply too much inertia to overcome in order to change the default to 25 cm. Changes should be made all at once since each incremental change will face just as much resistance as jumping directly to the desired result.

Los Angeles Philharmonic

The Los Angeles Philharmonic constructed a similar mockup in Dorothy Chandler Pavilion to prepare for the opening of Walt Disney Concert Hall across the street. In this case, the orchestra management—namely Ernest Fleischmann—was headstrong enough to force the orchestra on to 25 cm tall risers. Compared to their European counterparts, the American orchestra was accustomed to roomier accommodations on stage so the downstage edge of the semicircular risers was extended by about half a meter to provide more area. Some extra depth was also added upstage to fit a larger percussion ensemble. The result was neither too much nor too little floor space for the orchestra, and this layout, with only minor modifications resulting in the layout in Figure 14.1, has become the starting point for full sized concert halls.

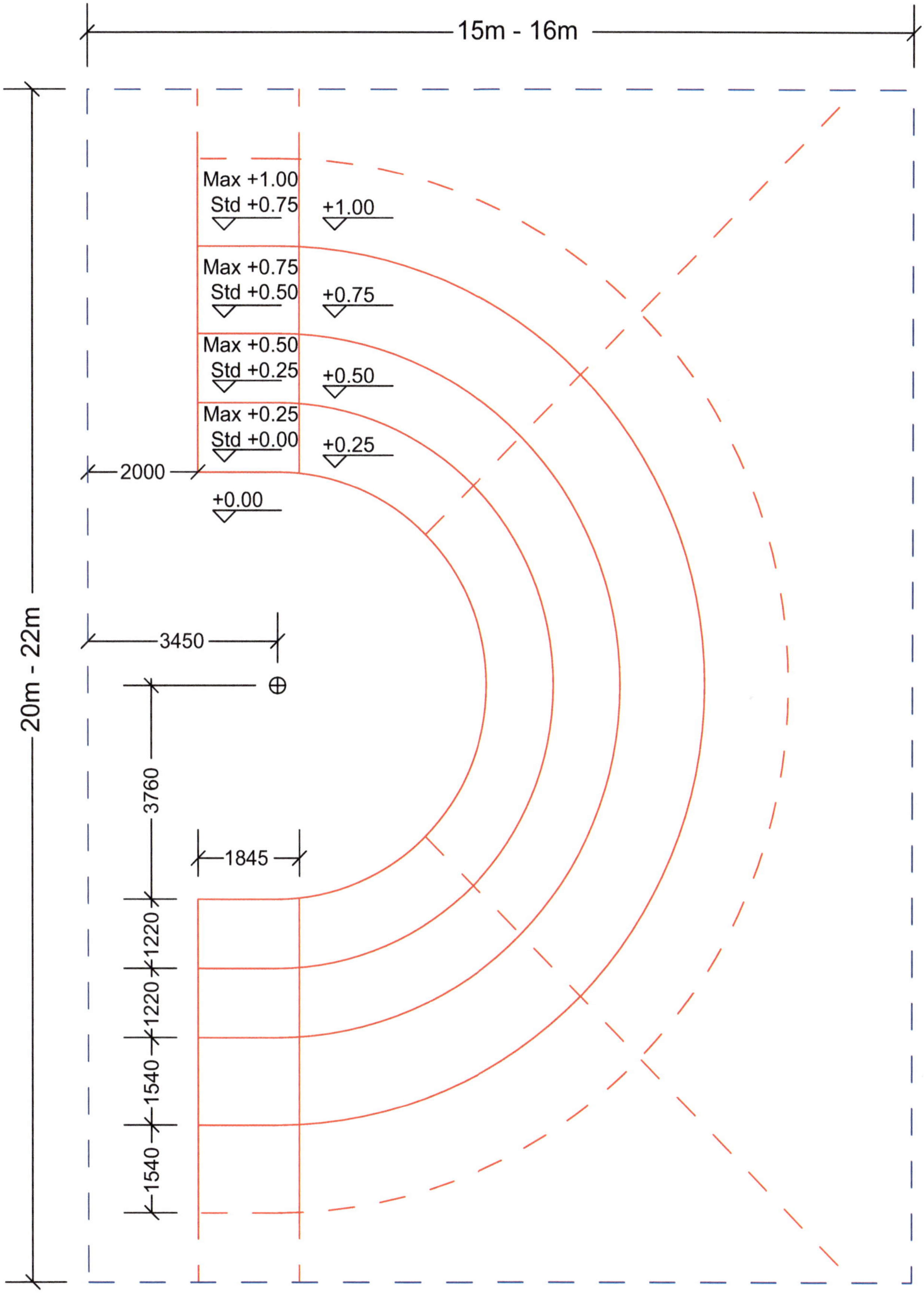

Fig. 14.1 Recommended riser dimensions

15. Stage Structure

The stage itself is a vital resonant body for the orchestra, in particular for instruments like cello and double bass which make a direct contact via their endpin. The analytical mathematics for the stage vibration would be prohibitively complex, so we are left relying on the trials and errors of previous experiences. In this way the development of the stage is much like the original development of the musical instruments themselves.

The first opportunity to test a variety of stage structures was in 1994 during the design of Kyoto Concert Hall. For the test, 14 podiums measuring 1.8×1.5 m were constructed with a variety of decking materials. A group of musicians from Kyoto Symphony and members of Nagata Acoustics evaluated comparisons of the podium as they listened to a selection of violin, cello, string bass, and timpani excerpts performed on each podium. The clear preference was a stage of 40 mm solid Hinoki wood supported by a system of sleepers and beams as shown in Figure 15.1. In the following years, the same preference was confirmed in similar tests with the New Japan Philharmonic Orchestra for Sumida Triphony Hall and with the Los Angeles Philharmonic for Walt Disney Concert Hall. Between the three listening tests, there was enough consensus to use this construction as the standard recommendation for all our subsequent projects.

Hinoki (*Chamaecyparis obtusa*) is a species of cypress that has been used for millennia as a high-quality building material in Japan. The tight grained wood is light and strong and making it a perfect material for temples, traditional Japanese bathtubs, and of course Noh and Kabuki stages. The association with professional performances is so close that the phrase *hinoki butai*—literally hinoki stage—is idiomatic for something like "main stage" or "center stage." Due to difficulties in obtaining hinoki outside of Japan, its cousin Alaskan Yellow Cedar (*Cupressus nootkatensis*), which is actually a cypress, is often used. Other members of the *Chamaecyparis* genus such as Port Orford Cedar (again a cypress) may be acceptable alternatives.

The sleepers and beams which support the decking must be arranged to enable the resonance of the stage. The sleepers can be thought to vibrate along with the decking and, if their spacing is too dense, they will increase the effective thickness of the vibrating membrane, encumbering the overall resonance. The beams are much sturdier and will become the nodal points for the vibration. Wide spacing is paramount or the stage will be too stiff. The material of the beams and sleepers is not as sensitive as the decking.

Y. Toyota et al., *Concert Halls by Nagata Acoustics*, https://doi.org/10.1007/978-3-030-42450-3_47

Stage structure testing with the Los Angeles Philharmonic

The location of instruments on the stage must also be considered when positioning the beams. They must be placed so that there is reduced chance of a cello or double bass endpin being inadvertently placed over a node (i.e., a beam), where they would receive little to no reinforcement from the stage. On the other hand, percussive instruments like timpani or the piano, which has its own soundboard, should be placed on nodes to maintain the desired percussive attack. For the piano, the leg under the bass keys would be the most important to locate over a beam. Next would be the point leg and finally the treble leg.

Since the beams run perpendicular to the sleepers which in turn run perpendicular to the decking, the orientation of the beams is fundamentally linked to the orientation of the decking. There is no real consensus as to the decking direction as seen in Table 15.1. Both have their advantages. At the front of the stage, decking—and therefore beams—running left-to-right would allow the piano to be easily centered for solo concerts. Decking running upstage-downstage would allow the piano to be easily moved out of the way of the podium in a concerto configuration. The seating arrangements of the cellos and string basses change with the orchestra, conductor, and repertoire making it difficult to predict the best orientation of the beams. This discussion is further complicated by the architect who must shape the visual impression, the theater consultant concerned with forced perspective, the stage manager who will be worried about the wear patterns and ease of rolling instruments on stage, the mechanical engineers who must design the machinery, and structural engineers who must make sure the stage meets building code.

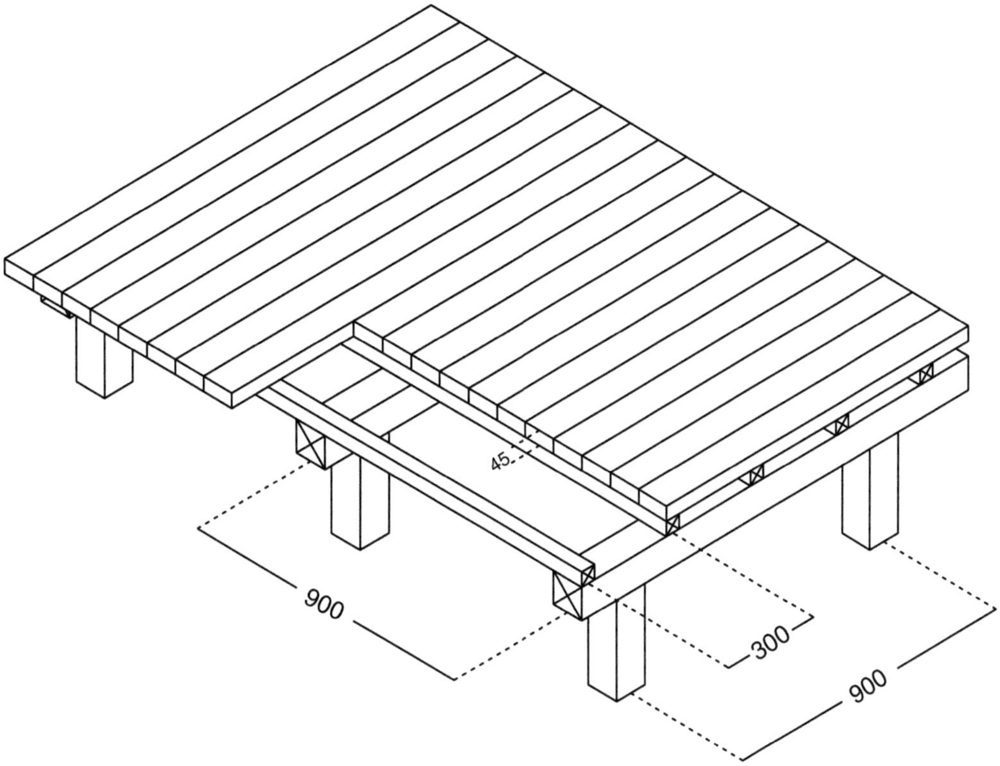

Fig. 15.1 Stage structure (in mm)

Table 15.1 Decking direction for historical halls

Left-right decking	Upstage-downstage decking
Birmingham Symphony Hall	Carnegie Hall, New York
Boston Symphony Hall	Dallas Symphony Hall
Gasteig, Munich	Leipzig Gewandhaus
Tokyo Bunka Kaikan	Muza Kawasaki Concert Hall
Suntory Hall, Tokyo	Sapporo Concert Hall
Vienna Musikvereinssaal	Walt Disney Concert Hall, LA

16. Variable vs. Fixed Acoustics

Adjustable acoustics sounds like some sort of magical system which can provide ideal acoustical environment for a wide range of programs in a single hall.

Usually, such a system changes the reverberation time by adjusting the amount of acoustically absorptive material exposed to the space. This can take the form of absorptive curtains, banners, mats or baffles, and panels which alternately expose reflective or absorptive surfaces. In more extreme cases, there have been trials which change the effective acoustical room air volume by opening or closing reverberation chambers which in turn changes the reverberation characteristics of the space.

The adjustable acoustics systems are added as a selling point in halls which can then claim to be tunable for different programs from solo recital to chamber music to symphonic orchestral music. The systems are also intended to adapt to the difference between empty and occupied conditions, as moving from rehearsal to concert.

However, there are several problems with these "almighty" systems. First, who is responsible for making the decision to change the acoustics? While the principal conductor for a resident ensemble may have enough time and consistency in ensemble and program to play with the adjustable acoustics, touring orchestras will be at the mercy of even the most well-intentioned technical crew. I have heard of several examples where the crew changed the adjustable acoustics from the "rehearsal" to "concert" mode without informing the conductor. This completely defeated the point of rehearsals since the hall acoustics had changed entirely, akin to playing with a new orchestra.

Many acoustically excellent halls—such as Vienna Musikvereinssaal, Amsterdam Concertgebouw, Boston Symphony Hall, Berlin Philharmonie, Suntory Hall, and Walt Disney Concert Hall—do not have an adjustable acoustics system but accommodate many different concert types. Instead, the discussion should focus on programs such as pop and rock events, lectures, and musicals which require amplified sound systems. In these non-acoustic situations, absorbing materials like curtains, banner, or acoustical baffles would be indispensable.

Y. Toyota et al., *Concert Halls by Nagata Acoustics*, https://doi.org/10.1007/978-3-030-42450-3_48

17. Sound System Design in Concert Halls

Introducing a sound system into concert halls which are dedicated to symphonic orchestral music is challenging since natural acoustics are often at odds with the requirements for amplification systems.

Sound systems were developed for large, often outdoor, spaces and in situations where it is necessary to produce enough sound power and add effects like artificial reverberation. Control over the directivity is of particular importance in achieving this goal of clear and loud delivery.

Taking these skills into a concert hall, the sound system designer is again trying to direct sound energy towards many different audience areas effectively. While mid- and high-frequency sounds are fairly directional, low-frequency sound is distributed broadly and spills out in all directions from the speaker. Adding loudspeakers for stereophonic effects would only be effective in a small area in front of the stage and distracting for those outside of this area. Amplified again by the room acoustics, it results in a sound which we would call "boomy" or "muddy," that is to say loud, bass heavy, and unclear.

The solution is to treat the loudspeakers as any other musical instrument. Placing the speaker on stage allows the sound system engineer to use the natural acoustics of a concert hall to their advantage instead of fighting against it.

Since the impression of space does not need to be artificially added in a concert hall, an omnidirectional mono speaker is most appropriate. A model with good sound reproduction and minimal distortion should be selected. While an omnidirectional loudspeaker would lack directivity control, this is comparable to most other instruments. Limiting the number of sound sources has the additional benefit of capping the overall sound power the system can produce.

Keeping the loudspeaker on stage would additionally improve the localization of sound, helping the amplified track blend with the other instruments. Hearing the location of the sound origin would be most natural for the entire audience. Placing the loudspeaker directly on the stage surface may place it too low to provide direct sound to the audience, so flying the loudspeaker would be possible, but it should still be located close to the level of stage, just above the musicians' heads, not close to the ceiling as is often the case. Flying the speaker also frees up space on stage from speaker stands and cables. Improving source localization on stage also benefits the differentiation with off-stage loudspeakers that can be selectively deployed for artistic and compositional effects throughout the room, in the same was as off-stage musicians performing on acoustic instruments.

Finally, the sound level must be kept low. When the mixing engineer cannot hear the music clearly in a conventional pop/rock venue, indoor or outdoor, their instinct is to raise the sound level. However, a concert hall is designed to faithfully transmit even the quietest solo violin to all the audience members. By lowering the level of amplification, the clarity is improved.

Using these principles facilitates the balance between amplified instruments or reproduced music and acoustic instruments by leveraging the room acoustical qualities of the space in similar manners. In many cases, such a system could also act as a public address system due to the improved intelligibility. Achieving a sound system in this style requires a unified goal among the client, architect, sound system designer, mixing engineer, loudspeaker manufacturer, and acoustician. This design concept must also be communicated to touring mixing engineers who will not be accustomed to the unique acoustics of the space. All parties must understand the fundamental differences between a concert hall and other venues.

Y. Toyota et al., *Concert Halls by Nagata Acoustics*, https://doi.org/10.1007/978-3-030-42450-3_49

18. Semi-staged Productions in Concert Halls

More and more, there is desire from venue managers to utilize their investment beyond the limits of symphonic orchestral music. Incorporating some multifunctionality has become somehow standard for newly constructed halls.

For example, semi-staged opera has become popular to include in concert hall programming. When played outside of a traditional opera house, the expectation of multiple, huge set pieces and large chorus is removed, meaning that the cost to the opera company is significantly reduced. Instead, abstract sets can be used on stage. A series of operas in Walt Disney Concert Hall over the past years have featured sets by architects Frank Gehry, Jean Nouvel, and Zaha Hadid. Suntory Hall has been continuing a tradition of a yearly opera performance for more than 20 years, including Tan Dun's opera *Tea*, commissioned in 2002. For these performances, a medium-sized orchestra is positioned either far downstage or over to one side of stage.

For recent concert halls in Russia—Mariinsky Concert Hall and Zaryadye Concert Hall—where opera is a particularly popular genre, opera was designed to be more than half of the program. In these halls, the inclusion of an orchestra pit is invaluable. These operatic performances have earned a positive acoustical reputation. The concert hall setting is generally better suited to amplify vocal performances since there are much fewer acoustically absorptive sets and theatrical curtains. In conjunction with moving the orchestra into the pit, vocalists may find it much easier to balance with the orchestra as compared to a traditional opera house. Furthermore, as long as the pit is incorporated into the design at an early stage, there would be no negative effects on non-opera acoustics.

Aside from opera, the inclusion of projectors has become increasingly common. The walls and ceiling of the concert hall can be used as the projecting surfaces to create magical art installations synchronized with the music. Even in halls like Shanghai Symphony Hall where the walls are neither flat nor white, projection mapping is still feasible, although this type of architecture encourages more abstract visualizations as opposed to concrete images or text.

Projectors do come with their own set of acoustical and architectural challenges. The machines and their ventilation systems are noisy, and must be located entirely within a concrete room with thick glazing to prevent disturbances to the musical performance. Furthermore, these systems can be expensive and will require increased maintenance and updating costs as the technology develops. While it is certainly possible to include projectors in surround-type halls like in New World Symphony or Shanghai Symphony Hall, more projectors are required to ensure that the visualization is visible from all seating positions. A typical end-stage, cinema layout can be created by using a retractable screen upstage, but this may reduce the number of saleable seats in the chorus and side seating areas.

However, the vineyard style hall also provides the opportunity to make room for off-stage ensembles, including within the classical repertoire. Circulation spaces or purpose-built platforms can be used to distribute small ensembles throughout the hall in a wide variety of locations. Some halls embraced this possibility as part of their design; notable examples include New World Symphony in Miami Beach, Philharmonie de Paris, or Elbphilharmonie in Hamburg.

Y. Toyota et al., *Concert Halls by Nagata Acoustics*, https://doi.org/10.1007/978-3-030-42450-3_50

Mariinsky Concert Hall. The orchestra is placed in the pit while stage is transformed for a production of *Figaro*

19. Concert Hall Organ Design

Adding an organ to a concert hall is quite common as it can be used both for solo recitals and as part of the orchestral repertoire.

Compared to the voluminous, stone churches for which organs were built, concert halls have shorter reverberation times and increased clarity. For this reason, it is especially important for the organ to be able to fully activate the available volume. Instead of pushing the organ back into a niche, it should be fully incorporated into the space; it must have room to breathe. The walls and ceiling around the organ must be heavy and continuous in order to radiate sound energy out into the room, similar to the requirements for the orchestra itself. The quality of the organ and its voicing is even more exposed, since the sound in a concert hall is clearer than in a church.

While solo organ recitals are possible in a concert hall, orchestral repertoire which has organ parts is the principal consideration. Placing the organ and console close to the stage improves the balance between organ and orchestra. Using a remote console on stage can help the organist communicate with the conductor and orchestra.

As long as the design is acoustically and architecturally balanced, the shape and position of the organ can be very flexible. The position of the organ does not necessarily need to be along the centerline, as it is in most concert halls and churches or cathedrals. Kyoto Concert Hall has an organ which is strongly off-center, which is especially noticeable since the room is otherwise symmetrical. The organ can be placed behind audience seats, as in Hamburg Elbphilharmonie, as long the organ is within the main volume of the hall rather than an adjacent subsidiary volume. Regarding shape, the Walt Disney Concert Hall organ is a unique collaboration between architect Frank Gehry and organ designer Manuel Rosales. The facade is integrated with the surrounding architecture, and, in spite of their curved shapes, composed entirely of functioning pipes.

Y. Toyota et al., *Concert Halls by Nagata Acoustics*, https://doi.org/10.1007/978-3-030-42450-3_51

Kyoto Concert Hall

Hamburg Elbphilharmonie

Walt Disney Concert Hall

20. A New Direction of Concert Hall Design

Two recent projects show a new direction for concert hall layouts, and a new possibility for shared experiences: Calderwood Hall at Isabella Stewart Gardner Museum in Boston and Pierre Boulez Saal at Barenboim–Said Academy in Berlin. In these halls, the stage is placed exactly in the center, surrounded equally on all sides by the audience.

The average distance from any seat to the stage is minimized in these configurations, greatly improving the visual and acoustical intimacy. Calderwood Hall features a remarkable 80% of seats in the front row, since only the ground floor has a second row of seats. Pierre Boulez Saal accommodates a larger audience with more rows in plan, but the extremely steep rake gives the impression of many front-row seats. Such a seating configuration is better suited to medium- to small-capacity halls with smaller stages, since larger ensembles have a natural orientation that competes with the omni-directionality of the space.

Without a clear directionality imposed by the architecture, the traditional performance orientation of the musicians is challenged. The best stage layout would the one chosen as if there were no audience at all. Chamber ensembles often face each other during rehearsal or recording sessions because of the improved communication between musicians, and therefore ensemble quality. Ideally, this tight communication would be transferred from rehearsal to the concert stage, even if not all musicians face all the audience. This method is used for Calderwood Hall where the square, symmetric seating layout is completely democratic with no indication of which direction is the front.

The piano, however, has a clear directionality, when the lid is opened in the usual manner. In Pierre Boulez Saal, it would be inappropriate to play piano with a lid. To counteract the psychoacoustical effect for the pianist of removing the lid, I suggested to Daniel Barenboim to think of the ceiling above the stage as the lid: it does the work of projecting the sound out to the audience. Barenboim has embraced this change and recommends this condition to visiting pianists as well. Should the lid be kept, an alternative is to rotate the piano during intermission, so as not to favor a single direction and to present viewer-listeners with different experiences within the same concert. Occasionally, larger ensembles with conductor performing in Pierre Boulez Saal also change orientation during intermission.

For solo vocal performances in a non-directional space, the challenge is even more pronounced than in a larger surround-style hall. In the opening concert series of Calderwood Hall, soprano Kiri Te Kanawa opted to rotate slowly during her performance. This response illustrates how a new environment can spark interesting ideas from performers.

By relinquishing a strict and imposed sense of directionality on the stage and in the hall, these "in-the-round" designs indeed encourage new approaches for performers and offer new experiences to the viewer-listeners. It is the task of architects and contributing designers to explore and expand the possibilities for other types of halls, even with larger seating capacities. More importantly, we hope they can foster new ways to present concerts by inspiring musicians, singers, composers, conductors, stage directors, dancers, visual and multimedia artists, venue managers, concertgoers, and even those who might not normally find themselves in a concert hall.

Y. Toyota et al., *Concert Halls by Nagata Acoustics*, https://doi.org/10.1007/978-3-030-42450-3_52

GPSR Compliance

The European Union's (EU) General Product Safety Regulation (GPSR) is a set of rules that requires consumer products to be safe and our obligations to ensure this.

If you have any concerns about our products, you can contact us on ProductSafety@springernature.com

In case Publisher is established outside the EU, the EU authorized representative is:

Springer Nature Customer Service Center GmbH
Europaplatz 3
69115 Heidelberg, Germany

Batch number: 10731528

Printed by Printforce, the Netherlands